ESO ASTROPHYSICS SYMPOSIA
Europcan Southern Observatory

Series Editor: Bruno Leibundgut

M. Kissler-Patig J.R. Walsh M.M. Roth (Eds.)

# Science Perspectives for 3D Spectroscopy

Proceedings of the ESO Workshop
held in Garching, Germany,
10-14 October 2005

Springer

*Volume Editors*

Markus Kissler-Patig
Jeremy R. Walsh

ESO - European Southern Observatory
Karl-Schwarzschild-Str. 2
85748 Garching
Germany

Martin M. Roth

Astrophysikalisches Institut Potsdam
An der Sternwarte 16
14482 Potsdam
Germany

*Series Editor*

Bruno Leibundgut

European Southern Observatory
Karl-Schwarzschild-Str. 2
85748 Garching
Germany

ISBN 978-3-642-09259-6 e-ISBN 978-3-540-73491-8



Springer is a part of Springer Science+Business Media
springer.com

Softcover reprint of the hardcover 1st edition 2007



Cover design: WMXDesign, Heidelberg

# Preface

"You will hold in your hands the proceedings of *the last conference ever* on integral-field spectroscopy." So the daring statement that we, organizers, made at the end of an extremely interesting and busy week of conference in October 2005. Indeed, the quality and diversity of the contributions in this book indicated that integral-field spectroscopy had evolved into a mature technique. Future books would not be dedicated to the technique as such but to the science achieved with '3D-spectroscopy,' that is a dedicated book after this one, assembling science results across so many fields, is unlikely to appear again (or have you seen a proceeding dedicated to 'long-lit spectroscopy' recently?).

In a sense this is very encouraging. The idea of this conference was an offspring of the Research and Training Network (RTN) 'Euro3D' sponsored by the European Commission. The goal of this network was to promote integral-field spectroscopy from an technique for experts to a common user utility for astronomical research. When we first talked about this conference the VLT, for example, was not yet equipped with any integral-field spectrograph. We were wondering whether anyone would actually attend the conference in 4 years' time and would have anything to report upon.

We were rewarded for our optimism and very positively surprised about the progress made in the years after that. A progress, of course, driven by the advent of so many integral-field spectrographs around the world, installed on 4m to 8–10m class telescopes. Many of these instruments are referred to in the present book. Yet, this is explicitly *not* a book dedicated to technical issues, but focusing on the science. Thus, the name of the conference, **Science Perspectives for 3D Spectroscopy**.

The reader will discover that integral-field spectroscopy is employed today to answer questions across the full spectrum of optical and near-infrared observational astronomy. The book is divided into nine parts following closely the program on the conference week: from solar system studies all the way to high redshift surveys. With 91 participants, 41 contributed talks, and 30 posters, these proceedings are a rich source state-of-the-art research with 3D spectroscopy.

We would like, of course, to use this foreword to thank the many people who contributed to the success of the meeting. Starting by the SOC (Jeremy

Allington-Smith, Santiago Arribas, Andy Bunker, Roland Bacon, Jonathan Bland-Hawthorn, Gerald Cecil, Roger Davies, Christophe Dumas, Pierre Ferruit, Reinhard Genzel, Andreas Quirrenbach, Gillian Wright), extending our big thanks to the LOC (Christina Stoffer, Karina Kjaer, Pam Bristow), and a special applause to Britt Sjoeberg for her brilliant idea to have the conference dinner in a Bavarian brewery.

ESO,
August 2006

*Markus Kissler-Patig*
*Jeremy R. Walsh*
*Martin M. Roth*

# Contents

## Part II Solar System(s)



## Part III Nearby Galaxies

# List of Contributors

**Abolmasov, Pavel**
Moscow State University
pasha@sao.ru

**Allington-Smith, Jeremy**
University of Durham
j.r.allington-smith@durham.ac.uk

**Amram, Philippe**
Laboratoire d'Astrophysique de Marseille philippe.amram@oamp.fr

**Arribas, Santiago**
STScI / IAC
arribas@stsci.edu

**Balkowski, Chantal**
Observatoire de Paris
chantal.balkowski@obspm.fr

**Bastian, Nate**
University College London - Department of Physics and Astronomy
bastian@astro.uu.nl

**Bik, Arjan**
ESO - Garching
abik@eso.org

**Bomans, Dominik**
Astronomical Institute, Ruhr-Univ. Bochum bomans@astro.rub.de

**Bouchet, Patrice**
GEPI, Observatoire de Paris-Meudon
Patrice.Bouchet@obspm.fr

**Bunker, Andrew**
University of Exeter
bunker@astro.ex.ac.uk

**Cecil, Gerald**
University of North Carolina
gerald@thececils.org

**Cesarsky, Catherine**
ESO - Garching
ccesarsk@eso.org

**Claire, Max**
Center for Adaptive Optics, UCSC
max@ucolick.org

**Covone, Giovanni**
Laboratoire d'Astrophysique de Marseille
giovanni.covone@oamp.fr

**Davies, Richard**
MPE
davies@mpe.mpg.de

**De Marchi, Guido**
European Space Agency
gdemarchi@rssd.esa.int

**Di Mille, Francesco**
Department of Astronomy
Padova University
dimille@pd.astro.it

**Dumas, Gaëlle**
CRAL - Observatoire de Lyon
gdumas@obs.univ-lyon1.fr

**Eisenhauer, Frank**
Max-Planck-Institut für extraterrestrische Physik, Garching
eisenhau@mpe.mpg.de

**Emsellem, Eric**
Centre de Recherche Astronomique de Lyon
emsellem@obs.univ-lyon1.fr

**Exter, Katrina**
IAC - also at LUDWIG
katrina@ll.iac.es

**Falcon-Barroso, Jesus**
Leiden Observatory
jfalcon@strw.leidenuniv.nl

**Ferruit, Pierre**
CRAL - Observatoire de Lyon
pierre.ferruit@obs.univ-lyon1.fr

**Foucaud, Sebastien**
University of Nottingham
sebastien.foucaud@nottingham.ac.uk

**Fuentes-Carrera, Isaura**
Instituto de Astronomia e Geofisica, USP (Brazil)
isaura@astro.iag.usp.br

**Gebhardt, Karl**
University of Texas
gebhardt@astro.as.utexas.edu

**Genzel, Reinhard**
Max Planck Institute for Extraterrestrial Physics genzel@mpe.mpg.de

**Gerssen, Joris**
Department of Physics, University of Durham
joris.gerssen@durham.ac.uk

**Goessl, Claus**
Universitätssternwarte München
cag@usm.uni-muenchen.de

**Grosbøl, Preben**
ESO - Garching
pgrosbol@eso.org

**Grupp, Frank**
Max Planck Institut für Extraterrestrische Physik
fug@usm.uni-muenchen.de

**Hammer, François**
GEPI, Observatoire de Paris-Meudon
francois.hammer@obspm.fr

**Hartung, Markus**
ESO - Chile
mhartung@eso.org

**Hopp, Ulrich**
University of Munich and MPE
hopp@usm.uni-muenchen.de

**Ivanov, Valentin**
ESO - Chile
vivanov@eso.org

**Jourdeuil, Emilie**
CRAL - Observatoire de Lyon
emilie@obs.univ-lyon1.fr

**Jungwiert, Bruno**
CRAL - Observatoire de Lyon
bruno@ig.cas.cz

**Kelz, Andreas**
Astrophysikalisches Institut Potsdam
akelz@aip.de

**Kissler-Patig, Markus**
ESO - Garching
jwalsh@eso.org

**Kjaer, Karina**
ESO - Garching
kekjaer@eso.org

**Koehler, Ralf**
MPE
rkoehler@mpe.mpg.de

**Krajnovic, Davor**
Oxford University
dxk@astro.ox.ac.uk

**Kuntschner, Harald**
ESO - Garching
hkuntsch@eso.org

**Lehnert, Matthew**
MPE
mlehnert@mpe.mpg.de

**Lemoine-Busserolle, Marie**
Institute of Astronomy (Cambridge)
lemoine@ast.cam.ac.uk

**Lopez, Rosario**
Departamento de Astronomia (Universitat de Barcelona)
rosario@am.ub.es

**Maillard, Jean-Pierre**
Institut d'Astrophysique de Paris (CNRS) maillard@iap.fr

**Malagnini Sicuranza, Maria Lucia** Dipartimento di Astronomia, Università degli Studi di Trieste
malagnini@ts.astro.it

**Marcelin, Michel**
OAMP
michel.marcelin@oamp.fr

**McDermid, Richard**
Leiden Observatory
mcdermid@strw.leidenuniv.nl

**Mendes de Oliveira, Claudia**
IAG/USP, Sao Paulo (Brazil)
oliveira@astro.iag.usp.br

**Mengel, Sabine**
ESO - Garching
smengel@eso.org

**Messineo, Maria**
ESO - Garching
mmessine@eso.org

**Mignoli, Marco**
INAF - Bologna Observatory
marco.mignoli@bo.astro.it

**Moiseev, Alexey**
Special Astrophysical Observatory
moisav@sao.ru

**Molla, Mercedes**
CIEMAT, Madrid
mercedes.molla@ciemat.es

**Monnet, Guy**
ESO - Garching
gmonnet@eso.org

**Monreal-Ibero, Ana**
Astrophysikalisches Institut Potsdam
amonreal@aip.de

**Mora, Marcelo**
ESO - Garching
mmora@eso.org

**Nesvadba, Nicole**
MPE
nicole@mpe.mpg.de

**Nuernberger, Dieter**
ESO - Chile
dnuernbe@eso.org

**Oosterloo, Tom**
Astron
oosterloo@astron.nl

**Peroux, Celine**
ESO - Garching
cperoux@eso.org

**Plana, Henri**
LATO - Universidade Estadual de Santa Cruz (Brazil)
plana@uesc.br

**Prugniel, Philippe**
Observatoires de Lyon & Paris
prugniel@obs.univ-lyon1.fr

**Pugliese, Giovanna**
ESO - Garching
gpuglies@eso.org

**Quirrenbach, Andreas**
Leiden Observatory
quirrenb@strw.leidenuniv.nl

**Relke, Helena**
MPE Garching
relke@usm.muenchen-uni.de

**Roth, Martin**
Astrophysical Institute Potsdam
mmroth@aip.de

**Smirnova, Aleksandrina**
Special Astrophysical Observatory (SAO)
alexiya@sao.ru

**Soucail, Genevieve**
Observatoire Midi-Pyrenées
soucail@ast.obs-mip.fr

**Sugai, Hajime**
Department of Astronomy, Kyoto University
sugai@kusastro.kyoto-u.ac.jp

**Swinbank, Mark**
University of Durham
a.m.swinbank@dur.ac.uk

**Tacconi-Garman, Lowell**
ESO - Garching
ltacconi@eso.org

**Tamburro, Domenico**
Max-Planck-Institute for Astronomy
tamburro@mpia.de

**Thatte, Niranjan**
University of Oxford
thatte@astro.ox.ac.uk

**Tsamis, Yiannis**
University College London
ygt@star.ucl.ac.uk

**van Breugel, Wil**
Univ. of California, LLNL
wil@igpp.ucllnl.org

**Vergani, Daniela**
IASF-MI
daniela@mi.iasf.cnr.it

**Walsh, Jeremy**
ESO - Garching
jwalsh@eso.org

**Weilbacher, Peter**
Astrophysikalisches Institut Potsdam
pweilbacher@aip.de

**Westmoquette, Mark**
University College London
msw@star.ucl.ac.uk

**Wisotzki, Lutz**
Astrophysikalisches Institut Potsdam
lwisotzki@aip.de

**Wright, Gillian**
UK Astronomy-Technology Centre
gsw@roe.ac.uk

# Part I

# 3D Instrumentation

# 3D Instrumentation

J. Allington Smith

Centre for Advanced Instrumentation. University of Durham, UK
j.r.allington-smith@durham.ac.uk

**Summary.** The main methods of obtaining "3D" data in optical/infrared astronomy are briefly reviewed before focussing on integral field spectroscopy (IFS). The major techniques are compared using a simple figure of merit: the specific information density. Although this analysis strongly favours image slicing generally, considerations of complexity, cost and signal/noise optimisation for a particular astrophysical study may produce different results. The advantage of using the *Advanced Image Slicer* paradigm to maximise both throughput and sampling efficiency within a small instrument volume is noted; together with the technical challenges. The range and diversity of IFS instrumentation currently available and planned is demonstrated. This shows that IFS is now a mature technique, central to the aspirations of all observatories and of great relevance to Extremely Large Telescopes.

## 1 Integral Field Spectroscopy and 3D Instrumentation

The term "3D spectroscopy" is often used to indicate any technique that produces spatially-resolved spectra over a two-dimensional field[1]. Integral field spectroscopy is that subset of 3D spectroscopy in which all the data for one pointing of the telescope is obtained simultaneously[2]

The oldest types of non-IFS instrumentation such as Fabry-Perot Interferometry (FPI; e.g. (18)), Imaging Fourier Transform Spectroscopy (IFTS; e.g. (31; 17)) and Hadamard transform spectrographs (e.g. (15; 16)) use the time domain (or a Fourier conjugate) to scan through wavelength space. This leaves them potentially sensitive to changes in the instrumental or sky background, but allows a wide field to be covered in one pointing. In contrast, IFS encodes all the spectral and spatial information in the same exposure resulting in a smaller field of view for a given detector format. IFS has the advantage of immunity to temporal variations in the sky background — especially when highly variable such as the OH lines longwards of 700nm — although these still affect calibrations. However wide-field applications requiring minimal spectral coverage often benefit from the use of the scanning devices; as

[1]The term *hyperspectral-imaging* is also used by the defence and earth-observation communities.

[2]Other types of 3D instrumentation in astronomy include, radio spectral line receivers working with scanning single-element antennae or arrays of antenna feeds or interferometer baselines.

do space applications where the sky background is very low and stable. Simpler examples of scanning systems are stepped-longslit spectroscopy where the position of the longslit is altered in 1D between exposures and narrow-line imaging where images are taken in multiple narrow passbands, usually at different times.

Another important class of 3D instrumentation is detector arrays with intrinsic energy resolution. In these, the energy of the incident photon is measured simultaneously with its position within the 2D detector surface. The obvious advantage is that no dispersive element is required which simplifies the optical system, resulting in the potential of better throughput. The energy is measured by counting the number of secondary particles or quasi-particles produced by each incident photon since this is proportional to its energy. The resolving power ($R \equiv \lambda/\delta\lambda$) thus depends on the size of the material's energy bandgap and the noise in the counting electronics. Such photon-counting detectors also experience non-linearity or saturation when the number of incident photons exceeds the counting rate of the electronics. The saturation may be caused either by the flux from the object or from the background.

One such device type, which is in regular astronomical use is based on Superconducting Tunnel Junctions (STJs; (24) and references therein). Currently, these are limited to $R < 30$, but could be improved through the use of different materials. They also suffer saturation effects from long-wavelength photons and so require exceptional IR-blocking filters. Other devices of this type using superconductors to provide small bandgaps are transition edge sensors (TES; e.g. (7)) and kinetic inductance devices (KIDs; e.g. (21)). Although these may well come to dominate astronomical spectroscopy they, like STJs, require further development to make arrays with sufficient numbers of spatial samples and resolving power for general use.

All 3D techniques produce a datacube of a scalar related to flux density as a function of spatial coordinates in the field and wavelength. To first order, the efficiency of all 3D techniques is the same. For example, an IFTS may provide a large number of spaxels at one time but require many timesteps to scan through the spectrum. Although IFS produces all the spectral information in one exposure, the field of view is necessarily limited so that a number of exposures with different pointings must be mosaiced to produce the same number of volume resolution elements (voxels). The same argument applies to stepped-longslit spectroscopy and multi-band imaging. To *second* order, the relative efficiency of the techniques differ significantly, depending on the details of how data from different exposures are combined, the variability of the background compared to fixed noise sources and the nature of the required data product.

3D techniques are generally preferable to slit spectroscopy for a number of reasons: (a) slit losses are eliminated; (b) accurate target acquisition is not required; (c) the actual target position can be recovered from the data by reconstructing an image — also an aid to accurate mosaicing; (d) errors in radial velocity due to differences in the barycentre of the slit illumination obtained from the object and from reference sources can be eliminated; (e) the global velocity field is recovered without bias imposed by the observer's choice of slit position and orientation; (f) atmospheric dispersion effects can be corrected without loss of light by manipulation of the datacube; (g) in poor or variable seeing, IFS is always optimally matched to the object PSF.

IFS brings a unique advantage in some cases, such as in crowded fields where the contiguous spectral and spatial information can be used together to overcome confusion (e.g for intermediate mass black holes detection in globular clusters); for the study of active galactic nuclei (e.g. Gerssen et al 2006, Garcia-Lorenzo et al. 1999); and the study of lensed galaxies to recover the source-plane radial velocity field (e.g. Swinbank et al. 2003).

## 2 The Different Methods of IFS

Fig. 1 summarises four techniques of IFS. See Allington-Smith (3) for more detail; the fourth of these is a recent idea proposed by Content (9) as a means to provide ~1 million spaxels to search for primaeval galaxies.

It useful to define a figure of merit (FOM): the specific information density (SID)

$$Q = \eta \frac{N_p N_q N_\lambda}{N_x N_y} \tag{1}$$

where $N_p$ and $N_q$ are the numbers of spatial resolution elements in orthogonal directions $p$ and $q$ in the field.

These quantities are related to the numbers of spaxels via $N'_p = N_p f_s$, $N'_q = N_q f_s$ where $f_s$ is the oversampling of the PSF by the *IFU*. $N_\lambda$ is the number of spectral resolution elements related to the number of spectral samples by $N'_\lambda = N_\lambda f_\lambda$ where $f_\lambda$ is the spectral oversampling by the *detector*. $N_x$ and $N_y$ are the numbers of pixels in the detector in orthogonal directions $x$ and $y$, and $\eta$ is the throughput of the IFU[3]. The coordinates, $x$ and $y$ are aligned with directions $p$ and $q$ on the sky in the sense that $p$ is the dimension that varies most rapidly along the slit which is defined by the detector $y$ direction. Although this is a useful FOM for comparing different IFS systems, a comparison of different 3D techniques requires a more general metric such as $Q/n_e t_e$ where $n_e$ is the number of separate exposures and $t_e$ is the duration of each.

The theoretical maximum SID is obtained when $N'_p N'_q N'_\lambda = N_x N_y$, for Nyquist sampling ($f_s = f_\lambda = 2$) as $Q_{max} = 1/8$. The case of super-sampling, by combining exposures with offsets of non-integral numbers of pixels, can

[3]Measured from the IFU's input to its output but including losses at the spectrograph stop, or, by comparison with the same spectrograph using a single slit of the same equivalent width.

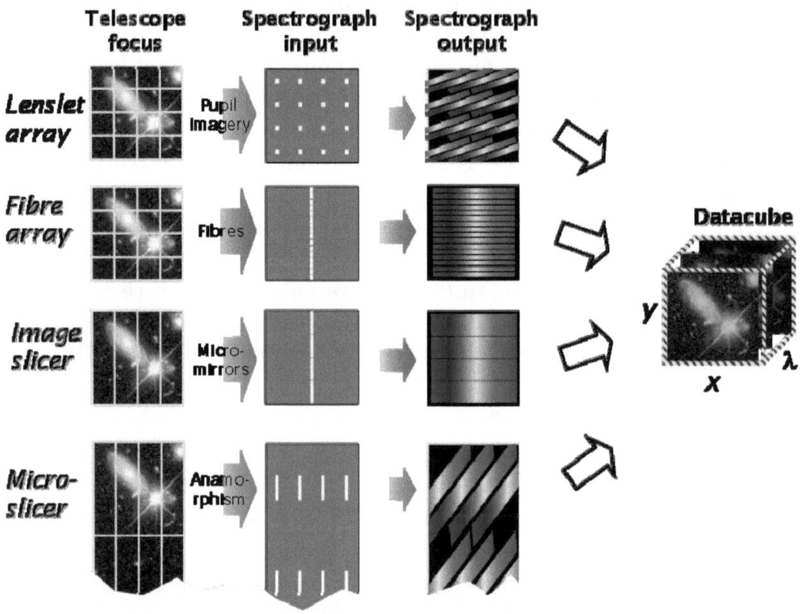


**Fig. 1.** A summary of the four main techniques of integral field spectroscopy.

be handled using this formalism by considering the combined observation to have been obtained with a virtual instrument with enhanced resolution. However the number of separate exposures will then need to be accounted for via the $Q/(n_e t_e)$ metric used for generalised 3D instruments.

The advantage of the SID as a FOM is that it is simply a measure of the *information content per detector pixel* and is thus appropriate for a discussion of cost-limited instrumentation whereas a FOM based on signal/noise is appropriate for discussion of specific astrophysical studies. SID makes no assumption about the contiguity of the spaxels within the field or the size of the field and so does not discriminate against sparsely-sampled IFS where contiguity is sacrificed for field (e.g. to measure the surface gravity over a face-on disk galaxy). Likewise, by defining the SID in terms of *resolution* elements rather than *sampling* elements, the statistical correlation between samples is naturally taken into account: it is only necessary to convert correctly between resolution elements and samples. The correlation may arise due to the, normally desirable, oversampling according to the Nyquist theory at an image conjugate, or due to crosstalk elsewhere in the instrument, e.g. at the slit, or be absent in the case of sparse sampling.

Expressions for the SID for each of the four techniques identified are given in Allington-Smith (3) in terms of instrumental parameters such as the spaxel-to-spaxel pitch on which the samples are imaged on the detector and the number and size of the gaps required to prevent cross talk between statistically-independent regions of the sky. Fig. 2 presents a comparison of the different techniques. The SID of each technique covers a wide range because of differences in the design of some types of instrument. For example, the Keck OSIRIS (19) has achieved an unusually high SID for a lenslet array IFS (e.g. Bacon et al. (6)) because the design strives to pack the pupil images as close together as possible. However, as noted above the SID does not take into account any influence that the close packing of the spectra might have on the recoverable signal/noise. This consideration also highlights the essential role played by the data reduction software in optimising scientific return.

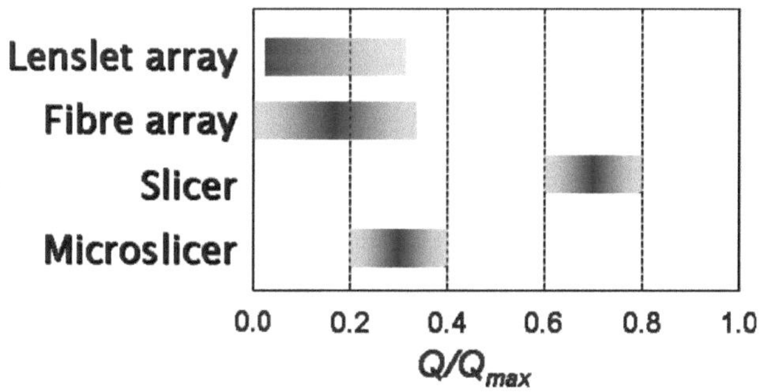


**Fig. 2.** A comparison of the Specific Information Density (normalised to the theoretical maximum) obtainable with existing IFS instrumentation. Although based on calculations of specific instruments, a greyscale is used to indicate the range — and barycentre — of the distribution generally available.

This comparison shows the near-optimal potential performance of image slicers (32). However, the other techniques are broadly similar to each other including the recent generation of lenslet-only devices which might naively be thought to be at a disadvantage because of the lack of reformatting. The new micro-slicer technique is well-suited to devices with very large numbers of spaxels, since it makes uses of easily-replicable anamorphic lenslet arrays rather than complex figured mirrors; and it is encouraging that this has the potential to reach moderately-high SID.

## 3 Image slicers

Results from early devices such as the GNIRS-IFU (2; 5) using the AIS principle (8) demonstrate excellent performance. Compared to the original design, the AIS brings the benefit of square spaxels, greater immunity to diffraction loss, smaller size and easier optical interfaces at the expense of greater complexity on the surface forms of the constituent mirrors (summarised by Allington-Smith (3)). The throughput (Fig. 3) is determined, apart from the small loss due to the non-unit reflectivity of the gold coating, exclusively by the roughness of the optical surfaces. For GNIRS, this was $\sigma = 15 - 20$nm.

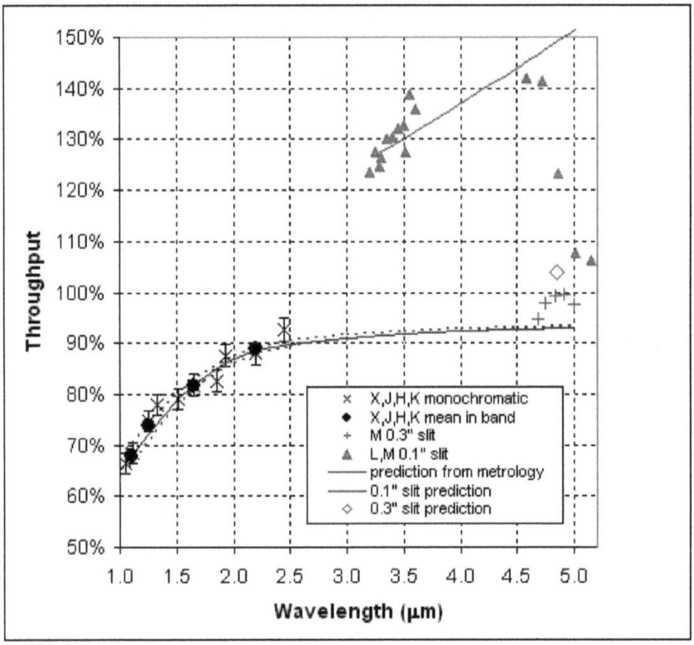


**Fig. 3.** The throughput of the GNIRS-IFU as a function of wavelength. The prediction assumes that the only loss mechanism is the reflectivity of the gold-coatings (almost independent of wavelength) and the total integrated scatter due to surface roughness. The prediction is calculated using measurements of the roughness as manufactured and refers to the part of the IFU which maps onto the extent of the GNIRS longslit, excluding a small damaged portion of one slice. The throughput represents the ratio of the throughput of the spectrograph with the IFU deployed and with the IFU replaced with a slit of twice the slice width. Results for a narrower slit are also shown after scaling to the same effective slitwidth to show how the IFU is better able to cope with diffracted light than the traditional slit, resulting in an apparent efficiency exceeding unity.

Although superior at all wavelengths to similar fibre-based IFUs in the optical, it is clear the smoother optics are required in the future especially if these devices are to be used at visible wavelengths as an alternative to lenslet and fibre systems. For example, KMOS (27; 29) requires $\sigma = 10$nm. This is probably achievable using the diamond-machined aluminium approach employed in GNIRS, but may need to be augmented by post-polishing. Trials suggest (J. Schmoll, priv. comm) that the initial surface may be improved by a factor two in this way. An alternative is to polish glass mirrors using traditional techniques which are then assembled and aligned individually to high accuracy (20) in contrast to the machined-metal approach which seeks to eliminate inter-facet alignment problems by making monolithic arrays of facets (26; 23; 11; 12). These two methods have their relative merits; the optimum choice being, as ever, dependent on the details of the astrophysical investigation and the environment of operation (e.g. terrestrial/space and warm/cold). For example, the IFU in the James Webb Space Telescope Near-infrared Spectrograph, which is to be built by SIRA and Durham CfAI, is baselined on the machined-metal approach.

## 4 Summary of current IFS projects

A recent conference held in Durham, 5-8 July 2005 ((4) and the many proceedings therein), included a large number of talks on IFS instrumentation ranging from those completed many years ago to those still on the drawing board. Ray Sharples provided a superb summary of the state-of-play (28). This has been adapted as Table 1 which shows current projects on 8-10m class telescopes and Table 2 which shows the new projects discussed in Durham.

**Table 1.** Integral field instruments on 8-10m class telescopes (adapted from Sharples (28)). (Instruments under construction are shown in **bold** font. Instruments which have a possible integral field mode, but which is not in the baseline capability, are shown in *italics.*)

| Telescope | Instrument |
|---|---|
| Gemini | GMOS(2), **NIFS***, GNIRS |
| VLT | FLAMES, SINFONI/SPIFFI, VIMOS, **KMOS, SWIFT, MUSE** |
| HET | **VIRUS** |
| Keck | OSIRIS |
| Magellan | IMACS |
| LBT | *MODS, LUCIFER* |
| Subaru | KYOTO-3D |
| SALT | *SCAM5* |
| GTC | **FISICA, FRIDA** |

Note: recently commissioned on Gemini-S (Peter McGregor, priv. comm).

**Table 2.** Instruments discussed at the Durham IFS classified by implementation method.

| Technique | Project |
|---|---|
| Lenslet systems | OSIRIS(Larkin) KYOTO-3D(Sugai) SNIFS(Copin) |
| Fibre-lenslet systems | GMOS(Murray) VIMOS(Lehnert/Bower) FLAMES(Lehnert) PMAS(Roth) CIRPASS(Parry) FALCON(Puech) |
| Slicer systems | SINFONI(Lehnert), GNIRS(Allington-Smith) MOMFIS(Prieto) FISICA(Eikenberry) SWIFT(Thatte) UIST(Ramsey-Howat) SNAP(Tilquin/Aumenier) FRIDA(Cuevas) NIRSPEC(Prieto) |
| Others | VIRUS(Hill), TF(Bland-Hawthorn), MEIFU(Content/Woodgate) |

Results from many of these instruments are described in these proceedings and the instruments themselves are described in the proceedings of the Durham conference with first authors as shown in parenthesis in Table 2 (except Lehnert and Thatte and Prieto for MOMFIS), so only general remarks will be given here.

Firstly, there are many projects: the value of IFS has at last been generally realised. Secondly, an examination of the literature, especially of these proceedings, shows that these have been used to generate exceptional astrophysical results: IFS is not a specialised technique but necessary for addressing major astrophysical questions. Thirdly, the methods employed cover almost the full range of techniques available. This reflects the comparison of techniques (above) which, aside from the slicer with its technological challenges, indicates little difference in terms of SID, and also that the most applicable FOM is a function of the blend of astrophysical studies to be addressed by a particular instrument, and may well produce different results from the SID used here.

## 5 IFS on Extremely Large Telescopes

One of the drivers for ELTs is resolved spectroscopy of distant stellar systems, exploiting the full diffraction limit of the telescope. The large apertures will allow IFS to reach its full potential by reducing photon-starvation. A number of lines of argument which take into account realistic AO performance and SNR limitations for 30m-class telescopes emphasize the 0.1" scale for spectroscopy but the uncorrected scale of 0.5" may also be relevant if the required developments in AO are not available initially.

These aims also place great emphasis on multiplexed IFS (22; 29) so that the ability of IFS to extract unambiguous and unbiased kinematic information, can be used to survey stellar groups by virial mass rather than light; and to overcome source confusion in dense systems. Another strong driver is that, at the diffraction limit, even modest fields of a few tens of arcminutes contain $10^9 - 10^{11}$ spaxels which cannot be sampled simultaneously. Instead, sub-fields of interest will be selected for high resolution imaging or IFS. This puts an even greater emphasis on performance since the total numbers of spaxels must be greatly increased to provide both a useful field for each IFU (say $(2/0.05)^2 \sim 2000$ spaxels) and a useful number of sub-fields (say 50, requiring $\sim 10^5$ spaxels in total). This favours the slicer approach since it provides the highest information density.

Linking the sub-fields to fixed IFUs or spectrographs requires new technology such as articulated arms (29; 25), or arrays of "beam-steering" mirrors (10). An alternative, already mature, technology is to use fibre-based IFUs with a pick-and-place robot (22).

## 6 Conclusions

This review has summarised the 3D techniques used in optical/infrard astronomy including integral field spectroscopy. The different methods of IFS have been compared using an appropriate figure of merit: the specific information density. While this analysis clearly favours image slicers, implementation requires the mastery of difficult skills. Fortunately good progress has been made. The advantages of the Advanced Image Slicer paradigm have been described. Despite this, the optimum choice of technique for any given astrophysical investigation depends on the details of the investigation and the optimisation of a figure of merit based on signal/noise.

It has been shown that IFS is highly relevant to ELTs, especially when multiplexed, both as a means to fulfill the primary science goals and as a practical means to sample the vast numbers of available spatial elements. Multiplexed-IFS places an even greater emphasis on performance than before to provide a useful multiplex gain without sacrificing the numbers of spaxels within each sub-field. This favours the use of the IFS technique which provides the highest specific information density: the slicer systems, and provides an extra spur to develop the technologies required to produce the optics.

## Acknowledgments

Thanks to Graham Murray, Jürgen Schmoll, Robert Content and Marc Dubbeldam for their huge and successful endeavours in producing the GMOS, IMACS and GNIRS IFUs. We acknowledge the support of the EU via its Framework programme through the Euro3D research training network HPRN-CT-2002-00305. GMOS and GNIRS are run by the Gemini Observatory which is operated by the Association of Universities for Research in Astronomy, Inc., under a cooperative international agreement.

## References

[1] Allington-Smith, J., and Content, R. 1998, PASP, 110, 1216-1234
[2] Allington-Smith, J., Dubbeldam, C.M., Content, R., Dunlop, C., Robertson, D.J., Elias, J, Rodgers, B. and Turner, J.E, 2004. SPIE 5492, 7-1-710
[3] Allington-Smith, J.R., 2006a. in "Integral field Spectroscopy: techniques and data production", New Astronomy Reviews, Elsevier.
[4] Allington-Smith, J.R., 2006b. in "Integral field Spectroscopy: techniques and data production", New Astronomy Reviews, Elsevier.
[5] Allington-Smith, J., Dubbeldam, C.M., Content, R., Robertson, D.J. and Preuss, W. 2006. MNRAS submitted.
[6] Bacon. R., Copin, Y., Monnet, G., Miller, B.M., Allington-Smith, J.R., Bureau, M., Carollo, C.M., Davies, R.L., Emsellem, E., Kuntschner, H., Peletier, R.F., Verolme, E.K. and de Zeeuw, T., 2001. MNRAS 326, 23-35.
[7] Benford, D. et al. 2002, Ninth International Workshop on Low Temperature Detectors. AIP Conference Proceedings, Volume 605, pp. 589-592 (2002).
[8] Content, R., 1997. Proc. SPIE, 2871, 1295
[9] Content, R. 2006. in "Integral field Spectroscopy: techniques and data production" in New Astronomy Reviews, Elsevier
[10] Cunningham, C. et al. 2004. SPIE 5382, 718.
[11] Dubbeldam, C.M, Robertson, D.J. and Preuss, W. 2004. SPIE 5494, 163
[12] Dubbeldam, 2006. in "Integral field Spectroscopy: techniques and data production" in New Astronomy Reviews, Elsevier
[13] García-Lorenzo, B., Mediavilla, E., Arribas, S., 1999. ApJ 518, 190.
[14] Gerssen, J., Allington-Smith, J., Miller, B.W., Turner, J. E. H., Walker, A. 2006. MNRAS 365, 29.
[15] Hanf, M., Kurth, S., Billep, D., Hahn, R., Faust, W., Heinz, S., Dötzel, W. and Gessner, 2004. SPIE, Vol 5445, pp. 128-131
[16] Hirschfeld, T. and Wyntjes, G., 1973. Appl. Opt. 12, 2876.
[17] Kauppinen, J., Partanen, J. Fourier Transforms in Spectroscopy pp. 271. ISBN 3-527-40289-6. Wiley-VCH, March 2001.
[18] Kulkarni, V., Woodgate, B., York, D., Thatte, D., Meiring, J., Palunas, P. and Wassell, E., 2006. ApJ, 636, 30
[19] Larkin, J. et al. 2006. in "Integral field Spectroscopy: techniques and data production" in New Astronomy Reviews, Elsevier
[20] Laurent, F. et al. 2006. in "Integral field Spectroscopy: techniques and data production" in New Astronomy Reviews, Elsevier
[21] Mazin, B., Day, P., LeDuc, H., Vayonakis, A., Zmuidzinas, J. 2002. SPIE 4849, 283
[22] Pasquini, L., Avila, G., Allaert, E., Ballester, P., Biereichel, P., Buzzoni, B., Cavadore, C., Dekker, H., Delabre, B, Ferraro, F., Hill, V.,

Kaufer, A., Kotzlowski, H., Lizon, J-L., Longinotti, A., Moureau, S. and Palsa, R.and Zaggia A., 2000. SPIE 4008, 129
[23] Preuss, W., 2006. in "Integral field Spectroscopy: techniques and data production" in New Astronomy Reviews, Elsevier
[24] Reynolds, A., Ramsay, G., de Bruijne, J., Perryman, M., Cropper, M., Bridge, C., and Peacock, A. 2005 AA, 435, 225
[25] Schmoll et al. 2006a. in "Integral field Spectroscopy: techniques and data production" in New Astronomy Reviews, Elsevier
[26] Schmoll et al. 2006b. in "Integral field Spectroscopy: techniques and data production" in New Astronomy Reviews, Elsevier
[27] Sharples, R. et al. 2004 SPIE, 5492 1179.
[28] Sharples R.M., 2006. in "Integral field Spectroscopy: techniques and data production" in New Astronomy Reviews, Elsevier
[29] Sharples R.M., 2006. in "Integral field Spectroscopy: techniques and data production" in New Astronomy Reviews, Elsevier
[30] Swinbank, A. M., Smith, J., Bower, R. G., Bunker, A., Smail, I., Ellis, R. S., Smith, Graham P., Kneib, J.-P., Sullivan, M., Allington-Smith, J., 2003. ApJ 598, 162.
[31] Thurman S. and Fenup J., 2005. Optics Express 13, 2160.
[32] Weitzel, L., Krabbe, A., Kroker, H., Thatte, N., Tacconi-Garman, L. E., Cameron, M., and Genzel, R. 1996. AA. Suppl., 119, p 531.

# 3D-NTT: A New Instrument for the NTT Based on Versatile Tunable Filter Technology

M. Marcelin[1] and the 3D-NTT collaboration[2]

[1] OAMP/LAM 2 place Le Verrier 13248 MARSEILLE cedex4 FRANCE
michel.marcelin@oamp.fr

[2] MARSEILLE: (LAM)Amram, Balard, Boissin, Boselli, Boulesteix, Buat, Cuby, Epinat, Gach, Joulié, Russeil, Surace; PARIS: (GEPI) Balkowski, Flores, Garrido, Hammer, Jagourel, Jégouzo, Proust, Sayède; CANADA: (LAE Montréal) Carignan, Chemin, Daigle, de Denus Baillargeon, Hernandez; AUSTRALIA: (AAO) Bland-Hawthorn, Jones; AUSTRIA: (Vienna) Zeilinger; BRASIL: (Sao Paulo) Fuentes-Carrera, Mendes de Oliveira & (Santa Cruz) Plana; DENMARK: (Copenhagen) Fynbo; FINLAND: (Ulu) Laurikainen; FRANCE: (CEA) Duc & (Strasbourg) Vollmer; GB: (Keele) Van Loon; GERMANY: (ESO) Pompei & (Bochum) Bomans, Dettmar; ITALY: (Padova) Bettoni, Buson, Giro, Marziani, Mazzei, Rampazzo; SPAIN: (IAA, Granada) Masegosa, Verdes Montenegro, Vilchez & (IAC, Canarias) Cepa; SWEDEN: (Uppsala) Bergvall, Östlin, Marquart; USA: (Tuscaloosa, Alabama) Sulentic

**Summary.** The 3D-NTT will be a visitor instrument for the NTT, built by GEPI (Paris) and LAM (Marseille) with the collaboration of LAE (Montréal, Canada) and AAO (Australia). It is a spectro-imager offering two modes: a low resolution mode (100-5000) with a Tunable Filter, and a high resolution mode (5000 - 40 000) with a standard scanning Fabry-Perot. A large variety of programmes may be led with such an instrument as has been shown recently (1997-2003) with the Taurus Tunable Filter on the AAT and WHT. In the frame of a large scientific coollaboration, gathering European teams as well as collaborators from other countries, we propose a Large Programme (IOGA) dedicated to the study of ionized gas in galaxies, at low and high z, to be undertaken with the 3D-NTT. Nearby IR galaxies (IRGs) are the key to understanding the formation and evolution of galaxies, as they are believed to be the local counter-part of the numerous luminous starburst galaxies at high z. To understand the nature, origin and evolution of IRGs, we propose to map the star formation and kinematics in 500 galaxies and obtain dust extinction, metal abundances and electron density maps in 50 of them for obtaining a reference sample with high spectral and spatial resolution (sub-kpc) to be compared with distant objects being observed on the VLT (GIRAFFE, SINFONI). We will take advantage of both modes proposed by the instrument: high resolution mode with scanning Fabry Perot (FOV 5.5' or 11'), and low resolution mode with Tunable Filter and larger field (FOV 20'). This mode will be used at high z to look for star forming galaxies around quasars with a range of intrinsic UV luminosity. Star forming galaxies exist in significant numbers around low power quasars and we want to check if this remains true around more distant and luminous quasars. This may be the first evidence that powerful UV fields can suppress or delay widespread star formation in galaxies and, as such, may provide important constraints on cosmological models.

# 1 The Instrument

The instrument will be a focal reducer attached at the Nasmyth focus of the NTT. It will offer a 20 arcmin field of view and two observing modes in the visible range.

## 1.1 The Low Resolution Mode

This mode will use a Tunable Filter (a Fabry-Perot with a low interference order) for which the transmitted bandwidth can be adjusted between 5 and 50Åand the central wavelength chosen everywhere in the visible domain.

This Tunable Filter is being built as a cooperative effort between the Laboratoire d'Astrophysique de Marseille, the Fresnel Institute and local private companies. The main difference, compared with Tunable Filters commercially available, is that it will cover a wide range of interference orders. This will be achieved thanks to long excursion piezo-actuators (400 $\mu$m), so that the same device can be used as a Tunable Filter or as a standard scanning Fabry-Perot. The spectral resolution will be adjustable between 100 and 5000 in the Low Resolution mode.

The detector will be a thin blue sensitive CCD, able to reach the [O II] line at 3727Å(interesting for metallicity measurements).

## 1.2 Comparison with other TF instruments

Table 1 summarizes the main characteristics of some TF instruments, compared with the 3D-NTT. Our instrument offers a large field of view, just behind the MMTF. However, the number of nights devoted to the TF mode will be larger and the [O II] line at 3727Åwill be seen by the 3D-NTT, which is not the case for the MMTF. OSIRIS will offer a large number of nights with a larger telescope but with a much smaller field, furthermore most of the observing time in the first few years will be devoted to the OTELO survey whose science goals do not overlap with ours.

**Table 1.** Comparison with other TF instruments

| Instrument | Telescope | Year | Diameter | FOV | nights/year |
|---|---|---|---|---|---|
| TTF | AAT/WHT | 1996-2003 | 3.9m | 9 arcmin | 50 |
| MMTF | Magellan | Dec. 2005 | 6.5m | 27 arcmin | 7 |
| OSIRIS | Grantecan | June 2006 | 10m | 6.7 arcmin | 100 (OTELO) |
| 3D-NTT | NTT | End 2008 | 3.5m | 20 arcmin | 37 (IOGA) |

A key point is that all of these TF instruments suffer from phase shift effects (since they are placed in the pupil) except for the 3D-NTT because our TF will be placed in the focal plane. This can be done since our focal reducer is a 2-stage reducer, with a first stage reducing the original aperture ratio of the Nasmyth focus of f/11 to f/7 so that the Tunable Filter can cover a 20 arcmin FOV with a reasonable size (100 mm diameter). The 2nd stage of the focal reducer offers a collimated beam in which is placed a standard scanning Fabry-Perot, used for the high resolution mode.

An example of monochromatic images of a galaxy obtained with a Tunable Filter (the TTF) can be seen in Fig.1 (left).

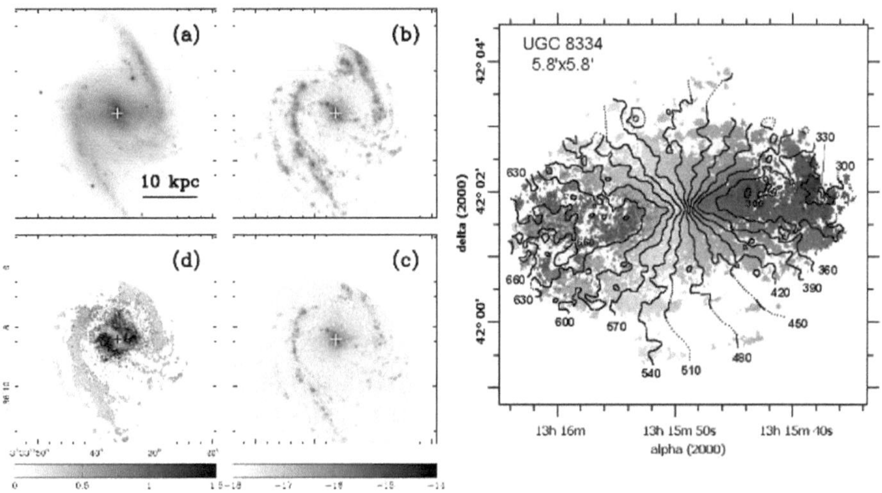


**Fig. 1.** Left(a to d): Red continuum, Hα, [N II] and [N II]/Hα images of the galaxy NGC 1365 obtained with the TTF (1). Right: Velocity field of the galaxy NGC 5055 (M63) observed with GHASP (2).

## 1.3 The High Resolution Mode

This mode will use a scanning Fabry-Perot in the collimated beam of the 3D-NTT focal reducer. The Tunable Filter will then be used as an interference filter selecting the spectral line to be scanned. Both interferometers will be scanned simultaneously, insuring that the line remains perfectly centered on the maximum of the transmission function of the whole device.

The spectral resolution will be adjustable between 5000 and 40000, well suited for mapping velocity fields (see Fig.1, right).

The detector for that mode will be a photon counting system with GaAs photocathode offering a high quantum efficiency. Because of the scanning process, such an Imaging Photon Counting System is more efficient than a CCD (3), especially when sky is veiled by cirrus clouds.

**Table 2.** Main characteristics of the two modes

| Field of view | Wavelength range | Interf. order | Detector | Pixel size |
|---|---|---|---|---|
| LR MODE | | | | |
| 20' x 20' | 350 to 850nm | 4 to 100 at Hα | CCD (4k x 4k with 15μm pix) | 0.29"/pix |
| HR MODE | | | | |
| 11' x 11' | 500nm to 800nm | 100 to 1200 at Hα | IPCS GaAs (1k x 1k with 17μm pix) | 0.64"/pix |
| or 5.5' x 5.5' | | | | or 0.32"/pix |

# 2 The Science

## 2.1 The Large Programme IOGA

IOGA is the acronym for: IOnized gas in star forming Galaxies through the Ages. One part of IOGA will be devoted to a survey of low z galaxies, observed with the 2 modes of the 3D-NTT, and a second part will be devoted to a survey of quasars around z=2, observed with the TF mode only.

- **First part of IOGA**
  It will be a survey of 500 IR galaxies, with the aim of :
  - mapping the SFR and kinematics (HR mode) for all of them
  - mapping the dust extinction, metal abundances and electron density for 50 LIRGS (LR mode)
  IOGA will provide a reference sample with high spectral and spatial resolution (sub-kpc) to be compared with distant objects observed with the VLT (GIRAFFE, SINFONI).
  The 500 galaxies will be selected from the QMW IRAS galaxy catalogue (4) up to a redshift 0.06. Most of them are IRGs (LIR $> 10^{10}$ L$\odot$) and some of them are LIRGs (LIR $> 10^{11}$ L$\odot$)
  The number of 500 is explained by statistical requirements: 5 mag bins x 5 Hubble types x 10 galaxies x 2 (bar and non-bar)= 500.
  The 50 galaxies that will be mapped in TF mode are LIRGs
- **Second part of IOGA**
  It will be a survey of a sample of 50 quasars at z ~2. The aim of this survey will be to look for ELGs around distant and luminous quasars. Although Lyman alpha emitters are common around nearby quasars (5), recent observations with the TTF suggest that it could be different around more distant and bright quasars (6). Indeed, powerful UV fields could suppress or delay widespread star formation in galaxies. This needs to be checked since it would put constraints on cosmological models. As a byproduct, we will be able to look for the Lyman alpha ionization cone around the observed quasars (7).

### 2.2 Other programmes

Among the many possible observing programmes that can be performed with such an instrument, one can mention the following:

- **Follow up of other surveys at different wavelengths** like:
  Line maps (SINGS/Spitzer, AMIGA).
  Accretion and galaxy evolution in poor groups.
- **Nearby galaxies**
  Low-z super-wind galaxies and shocked gas.
  Galactic winds of dwarf irregular galaxies.
  Mapping the large scale potential wells of galaxies using PN.
  Search for warm ionized gas around nearby radio galaxies.
  Identification and Mapping of star forming galaxies.
- **AGN**
  Seyfert 2 and LINERS as well as filamentary ionized structures in clusters.
  Study of ionization cones and unification models of AGN.
- **Clusters**
  Ionized gas in intracluster medium.
- **Distant galaxies**
  Search for high-redshift gravitationally lensed galaxies.

### 2.3 Data reduction and Data Base

The reduction of observations in Tunable Filter mode will benefit from the experience of the TTF (8).

High Resolution mode observations will be reduced with the ADHOCw software (downloadable from the Web at http://www-obs.cnrs-mrs.fr/adhoc/adhoc.html). This software has been used extensively for the reduction of the GHASP survey of velocity fields of 200 nearby galaxies (2).

A Data Base for Fabry-Perot observations is being prepared (http://fpdatabase.obspm.fr/) and will be used for the 3D-NTT data. The data will be available for public access one year after the observations.

## References

[1] S. Veilleux, P.L. Shopbell, D.S. Rupke et al: AJ **126**, 2185 (2003).
[2] O. Garrido, M. Marcelin, P. Amram et al: MNRAS, **362**, 127 (2005).
[3] J-L. Gach, O. Hernandez, J. Boulesteix et al: PASP **114**, 1043 (2002).
[4] M. Rowan-Robinson, W. Saunders, A. Lawrence, K. Leech: MNRAS **253**, 485 (1991).

[5] J.M. Barr, J.C. Baker, M.N. Bremer et al: AJ **128**, 2660 (2004).
[6] P.J. Francis and J. Bland-Hawthorn: MNRAS **353**, 301 (2004).
[7] M. Weidinger, P. Møller, J.P.U. Fynbo: Nature **430**, 999 (2004).
[8] D.H. Jones, P.L. Shopbell, J. Bland-Hawthorn: MNRAS **329**, 759 (2002).

# The Integral Field Unit of the Near Infrared Spectrograph for JWST

S. Arribas[1-3], P. Ferruit[4], P. Jakobsen[3], T. Boeker, A. Bunker, S. Charlot, D. Crampton, M. Franx, M. García-Marín, R. Maiolino, G. de Marchi, H. Moseley, B. Rauscher, M. Regan, H-W. Rix and J. Valenti

[1] Space Telescope Science Institute, Baltimore, USA arribas@stsci.edu
[2] Instituto de Astrofsica de Canarias, La Laguna, and CSIC, Spain,
[3] European Space Agency, ESTEC, The Netherlands, pjakobsen@rssd.esa.int
[4] Centre de Recherche Astronomique de Lyon, France, ferruit@cral.fr

**Summary.** The Near Infrared Spectrograph (NIRSpec) on board of the James Webb Space Telescope (JWST) will be equipped with an Integral Field Unit (IFU) which will allow the two-dimensional spectral characterization of astronomical objects with unprecedented depths, especially in the 2-5 micron wavelength range. In particular, NIRSpec-IFU observations of high-z galaxy populations will permit the capture of their often complex two-dimensional physical and kinematical structure, a pre-requisite for carrying out a number of studies. These types of observations will also help to understand possible observational biases of large 1D (i.e. slit) surveys to be carried out in multi-object mode. Here we describe the main characteristics of the NIRSpec-IFU, as well as its expected sensitivity and performance for observations of high-z galaxy populations.

## 1 Introduction

The James Webb Space Telescope (JWST) is a collaborative effort between NASA, the European Space Agency (ESA) and the Canadian Space Agency (CSA) to develop a large ($\sim$ 6.5 m), near- and mid-infrared (0.6-28 $\mu$m) optimized space telescope. The scientific objectives of JWST fall into four main themes (4): i) The End of the Dark Ages: First Light and Reionization, ii) The Assembly of Galaxies, iii) The Birth of Stars and Protoplanetary Systems, and iv) Planetary Systems and the Origins of Life.

To reach its science goals JWST will be equipped with four instruments: FGS, the guider camera (equipped with a Tunable-Filter capability operating between 0.8 and 5 $\mu$m); NIRCam, a near-infrared (0.6-5 $\mu$m) camera; MIRI, a mid-infrared (5-28 $\mu$m) instrument (camera and spectrograph); and NIRSpec, a near-infrared (0.6-5 $\mu$m) spectrograph.

NIRSpec is a complex instrument able to operate at three spectral resolutions (R=100, 1000, 2700) and several science modes. A Multi-Object Spectroscopic (MOS) mode will allow to obtain the spectra of a large number ($>$ 100) of objects in one single exposure. The target selection over a 3.6' x 3.4' field of view will be performed using a fully-addressable array of micro-shutters developed by NASA at the Goddard Space Flight Center. In addition to the MOS mode, NIRSpec will have an Integral Field Spectroscopic mode (see next section) and a set of five classical slits. Figure 1 shows the location of the different input apertures at the (*slit*) focal plane. NIRSpec is under the responsibility of ESA and it is developed by EADS-Astrium.

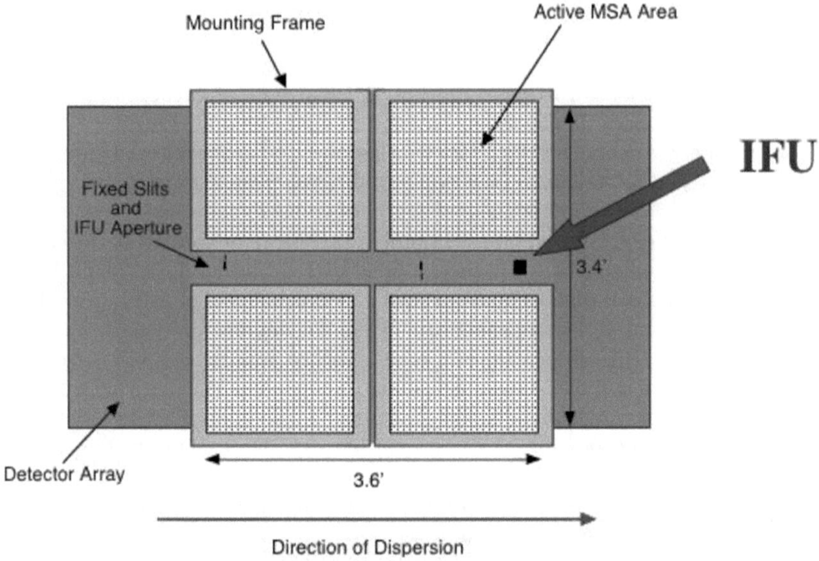


**Fig. 1.** Input apertures at the slit focal plane of NIRSpec. Note that the optics of the IFU transforms the input square FoV (marked in the figure) into a pseudo-slit.

## 2 The Integral Field Unit of JWST-NIRSpec

### 2.1 Science Motivation

A major theme for the NIRSPec-IFU will be the exploration of the physical and kinematic structures of high-z galaxy populations in the early universe. The fact that important optical diagnostic features will shift into the NIRSpec spectral domain at high-z , will allow to extend current observations of local galaxies up to early galaxies populations. This gives the possibility to follow, observationally, the evolution of fundamental galaxy properties along cosmic time and, therefore, strongly constrain models of formation and evolution of galaxies. The two-dimensional spectral information provided by NIRSpec-IFU is particularly important because high-z galaxies often show complex 2D structures (differential extinction effects, decoupled ionized and stellar structures, very complex velocity fields due to interactions and mergers, etc), and the sole use of slit spectroscopy can produce misleading results and/or difficult interpretation (see (2), and references therein).

NIRSpec-IFU observations will allow to evaluate the importance of orbital ordered motions (i.e. *rotation*) and to determine dynamical masses from spatially resolved kinematics for those systems dominated by rotation. The evolution of dynamical masses with redshift is an important observational information to constrain models, and complement the methodologically different estimates of stellar masses based on stellar population models. Similarly, IFS data will permit to directly determine if the ionized gas and stellar components are or not coupled and, therefore, infer more accurate star formation rates (SFR). Other related studies requiring two-dimensional spectral information include the study of large scale outflow motions and their role in the enrichment of the intergalactic medium, and the study of the two-dimensional stellar populations of galaxies as a tool to understand their past history. All these NIRSpec-IFU observations will also help to understand possible observational biases of large 1D (i.e. slit) surveys to be carried out in multi-object mode using the Micro-Shutter Array.

Other many programs including a number of *galactic* studies (e.g. characterization of proto-planetary disks, PNs, etc) will benefit from this capability. Note that the performance of this IFU will be unique, especially in the spectral range 2-5 $\mu$m (see below).

## 2.2 Basic Characteristics and Requirements

The main characteristics of the integral field unit of NIRSpec are summarized in Table 1. This table is mainly based on technical requirements. This unit will be developed by the industrial company SIRA, and manufactured by the Center for Advance Instrumentation, Durham, UK.

**Table 1.** Main characteristics and requirements of the NIRSpec Integral Field Unit

| | |
|---|---|
| Approach | Advance Image Slicer |
| Number of Slices | 30 |
| Field of View | 3" $\times$ 3" |
| Sampling / Spaxel | 0.1" $\times$ 0.1" |
| Spectral performance | All NIRSpec spectral configurations |
| Throughput | $>$ 50 per cent |
| Cross-talk | $<$ 5 per cent |
| Diffraction losses | $<$ 15 per cent |
| Optical quality | OTE + NIRSpec + IFU diffraction-limited at 3.4 $\mu$m |
| Mechanics | Entirely passive (no extra moving parts) |
| Operations | Point and shoot |

### 2.3 Expected performance

#### Sensitivity

In figure 2 we present the expected sensitivity curves of the NIRSpec - IFU for a point source at a resolution of R=100 (continuum), and R=2700 (unresolved emission lines). The case R=1000 is not shown but, for the case of unresolved lines, follows close to R=2700. These curves have been obtained assuming the IFU requirements.

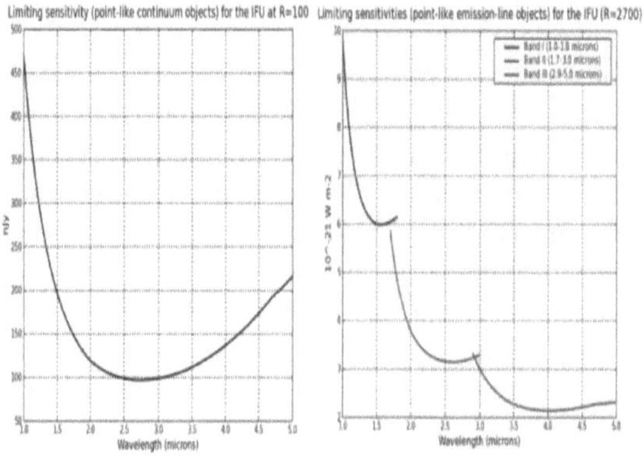


**Fig. 2.** Expected sensitivity limit curve for NIRSpec-IFU observations of a point source at R=100, and a S/N=10 in $10^4$ sec.

#### Comparison with ground-based IFUs

To evaluate to full potential of this IFU, apart from the sensitivity curves shown above, it is important to bear in mind that JWST-NIRSpec will cover the full range 0.6-5 $\mu$m continuously, and will be affected by a very low level of background. In fact, at difference of ground observations, NIRSPec will be dominated by detector noise in most of the cases. Therefore, though in the range 0.6-2.2 $\mu$m NIRSPec-IFU may be rivaled by other ground instruments attached to larger aperture telescopes (assuming observations through the atmospheric transmission windows and between the OH emission lines), this unit is expected to have a unique performance in the range 2.2-5 $\mu$m. Note that, in the case of 2D spectroscopy, the observation between the telluric emission lines for the whole FoV, may be more difficult if the object is affected by a strong velocity field.

#### Simulated Observations

In order to have a more direct idea of its capabilities, Figure 3 gives the expected exposure times when observing with NIRSpec-IFU a sample of Lyman

Break galaxies at $< z >\sim 2.3$ (black symbols). These galaxies have been already observed by (1) and references therein, with Keck and VLT. The exposure times assume the expected sensitivity of the instrument (3), and were computed at the full spatial and spectral resolution of the IFU (i.e. 0.1", R=2700), for an average S/N of 10 within the half-light area in H$\alpha$. In this plot we can see that these objects can be observed with exposure times ranging between a few minutes and a few hours. Then the observed properties of these galaxies were projected at higher redshift ($< z >\sim 4$ and 6) and the exposure times estimated for the conditions mentioned above. No evolution in galaxy size or surface brightness was taken into account, which led to conservative estimates for the observing times. The main conclusion of this plot is that NIRSpec-IFU will be able to observe, at its full spectral and spatial resolution, objects similar to the star forming galaxies considered here all the way down up to z=6, or even larger.

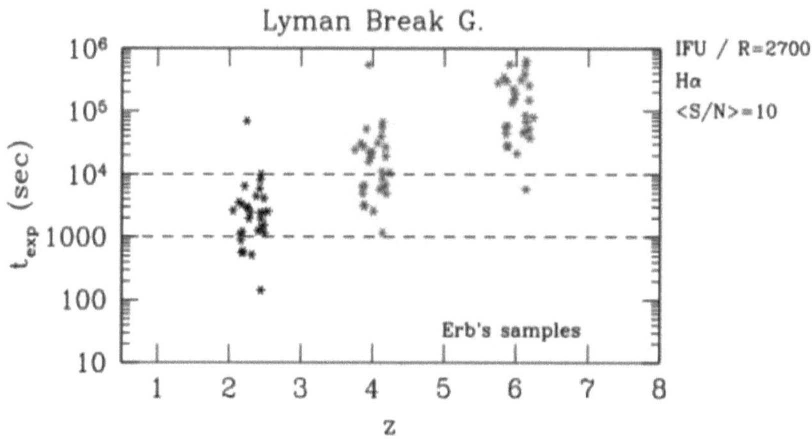


**Fig. 3.** Estimated exposure times when observing LBGs at different redshifts (see text).

Figure 4 shows another type of simulated observation for a more massive star forming galaxy class, the ultraluminous infrared galaxies (ULIRGs). Here we have made use of actual integral field observations at optical wavelengths of IRAS 15206+3342 carried out with the INTEGRAL system at the 4.2 m. WHT. The actual angular size and surface brightness distribution of the H$\alpha$ map (top panel) have been modified to the corresponding values if this object were located at z=2 (bottom panel). Making use of the expected sensitivity curves and tables for the instrument (3), the new surface brightness in the different parts of the image can be translated into S/N for a given exposure time. The case presented here corresponds to a spectral resolution R=2700. An exposure time of $10^4$ sec. leads to a S/N larger than 30 in the inner regions. The color palette also indicates the S/N in other parts of the image.

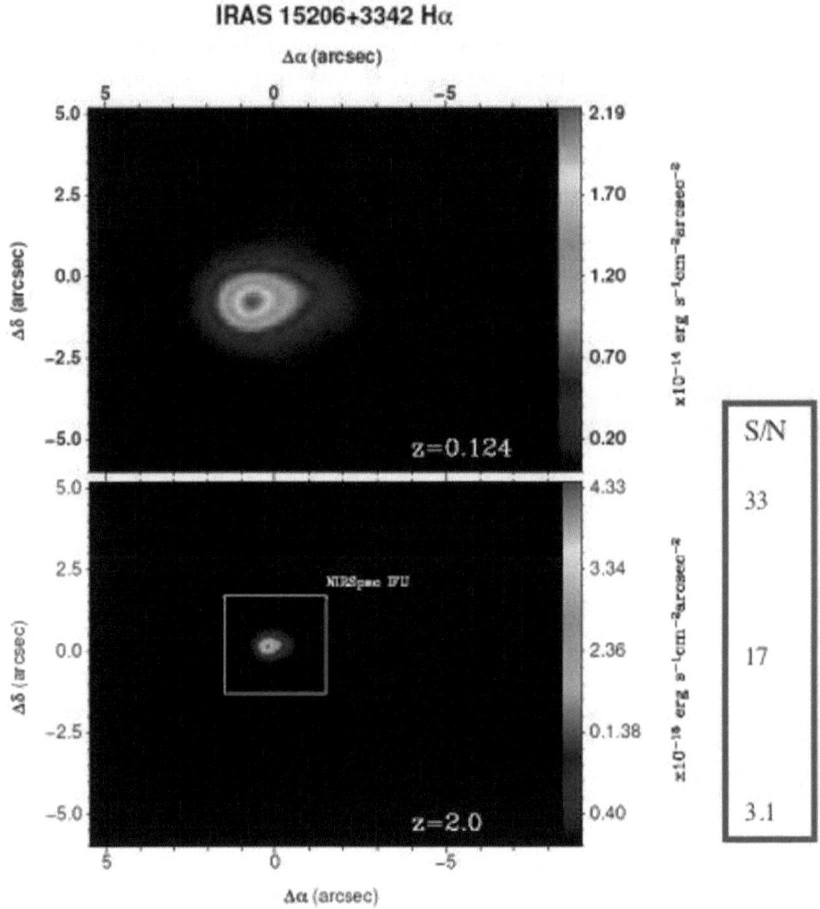


**Fig. 4.** Simulated observations of IRAS 15206+3342 in Hα, artificially shifted up to z=2. (see text)

## References

[1] D.K. Erb et al: ApJ. **612**, 122 (2004)
[2] M. García-Marín et al. , these proceedings (2006)
[3] P. Jakobsen: NIRSpec sensitivity tables. (Priv. Communication, 2005)
[4] J. Mather et al.: JWST Science Requirements Document. JWST-RQMT-002558 (2005)

# Scientific Prospects of the 3D-NTT Instrument

D. J. Bomans[1] and the 3D-NTT collaboration[2]

[1] Astronomical Institute of the Ruhr-University Bochum, Universitätsstr. 150, D-44780 Bochum, Germany
bomans@astro.rub.de

[2] M. Marcelin[2], P. Amram[2] , C. Balkowski[3], J. Bland-Hawthorn[4], A. Boselli[2], J. Boulesteix[2], V. Buat[2], C. Carignan[5], R.-J. Dettmar[1], P.A. Duc[3], H. Flores[3], F. Hammer[3], H. Jones[4], E. Laurikainen[6], C. Mendes de Oliveira[7], G. Östlin[8], E. Pompei[9], R. Rampazzo[10], J. Sulentic[11], L. Verdes Montenegro[12], W. Zeilinger[13], N. Bergvall[14], D. Bettoni[10], J. Cepa[15], D. Buson[14], J. Fynbo[16], P. Marziani[10], J. Masegosa[12], P. Mazzei[10], H. Plana[7], D. Russeil[2], J. Van Loon[17], J. Vilchez[15], B. Vollmer[18], L. Chemin[5], O. Daigle[1,5], M.-M. de Denus Baillargeon[5], B. Epinat[2,3], I. Fuentes Carrera[19], O. Garrido[2,3], O. Hernandez[5], T. Marquart[14], J. van Eymeren[1]
1) Ruhr-Univ. Bochum, 2) Marseille Obs., 3) Paris Obs., 4) AAO, 5) LAE Montreal, 6) Oulu Univ., 7) Sao Paulo Univ., 8) Stockholm Univ., 9) ESO, 10) Padova Obs., 11) Univ. Alabama, 12) IAA Granada, 13) Vienna Univ., 14) Uppsala Univ., 15) IAC La Laguna, 16) Copenhagen Univ., 17) Keele Univ., 18) Strasbourg Univ., 19) UNAM Mexico City

## 1 3D-NTT

We propose to develop an imaging spectrometer (3D-NTT) which is well optimized for the detection and kinematical study of the ionized gas in galaxies with the ESO/NTT. We plan to have a low resolution mode, with a Tunable Filter for detection of sources and large-scale mapping in a 20' field (with 0.29" pixels) and CCD detector, and a high resolution mode, with scanning Fabry-Perot for obtaining detailed velocity fields over a smaller field (11' with 0.64" pixels or 5.5' with 0.32" pixels) with a new photon-counting detector. For the high-redshift studies and for abundance work at low redshift (needing the [OII] line at 3727Å), the 3D-NTT will operate with a greatly improved blue sensitivity compared to previous Tunable Filter/Fabry-Perot instruments. This instrument will exploit the excellent seeing and photometric conditions offered by the NTT.

A more detailed description of 3D-NTT, a comparison of 3D-NTT to other instruments, and our project plans are described in more detail in Marcelin et al. (this volume).

## 2 Science Program

With this instrument we propose a large program, called IOGA (IOnized gas in starforming Galaxies through the Ages), defining a core science framework for the 3D-NTT instrument.

### 2.1 Large Project part A: Local IR galaxies

This part of IOGA would use both Fabry-Perot and Tunable Filter mode of 3D-NTT. We plan to observe $\sim$ 500 galaxies up to a redshift z$=$ 0.06 selected from the QMW IRAS galaxy catalog in order to study the kinematics of LIRGs and IRGs ($L_{ir} > 10^{10}$ $L_{sol}$) in the nearby universe (with the high resolution Fabry-Perot mode) and to identify the physical processes that lead to intense star formation. This will also provide the local comparison sample for high redshift starburst galaxies.

### 2.2 Large Project part B: Radiative feedback of QSO

This part of IOGA would use the Tunable Filter mode. We plan to look for star forming galaxies around a sample of 45 quasars at z$\sim$ 2. The main goal will be to test the dependency of the luminosity of the quasars on the presence or absence of wide spread star formation around them, and therefore radiative feedback very similar to the epoch of galaxy formation.

### 2.3 GO-type Programs

Furthermore, the unique capabilities of 3D-NTT will provide possibilities for a wide variety of highly competitive GO type programs. For example: Identification and mapping of star forming galaxies, Seyfert 2 and LINERS as well as filamentary ionized structures in clusters; Study of low-z super-wind galaxies and of shocked gas; Mapping the large scale potential wells of galaxy using planetary nebulae; Study of ionization cones and unification models of AGN; Search for high-redshift gravitationally lensed galaxies; Search for warm ionized gas around nearby radio galaxies; Study of halo gas in dwarf galaxies and massive spiral galaxies.

## 3 Conclusions

3D-NTT would provide a unique survey instrument for many areas of astrophysics, optimally suited to produce local samples for the comparison of high-z kinematic and star formation studies, and study emission line objects at high-z in its own right. Many more projects, especially concerning the better understanding of the kinematics and the feedback in local galaxies, and all kinds of galactic and extragalactic emission line objects would also be possible. We only could give a (certainly personally biased) collection of science prospects for 3D-NTT.

# Handling IFU Datasets in the Virtual Observatory

I. Chilingarian[1,2], F. Bonnarel[3], and M. Louys[3] and J. McDowell[4]

[1] CRAL Observatoire de Lyon, France
[2] Sternberg Astronomical Institute, Russia
[3] CDS Observatoire de Strasbourg, France
[4] Harvard-Smithsonian Center for Astrophysics, USA

**Summary.** We propose a formalised description of IFU datasets sufficient for providing data discovery, access and processing in the Virtual Observatory framework.

**Key words:** virtual observatory, data archives, data modelling, IFU data

## 1 Characterisation Data Model of IVOA

The Virtual Observatory is a joint effort of astronomers and computer scientists all over the world, intended to provide easy and effective access to the astronomical data and software tools and services for processing and analysing them. The key point of the VO development is to achieve interoperability between different archives and tools (see (1)). An important topic related to interoperability is to build a data model for describing any observable astronomical dataset (Observation DM, (2)): standardized self sufficient and self-consistent description of their metadata, that should be sufficient for making any sort of data analysis. This work is being carried out by the IVOA (International Virtual Observatory Alliance, (1)) Data Models working group.

One of the important and basic subclasses of Observation DM is Characterisation DM, giving physical insight to the dataset. This data model has a multi-layered structure, describing each of the following axes: spatial, temporal, spectral, observable (flux). Every layer corresponds to increasing level of granularity of description. Every axis on every level is characterised by the following properties: Coverage, Resolution, Sampling Precision. The localisation and "volume" in the multidimensional space are given by the Coverage class. The "Resolution" gives the actual extent of each volume element in the parameter space, and finally the "Sampling Precision" class describes how the parameter space is scanned by the data. The rawest level of description provides only a single approximate position in the parameter space. The fourth (maximum) level defines a full sensitivity function, giving the absolute transmission factor for each voxel.

For example, the first level of characterisation for Spatial Coverage ("Location") corresponds to the approximate position of the observation on the sky. The second level ("Bounds") is a bounding box. The third level ("Support") tells whether the area, covered by the dataset is contiguous and provides the precise boundaries as a polygon (or union of polygons). We refer to the detailed description of the Characterisation Data Model given in the IVOA Note (3).

## 2 Characterising IFU datasets

The Characterisation Data Model has enough capabilities for describing complex datasets such as IFU data. We propose to limit the description of the datasets with the second level of granularity (Bounds) due to the complexity of resulting information, and provide further levels for individual spectra (spaxel-related).

The way of computing spatial characterisation metadata for IFU datasets is shown in Fig 1. Complete example of the characterisation metadata for a "live" dataset obtained with the Multi-Pupil Fiber Spectrograph at the 6-m telescope of SAO RAS is given in Appendix A of (3) (XML format) and Appendix B of (3) (VOTable format).

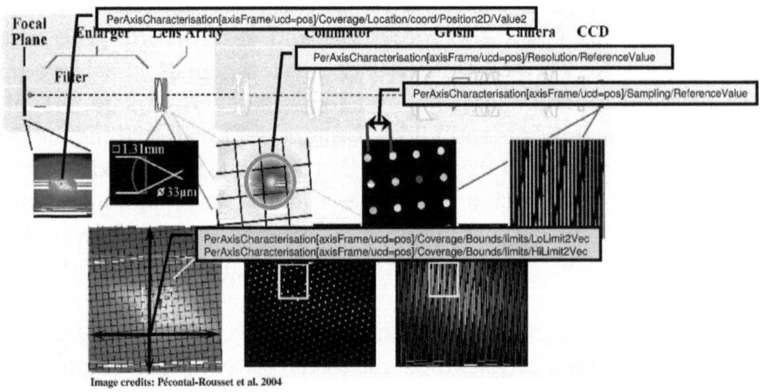


**Fig. 1.** Computing characterisation metadata of the spatial axis for IFU datasets

Euro3D FITS format, developed by the Euro3D consortium (4), provides a structure where the characterisation metadata up-to the second level of granularity could be put. Putting the third and fourth levels requires minimal modifications to the format.

## 3 Accessing IFU datasets

IVOA Simple Spectral Access Protocol (SSAP, (5)) was developed for providing data discovery and access for 1D spectra. However, the present version of it (v0.9) is versatile enough for dealing with almost any sort of astronomical data, thanks to the mechanism of handling any service-specific parameters and referencing to service-specific data model and data format.

We propose to use SSAP for accessing IFU datasets. The first two granularity levels of characterisation, given in the SSAP query response, allow to process elaborated queries for selecting the datasets that astronomer is interested in. The data including characterisation metadata can be delivered in the Euro3D format.

## 4 Summary

Presently all the necessary infrastructural components exist for publishing IFU data in the Virtual Observatory: building VO-compliant science-ready 3D spectroscopic archives. First such archives are expected to appear in the first semester of 2006.

Characterisation metadata on the highest granularity level provides a possibility to carry out almost any, even very elaborated sort of data analysis, such as studies of internal kinematics and stellar populations in galaxies. Thus, defining the data model for 3D spectroscopic datasets is a crucial point toward development of VO services for scientific data analysis (virtual instrumentation).

## 5 Acknowledgements

We appreciate the organizing committee support provided to IC for attending the workshop, and support provided by VO-France. IC is supported by the INTAS PhD fellowship 04-83-3618. We also thank IVOA Data Modelling working group for fruitful discussions concerning characterisation of 3D datasets.

## References

[1] International Virtual Observatory Alliance web-site: `http://www.ivoa.net/`
[2] Observation Data Model; J. McDowell et al., 2004, IVOA DM WG Internal Working Draft
[3] IVOA Data Model Working Group note on the Characterisation Data Model, made on 2005/11/29, editors: J. McDowell et al.
[4] M. Kissler-Patig et al., AN, 2004, v.325, p.159
[5] SSAP v. 0.90; M. Dolensky et al., 2005, IVOA Working Draft

# Simulations of High-$z$ Starburst Galaxies for the JWST-MIRI Integral Field Unit

M. García-Marín[1], N. P. F. Lorente[2], A. Glasse[3], L. Colina[4] and G. Wright[5]

[1] DAMIR/IEM/CSIC maca@damir.iem.csic.es
[2] UK-ATC npfl@roe.ac.uk
[3] UK-ATC achg@roe.ac.uk
[4] DAMIR/IEM/CSIC colina@damir.iem.csic.es
[5] UK-ATC gsw@roe.ac.uk

**Summary.** We present the first realistic simulations of how high-redshift star forming galaxies will be observed by MIRI (Mid IR Instrument). The simulations presented have been developed using the new tool Specsim, and are part of an ongoing project aimed at studying a representative sample of Ultraluminous Infrared Galaxies (ULIRGs) with integral field spectroscopy (IFS) data.

## 1 MIRI: The Mid-IR instrument for the JWST

MIRI is the Mid-IR instrument for the James Webb Space Telescope (JWST). With an operational range between 5 and 28 $\mu$m, it will be crucial for achieving the scientific goals of the JWST. Here we will focus our attention on the MIRI-MRS (Medium Resolution spectrometer) (6), formed by four individual integral field units (IFU, Fig. 1 Left) with concentric FoV on the sky, whose main characteristics are detailed in Table 1. The detection limit for the MIRI-MRS (10-$\sigma$ in $10^4$ sec) will be $1\times10^{-20}$ W m$^{-2}$ at 10 $\mu$m (Fig. 1 Right).

**Table 1.** Summary of MIRI IFU parameters

| IFU | N. Slices | $\lambda_{range}$ ($\mu$m) | R$_{spectral}$ | Slice Width (arcsec) | FoV (arcsec$^2$) |
|---|---|---|---|---|---|
| 1 | 21 | 4.87-7.76 | 2450-3710 | 0.176 | 3.7×3.7 |
| 2 | 17 | 7.45-11.87 | 2480-3690 | 0.277 | 4.7×4.51 |
| 3 | 16 | 11.47-18.24 | 2510-3730 | 0.387 | 6.20×6.13 |
| 4 | 12 | 17.54-28.82 | 2070-2490 | 0.645 | 7.74×7.93 |

## 2 Specsim: The MIRI-MRS simulator

Specsim (4) is an IDL based simulator that generates FITS frames which are a good approximation of the ones that will be taken by the MRS in orbit.

The simulator allows one to model astronomical targets, and creates a data cube with the source, background and noise information (Fig. 2 Left, top). Once the target has been produced, the program simulates the effects of the image slicers, the dispersion, the dichroics, the flat field response and the cosmic ray effect among others. The result is a detector *raw* image (Fig. 2 Left, bottom).

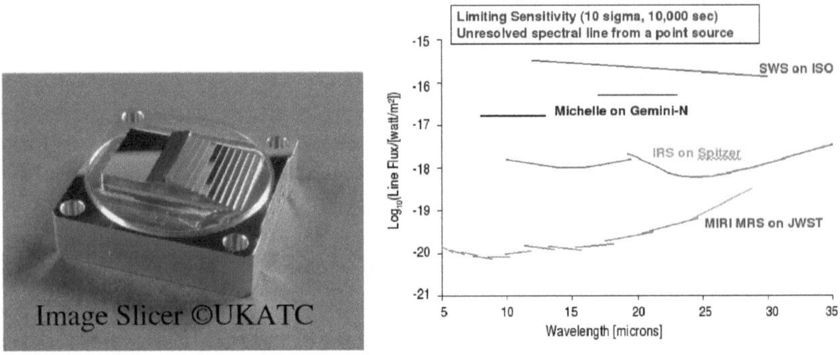


**Fig. 1.** *Left:* Prototype image slicer for MIRI. *Right:* Sensitivity for the MIRI-MRS compared with other instruments.

## 3 Modelling galaxies at high redshift: Technique and results

The simulation of galaxies with Specsim requires three main inputs: the morphology of the source, the spectral model of the continuum and information about the emission lines. In the case of the morphology, the program allows one to model a galaxy using a set of primitives (extended, point sources) for each astronomical target in the FoV. For the continuum definition we have specifically included three options for the galaxies: STARBURST99 models (3), (5), Bruzual & Charlot models (1) and Lagache, Dole & Puget phenomenological model for galaxies (2) that includes PaH lines. Finally, the emission lines are modeled using resolved (Gaussian) or unresolved profiles adapted for each individual case. The emission line information (e.g. EW, star formation rate) can be used for deriving the continuum age and mass.
In this case the goal is to simulate galaxies at high redshift, thus once the continuum and the emission line information have been selected, cosmological effects (i.e. enlargement and dimming of the spectra; $H_0$=70 km $s^{-1}$ $Mpc^{-1}$, $\Omega_\Lambda$=0.7, $\Omega_M$=0.3) need to be applied for each redshift.
The capability of MIRI of detecting the typical emission lines from galaxies has been studied; our preliminary determinations show that for intermediate

redshifts (z=2, 3) a galaxy with a luminosity of L(Pa$\alpha$)=1$\times$10$^{42}$ erg s$^{-1}$ (i.e. SFR$\simeq$60 M$_\odot$ year$^{-1}$) will be detected at 10$\sigma$ in $\sim$3$\times$10$^3$ s. At higher redshifts, like z=7, H$\alpha$ will be one of the most intense lines observable by MIRI, and a galaxy with L(H$\alpha$)=1$\times$10$^{43}$erg s$^{-1}$ (i.e. SFR $\simeq$80 M$_\odot$ year$^{-1}$) will be detected with 10$\sigma$ in $\sim$4$\times$10$^3$ s. Another aspect that we have analysed is how well MIRI will recover the spatial structure of high redshift galaxies (see Fig. 2, Right). As a reference, the IFU1 at 5 $\mu$m will be able to resolve galaxies and regions with intense stellar formation separated by $\sim$ 1.5 kpc at z=2 and $\sim$1 kpc at z=7; thus it will be possible to study aspects like the distribution of the ionised gas and the kinematics of those systems. The dithering method it is recommended to ensure that the PSF is fully well sampled and to minimise effects like a source falling between two IFU slices.

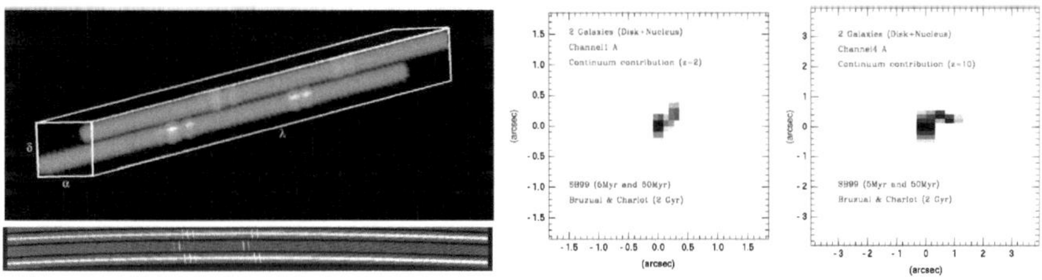


**Fig. 2.** *Left Top:* Data cube with two extended sources (with a point source as nucleus) simulating two galaxies with H$\alpha$+[N II] and [S II] at z=7. *Left Bottom:* Detail of the final *raw* MIRI MRS image generated by Specsim. *Right:* Simulated interacting galaxy (two discs of semi-minor and semi-major FWHM axes of 1 kpc and 2 kpc, with the nuclei located 2.4 kpc apart). At redshift 2 (IFU 1) it is possible to separate the two individual galaxies, while at redshift 10 (IFU 4) it is impossible to resolve the two individual components.

**Acknowledgments. MIRI draws on the expertise of the following organisations:** Research Center, USA; Astron, Netherlands Foundation for Research in Astronomy; CEA Service d'Astrophysique, Saclay, France; Centre Spatial de Liége, Belgium; Consejo Superior de Investigacones Científicas, Spain; Danish Space Research Institute; Dublin Institute for Advanced Studies, Ireland; EADS Astrium, Ltd., European Space Agency, Netherlands; UK; Institute d'Astrophysique Spatiale, France; Instituto Nacional de Técnica Aerospacial, Spain; Institute of Astronomy, Zurich, Switzerland; Jet Propulsion Laboratory, USA; Laboratoire d'Astrophysique de Marseille (LAM), France; Lockheed Advanced Technology Center, USA; Max-Planck-Insitut für Astronomie (MPIA), Heidelberg, Germany; Observatoire de Paris, France; Observatory of Geneva, Switzerland; Paul Scherrer Institut, Switzerland; Physikalishes Institut, Bern, Switzerland; Raytheon Vision Systems, USA; Rutherford Appleton Laboratory (RAL), UK; Space Telescope Science Institute, USA; Toegepast-Natuurwetenschappelijk Ondeszoek

(TNO-TPD), Netherlands; UK Astronomy Technology Centre (UK-ATC); University College, London, UK; Univ. of Amsterdam, Netherlands; Univ. of Arizona, USA; Univ. of Cardiff, UK; Univ. of Cologne, Germany; Univ. of Groningen, Netherlands; Univ. of Leicester, UK; Univ. of Leiden, Netherlands; Univ. of Leuven, Belgium; Univ. of Stockholm, Sweden, Utah State Univ. USA

## References

[1] Bruzual, G. & Charlot, S. 2003, MNRAS, 344, 1000
[2] Lagache, G., Dole, H., Puget, J.L. et al. 2004, ApJS, 154, 112
[3] Leitherer, C. et al. 1999, ApJS, 123, 3
[4] Lorente, N. P. F., Glasse, A. & Wright G. 2005, astro-ph/0511036
[5] Vazquez G.A. & Leitherer, C. 2005, ApJ, 621, 695
[6] Wright G. et al. 2004, SPIE, 5487, 653

# IFUs: Disentangling the Light from Neighbour Fibers

A. Kanaan[1], C. Mendes de Oliveira[2], C. Strauss[3], B. V. Castilho[4] and F. Ferrari[5]

[1] Dep. de Física - Universidade Federal de Santa Catarina `kanaan@astro.ufsc.br`
[2] Instituto Astronômico e Geofísico - Universidade de São Paulo `oliveira@astro.iag.usp.br`
[3] Instituto Nacional de Pesquisas Espaciais `strauss@astro.iag.usp.br`
[4] Laboratório Nacional de Astrofísica `bruno@lna.br`
[5] Universidade Estadual do Rio Grande do Sul `ferrari@if.ufrgs.br`

Integral Field Unit (IFU) spectrographs allow continuous space coverage of a small part of the sky by using a microlens array at the telescope focal plane. Each microlens images the telescope pupil onto an optical fiber, which is in turn taken to a bench spectrograph. A larger number of microlenses allows for higher spatial coverage and/or spatial resolution.

A larger number of fibers also implies in larger spectrographs, this means higher costs and higher telescope payloads. As the fibers are brought closer together at the spectrograph slit, the smaller the spectrograph becomes, and bigger becomes the contamination problem.

We have been working to determine how close together can be the fibers at the slit plane of a spectrograph, and yet have an acceptable amount of contamination. To separate the signal of a fiber from its neighbor, we devised the following strategy: a- measure the width and position of the spectrum of every fiber at all wavelengths as part of our calibration process, using non-linear Gaussian fits (see Figure 1); b- when observing our science objects we use the already determined positions and widths and solve only for intensity, using a linear Gaussian fit. This is currently done for Gaussian profiles (spatially), but could be implemented for other shapes, including numerical tables. We note that the mask procedure is always used, whether we extract our spectra using profile fitting or simple aperture extraction.

We have tested our method by running simulations changing fiber spacing from 1 to 5 pixels and FWHM of the spatial profile from 1 to 5 pixels. Our tests roughly show that for equal values of spacing and FWHM simple aperture extraction gives a contamination of 15% between neighbors, while this number falls to 0.1% for Gaussian fits. The quality of the Gaussian fits depend on SN and we conclude that one can easily push as far as 1 pixel separation and 2 pixels FWHM at a SN=5, or equivalently to 2 pixels separation and 4 pixels FWHM also for SN=5. Simple aperture extraction at the same parameters produces a 50% contamination, while Gaussian fits produce 2% contamination. Figure 2 presents contour lines of contamination

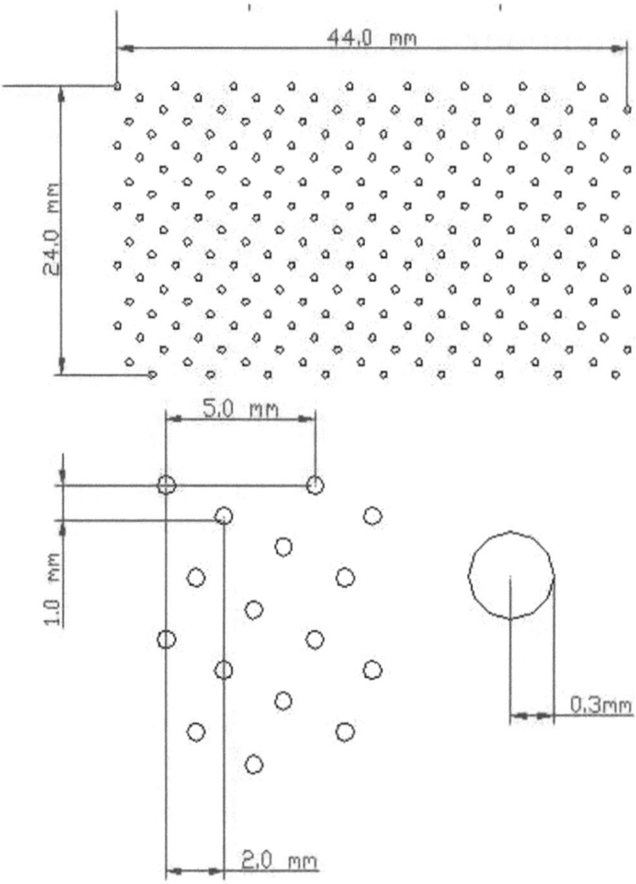


**Fig. 1.** This figure shows the current mask used with the Eucalyptus spectrograph at the 1.6m telescope in Brazil. The holes are arranged in a way that we need to move the mask to 5 successive positions and take one spectrum at each position in order to have one spectrum of each and every fiber. The mask holes are 0.3mm in diameter, compared to 1mm of the microlens size. The hole size being much smaller than the microlens insures that we don't contaminate a neighbor microlens thanks to bad alignment.

as a function of fiber separation and spectrum FWHM for Gaussian fits. Figure 3 is the same as Figure 2 for aperture extraction.

We have validated our results analyzing data obtained with the Eucalyptus spectrograph (in use at the 1.6m telescope in Brazil) at two samplings, spacing=FWHM=3 pixels and spacing=FWHM=5 pixels.

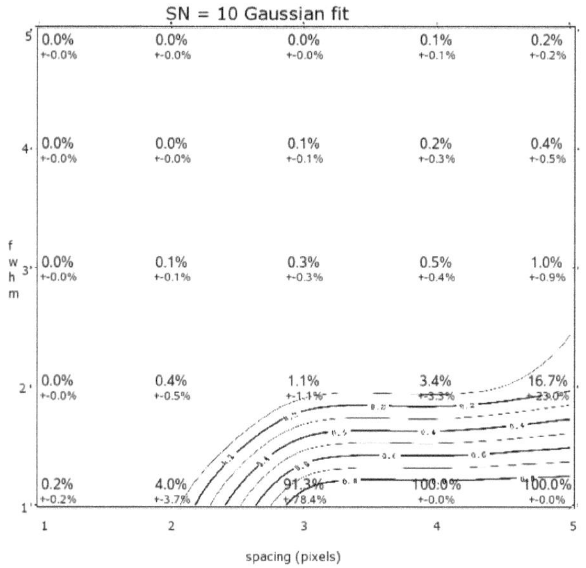


**Fig. 2.** This figure shows contour lines of contamination. The iso-contamination lines are equally spaced in intervals of 20%. In the x-axis is the spacing between fibers. In the y-axis the FWHM. The numbers superposed to the plot show the average contamination. The smaller numbers below are the scatter in contamination measured in Monte-Carlo simulations.

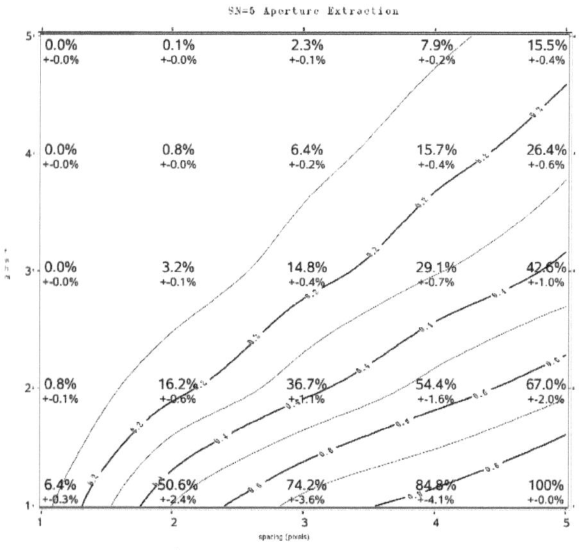

**Fig. 3.** Same as previous figure for aperture extractions. Note how contamination is much higher.

# OSIRIS – a New Integral-Field Spectrograph at Keck Observatory

A. Quirrenbach[1,2], J. Larkin[3], M. Barczys[3], T. Gasaway[2], C. Iserlohe[4], A. Krabbe[4], M. McElwain[3], I. Song[3], J. Weiss[3] and S. Wright[3]

[1] Sterrewacht Leiden, P.O. Box 9513, NL-2300RA Leiden, The Netherlands
[2] University of California, San Diego, USA
[3] University of California, Los Angeles, USA
[4] Universität Köln, Germany

## 1 Scientific Capabilities of OSIRIS

OSIRIS (OH-Suppressing InfraRed Integral-field Spectrograph) is a new facility instrument for the Keck Observatory. It provides three-dimensional near-infrared spectroscopy at the resolution limit of the Keck II telescope, which is equipped with adaptive optics and a laser guide star. The innovative capabilities of OSIRIS enable many new observing projects. OSIRIS can perform detailed studies of the stellar content and dynamical properties of galaxies in the early Universe. In quasars, radio galaxies, and nearby active galactic nuclei, OSIRIS can elucidate the relation of the central black hole to the properties of the host galaxy, and the mechanism feeding gas into the central engine. In the center of our own Galaxy, it is possible to investigate the properties of the massive black hole and the stars in its immediate vicinity. OSIRIS can perform spectroscopic observations of young stars and their environment, and of brown dwarfs. Imaging spectroscopy of the giant planets, their moons, asteroids, and Kuiper Belt objects can shed new light on meteorology, mineralogy, and volcanism in the Solar System. For more details on the science case of OSIRIS see Quirrenbach et al. (2002).

## 2 The Instrument

OSIRIS is an integral-field spectrograph with a TIGER-type design (Bacon et al. 1995). A detailed description of OSIRIS has been given by Larkin et al. (2002). The spectral resolution is $R \approx 3800$; the spectra are sampled at the Nyquist frequency by the detector. In its "wide-band" mode, OSIRIS takes a 1700-pixel spectrum providing full coverage of one of the infrared bands ($z$, $J$, $H$, or $K$) for each one of 1024 spatial pixels arranged in a $16 \times 64$ rectangular grid. A second "narrow-band" mode is available, in which shorter (400-pixel) spectra covering only a partial band are taken for $48 \times 64$ spatial pixels. Four pixel scales ($0\farcs02$, $0\farcs035$, $0\farcs05$, and $0\farcs1$ per pixel) provide a field-of-view ranging from $0\farcs32 \times 1\farcs28$ to $1\farcs6 \times 6\farcs4$ for the wide-band mode, and from $0\farcs96 \times 1\farcs28$ to $4\farcs8 \times 6\farcs4$ for the narrow-band mode. OSIRIS uses a

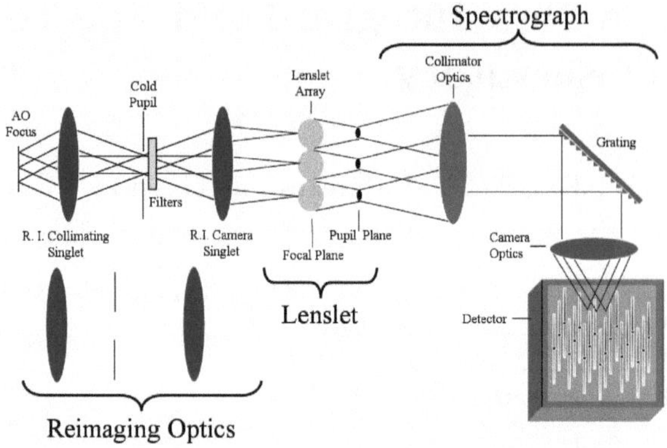


**Fig. 1.** Schematic drawing of OSIRIS. Light from the telescope enters from the left through the adaptive optics system. The image scale is selected by moving one of four sets of reimaging optics into the beam. The spectral bandpass is defined by a filter. The lenslet array produces a pupil image for each spatial pixel in the field.

fixed grating blazed at 6.5 $\mu$m, so that the 3rd to 6th orders cover the desired wavelength range. A filter wheel is used to select among the available bands. The detector is a 2048 $\times$ 2048 pixel Hawaii 2 array.

In addition to the spectrograph, an imaging camera is integrated into the OSIRIS dewar. It provides a 20$''$ field with a pixel scale of 0$''$.02 and is equipped with 18 filters. The centers of the imaging and spectroscopic fields are separated by 20$''$. The camera is useful as a finder for faint objects, and it can perform parallel observations during spectroscopic exposures. The typical integration times of the spectrograph are sufficiently long to get good signal-to-noise on "random" stars within the imager field-of-view. The imager can therefore be used for monitoring of the point spread function, and it provides fiducials for the spatial registration of individual spectroscopic data cubes.

## 3 Data Products, Calibration, and Data Reduction

Because the only moving parts in OSIRIS are the reimaging lens assemblies (providing the different pixel scales), the filter wheel (defining the bandpass), and the mask slide (selecting the illuminated lenslet rows), there are only 92 discrete data products (4 pixel scales $\times$ 23 wavelength bands). The spectrograph is designed to be very stable, which is possible because it is located on the Nasmyth platform and thus does not experience a variable gravity vector. It is thus expected that it will not be necessary to perform calibrations each night, but that one can use a data base of calibration observations for all 92 data products, which will only have to be updated occasionally.

Important steps in the data reduction and calibration comprise correction of detector non-linearity, removal of bad pixels and fixed-pattern noise,

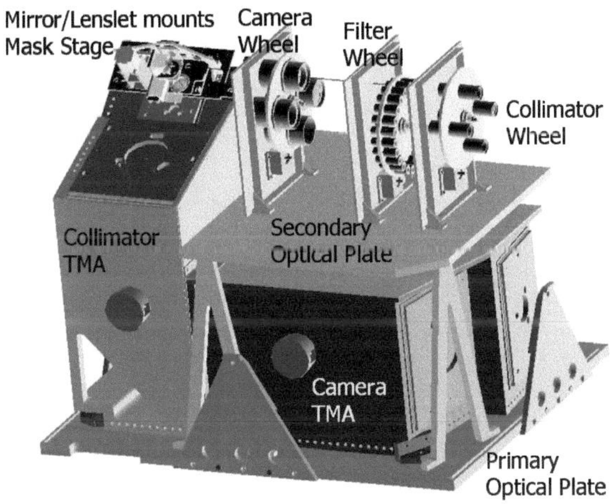


**Fig. 2.** CAD drawing of the mechanical design of OSIRIS. Baffles, shields, supports, vacuum vessel and frame are not shown. Light enters from the top right. The detector assembly is hidden by the TMA housings.

sky subtraction, spatial rectification, division by a flat field, interpolation of missing pixels, wavelength calibration, assembly of the data cube, correction for the instrumental and telluric transmission, and for dispersion (see Krabbe et al. 2004 for details). A critical step in the data reduction is the spatial rectification, in which the flux received in each detector pixel is associated with the lenslets. Since the individual spectra are separated by only two pixels on the detector, cross-talk between them has to be taken into account properly.

We have developed a procedure that distributes the light received at each detector pixel among all lenslets that contribute to the illumination of that pixel in a consistent way (Quirrenbach et al. 2006). This algorithm uses a set of 16 calibration exposures, in which all rows of lenslets but one are covered by the lenslet mask mechanism. Matrices describing the mapping between the lenslet array and the detector can be populated from these calibration images. For each scientific exposure, the computation of the spectrum incident on each lenslet from the observed data consists of an inversion of these matrix equations, which is achieved with an iterative algorithm.

## References

[1] Bacon, R., et al. (1995). A&AS **113**, 347-357
[2] Krabbe, A., et al. (2004). SPIE Vol. 5492, p. 1403-1410
[3] Larkin, J.E., et al. (2002). SPIE Vol. 4841, p. 1600-1610
[4] Quirrenbach, A., et al. (2002). SPIE Vol. 4841, p. 1493-1502
[5] Quirrenbach, A., et al. (2006). New Astron. Rev., in press

# A User–oriented Comparison of the Techniques for 3D Spectroscopy

S. di Serego Alighieri

INAF – Osservatorio Astrofisico di Arcetri, Largo E. Fermi 5, Firenze, Italy
sperello@arcetri.astro.it

## 1 Introduction

3D spectroscopy attempts to get closer to the fundamental goal of astronomical observing techniques, which is to **record the direction, wavelength, polarization state and arrival time for every incoming photon over the largest field of view**. In fact using 3D spectroscopy, the wavelength and the incoming direction in a 2D field of view are recorded in a (x,y,$\lambda$) data cube, in contrast with standard techniques which either do imaging over a 2D field, or spectroscopy along a 1D slit. There are two main ways of doing 3D spectroscopy: the best one is to simultaneously record both direction and wavelength, while in the other case these are not recorded at the same time, but scanning in one of the 3 dimensions is required. Clearly the latter throws away some of the incoming photons, therefore requiring longer exposure times, and has problems with variable observing conditions. Nevertheless it can be useful for some particular applications.

## 2 Simultaneous Techniques

**Integral field spectroscopy (IFS)** rearranges over a 2D detector the spectra coming from every pixel in a 2D field of view. Therefore it provides a straightforward way to fill the (x,y,$\lambda$) data cube. The rearrangement of the spectra can be done with microlens arrays (e.g. SAURON (2)), fibre–lenslet arrays (e.g. IFS in GMOS (1)), or image slicers (e.g. SINFONI (8)). It is becoming a popular technique, since most modern spectrographs on large telescopes have IFS capability. IFS in general has the advantage of providing great flexibility in the choice of the spectral resolution and wavelength range and of being easily fed by adaptive optics systems. The main disadvantage is the limited field of view, since the number of spatial elements is limited by the number of pixels along one side of the detector. This drawback is at least partially overcome by an obvious development of IFS: the field of view, particularly for IFS using fibre–lenslet arrays, can be separated in several disjoint regions, for example to cover several galaxies in a cluster. Examples of this development are the multiple IFS of GIRAFFE (9), and the even more flexible programmable IFS concept (5). Because of its flexibility, IFS is suited to a large number of applications, from kinematical studies of the Galactic centre to stellar population and kinematical studies of distant galaxies (10).

Also **slitless spectroscopy** is capable of simultaneously recording a (x,y,$\lambda$) data cube: originating from the objective prism technique, used on Schmidt telescopes for more than fifty years, it is easily implemented in modern imaging (focal reducer) spectrographs by removing the slit. Therefore it records spectra of all objects over the whole field of view, which can be quite large, like the 14'x14' field of VIMOS (12). The disadvantages are the high sky background, since on every detector pixel the sky is integrated over the whole wavelength range, and the overlap of spectra in the dispersion direction. Still this technique is particularly useful for surveys and searches of special objects, when the sky background is very low, like in space or in small atmospheric windows. For example it has been successfully used with ACS on the HST for GRAPES, a spectroscopic survey of the Hubble Ultra Deep Field down to an AB magnitude limit of $z = 27.2$, leading to the discovery of a large number of emission line objects over a huge redshift range, like AGN and Lyman $\alpha$ galaxies (15). In this case the effects of the spectra overlap has been substantially reduced by taking spectra at various position angles. An example of the use of slitless spectroscopy from the ground is the search for Lyman $\alpha$ emitters at z=6.5 in the atmospheric window centred at 915 nm. In this case the spectral range can be limited to the 20 nm width of the window by using a narrow-band filter. Therefore both the sky emission and the spectra overlap are greatly reduced and very faint emission line objects can be found down to a line flux of $2 \times 10^{-17} erg\ cm^{-2} s^{-1}$ (11).

**Energy–resolving detectors** are imaging arrays where each pixel has some energy resolution. Therefore these are true 3D devices capable of simultaneously recording the (x,y,$\lambda$) data cube, and no spectrograph is necessary. Being mostly photon-counting detectors, they also have a very good temporal resolution. Their main disadvantages are the very limited spectral resolution and field of view. Two different technological approaches are being explored in the optical range: the Superconducting Tunnel Junctions (STJ (14)) and the superconducting transition-edge sensors (6). Currently STJ detectors using tantalum metal films have a good quantum efficiency in the optical range (about 70%), but have a resolution $\frac{\lambda}{\Delta\lambda}$ of only about 20 and a total number of pixels of about 100. The latter could reasonably be increased to 10000. High speed energy-resolved observations of rapidly variable stars and optical pulsars have been obtained with STJ detectors, and their use as order sorters in an intermediate resolution spectrograph (not 3D) has been investigated (7).

Although purists may not consider **multi–object spectroscopy** as true 3D spectroscopy, since it does not completely cover a 2D field of view, it does however produce spectra of many objects in a large field, and very suitably fulfils the needs of many applications, making it the most popular 3D technique. Practically all telescopes have MOS instruments, using either a fibre positioner coupled to a spectrograph, movable slitlets, or a multi-aperture plate. The latter implementation has advantages in terms of better sky subtraction and throughput than fibres, and a larger number of objects and better positioning flexibility than slitlets. A good example is VIMOS on the VLT (12), which is capable of simultaneously recording spectra of 1000 objects over a 14'x14' field of view. The disadvantages are that it requires prior imaging (and mask preparation), that objects have to be preselected (not good for object searches), and that it is not capable of a complete 2D coverage of extended objects.

## 3 Scanning Techniques

**Tunable imaging filters** cannot simultaneously record the data cube, but require scanning in wavelength. The most used in astronomy is the Fabry–Perot filter, which uses interference between two glass plates (4). They have very good imaging capability, a large field of view and good spectral resolution. They suffer from the so–called phase problem: the central wavelength is not constant over the field of view. Therefore reconstructing the (x,y,$\lambda$) data cube is not straightforward. Fabry–Perot filters have been used for a large number of applications mostly on nearby galaxies and nebulae.

**Imaging Fourier Transform Spectroscopy (IFTS)** is a special technique using the interference of two optical beams. Although it requires several exposures by scanning a movable mirror, and the reconstruction of the (x,y,$\lambda$) data cube is not straightforward, but requires heavy computation, nevertheless the scanning does not imply any loss of photons, which are all recorded over the full field of view and wavelength range (3). A disadvantage compared to the simultaneous 3D techniques, like the IFS, is that the readout noise affects the final data cube not just once, but a number of times equivalent to the number of spectral elements. Also each spectral element suffers the sky noise of the whole bandpass. Therefore IFTS is competitive when a reduced number of spectral elements is required over a large field, as, for example in kinematic studies of the Galactic centre. One of the few examples of IFTS used in astronomy is BEAR on the CFHT (13).

**Scanning long–slit spectroscopy** does not require a new instrument, but uses a normal long-slit spectrograph. It does not simultaneously fill the data cube, but can be used for very elongated objects, like edge-on galaxies, when only coarse information is required in the second spatial direction.

## 4 Selection of the Most Suitable Technique

Although advantages and disadvantages can be found (see Table 1), it is hard to say which technique is best. One has rather to find the technique which most efficiently fills the (x,y,$\lambda$) data cube for each specific application. In most cases the data cube is largely empty. However it is exactly in these empty spaces that one can make serendipitous discoveries.

**Table 1.** Synopsis of the techniques for 3D spectroscopy

| Technique | Advantages | Disadvantages | Applications |
|---|---|---|---|
| IFS | Simultaneous (x,y,$\lambda$)<br>Spectral flexibility<br>Disjoint regions possible | Limited f.o.v. | Galactic centre,<br>distant galaxies, etc. |
| Slitless spectr. | Simultaneous (x,y,$\lambda$)<br>Normal spectrograph<br>Large f.o.v. | High sky background<br>Spectra overlap | Surveys<br>Searches for objects |
| Energy–res. det. | Simultaneous (x,y,$\lambda$)<br>No spectrograph<br>Good temporal res.<br>Good efficiency | Very limited $\frac{\lambda}{\Delta\lambda}$<br>Very limited f.o.v.<br>Under development | Rapid variables<br>Optical pulsars |
| MOS | Normal spectrograph<br>Spectral flexibility<br>Very large f.o.v. | 2D field not covered<br>Prior imaging<br>Mask preparation | Large redshift surveys |
| Tunable im. filters | Large f.o.v.<br>Good spectral res. | Scanning in $\lambda$<br>Variable central $\lambda$ | Nearby galaxies<br>Nebulae |
| IFTS | No loss of photons | Scanning required<br>High sky background<br>Heavy computations | Galactic centre |
| Long slit scanning | Normal spectrograph | Scanning required | Very elongated objects |

## References


[1] J. Allington–Smith, G. Murray, R. Content et al: PASP **114**, 892 (2002)
[2] R. Bacon, Y. Copin, G. Monnet et al: MNRAS **326**, 23 (2001)
[3] C.L. Bennett: A.S.P. Conf. Ser. Vol. 195, p. 58 (2000)
[4] J. Bland–Hawthorn: A.S.P. Conf. Ser. Vol. 195, p. 34 (2000)
[5] J. Bland–Hawthorn, A. McGrath, W. Saunders et al: Proc. SPIE Vol. 5492, p. 242 (2004)
[6] B. Cabrera, R.M. Clarke, P. Colling et al: Appl. Phys. Let. **73(6)**, 735 (1998)
[7] M. Cropper, M. Barlow, M.A.C. Perryman et al: MNRAS **344**, 33 (2003)
[8] F. Eisenhauer, R. Abuter, K. Bickert et al: Proc. SPIE Vol. 4841, p. 1548 (2003)
[9] H. Flores, M. Puech, F. Hammer et al: A&A **420**, L31 (2004)

[10] S. Gillessen, R. Davies, M. Kissler–Patig et al: The Messenger **120**, 26 (2005)
[11] J.D. Kurk, A. Cimatti, S. di Serego Alighieri et al: A&A **422**, L13 (2004)
[12] O. Le Fèvre, M. Saisse, D. Mancini et al: Proc. SPIE Vol. 4841, p. 1670 (2003)
[13] J.P. Maillard: A.S.P. Conf. Ser. Vol. 195, p. 185 (2000)
[14] M. Perryman, C. Foden, A. Peacock et al: Nuc. Inst. Meth. A **325**, 319 (1993)
[15] N. Pirzkal, C. Xu, S. Malhotra et al: ApJS **154**, 501 (2004)

Part II

# Solar System(s)

# The Potential of Integral Field Spectroscopy Observing Extrasolar Planet Transits

S. Arribas[1−4], R. L. Gilliland[1], W. B. Sparks[1], L. López-Martín[2], E. Mediavilla[2] and P. Gómez-Alvarez[2]

[1] Space Telescope Science Institute, Baltimore, USA arribas@stsci.edu
[2] Instituto de Astrofísica de Canarias, Tenerife, Spain
[3] European Space Agency, ESTEC, The Netherlands
[4] Consejo Superior de Investigaciones Cientificas, Spain

**Summary.** We explore the use of Integral Field Spectroscopy (IFS) for observing extrasolar planet transits. We show that IFS can improve the photometric accuracy in high S/N time series observations with respect to standard slit spectrophotometry. Ground based observations of HD209458b obtained with WHT + INTEGRAL during a transit have led to an accuracy in relative spectrophotometry (i.e line / continuum) not far from that expected by photon noise. This result indicates that these types of ground based observations can constrain the characterization of the transmission spectrum of extrasolar planets.

## 1 Introduction

The transit method (TM) has recently gained much attention due to its potential for discovering earth-like planets. This method allows one a detailed characterization of the star/planet system and offers the possibility of studying the existence of planetary satellites, rings, and atmospheric features (see, for instance, Deeg 2002, and Deming et al 2005a and references therein).

The success of the TM largely relies on the (spectro) photometric accuracy for which the light curve can be obtained. For some science applications (e.g. detection of planetary atmospheric features) only *relative* spectrophotometry (i.e. band1/(nearby)band2 and in/out transit) is required, which reduces drastically the systematic correlated errors. Experiments with the HST (e.g. Charbonneau et al. 2002; Gilliland and Arribas 2003) have demonstrated that accuracies of $\sim 10^{-5}$ can be reached from space, though such a high accuracy is not always required (Vidal-Madjar et al. 2003, 2004).

Pioneering studies from the ground include those of Coustenis et al. (1998), Rauer et al. (2000), Bundy & Marcy (2000), Moutou et al. (2001), and Brown et al. (2002). More recently Deming et al. (2005b) and Narita et al. (2005) have studied the transmission spectrum of HD209458 from the ground. All these ground based observations have gained importance since STIS–HST is no longer operational.

In this paper we summarize the advantages of Integral Field Spectroscopy (IFS) when observing extra-solar transits, and comment on the IFS observations of HD209458. Further details are given in Arribas et al. (2006).

The reader may be also interested in other related applications of IFS (e.g. Arribas, Mediavilla, Fuensalida, 1998; Sparks and Ford, 2002).

## 2 Proposed Method: The potential of IFS in observations of planet transits

Here we want to highlight three main advantages of IFS when performing high S/N time series spectrophotometric observations:

i) Ability to gather photons during the transit: S/N enhancement.

Gilliland, Goudfrooij, and Kimble (1999) proposed to increase the photometric accuracy of very high S/N observations by using the spectroscopic mode of STIS. By distributing the star light along the dispersion direction it was possible to increase the total number of photons collected before the detector reaches the saturation limit. This simple idea enhances the duty cycle of the instrument (i.e. total number of photons by unit of available time, including overheads), which translates to an increase in the S/N when the photon noise is the major source of noise.

IFS allows one to extend this idea. Thanks to the 2D $\rightarrow$ 1D conversion done by a fiber bundle (or by an image slicer), the star light in an integral field spectrograph is not only spread along the dispersion direction but also across dispersion. This technique allows us to use a larger area of the detector, boosting the limit imposed by the photon-noise per exposure. Note that to optimize the use of the detector, the image of the star at the focal plane should be well oversampled by the Integral Field Unit (IFU). This can be obtained in different ways: (using very small spaxels, by defocusing, or even making IFS of the pupil. In contrast to slit-spectroscopy, this does not imply a degradation in the spectral resolution of the observations.

ii) Ability to track instabilities.

Stability is crucial for transit observations. Of course, beyond some limits it is not possible to control the stability of the environment and/or the observational set-up. When uncontrolled instabilities dominate the photometric accuracy of the observations, a good record of them gives the possibility to de-correlate the photometric signal. This provides another important advantage of the IFS. Since the spectra from which the photometric information is obtained, the image of the object (PSF), as well as other variables tracking the stability of the instrument are extracted from the same data cube, it is possible to remove photometric variations induced by PSFs or instrument instabilities.

iii) Autocalibrations.

Another advantage of IFS for these type of observations is a consequence of the large number of spectra collected simultaneously, which can be used to *autocalibrate* the data themselves from detector and background signatures (see Section 3.3 for a practical realization).

## 3 Ground-based IFS Observations of HD209458

We performed the observations with the 4.2m WHT during 2004 August 17/18 using the INTEGRAL system (Arribas et al. 1998). For this analysis we have used more than 5200 spectra (from a potential of ~ 30000), obtained in 142 exposures during a period of more than 7 hours. Details about the observations and reductions can be found in Arribas et al. (2006).

### 3.1 Auxiliary variables

We studied the variation of a set of auxiliary variables along the night. Except for the airmass, these variables were obtained from the reduced spectra and images and, therefore, extracted from the data cube. Table 1 lists the set of selected variables.

**Table 1.** Decorrelation Parameters

| # | Variable | Coefficient | Error | Correlation | Conservation Factor |
|---|---|---|---|---|---|
| 1 | Airmass | 1.074 | 5.41E-04 | 0.76 | 0.88 |
| 2 | Total flux | 8.69E-04 | 1.33E-10 | −0.14 | 0.88 |
| 3 | Background | 7.74E-02 | −6.61E-07 | −0.48 | 0.75 |
| 5 | Spectra Shift (across disp.) | 1.60 | 3.60E-04 | −0.29 | 0.32 |
| 6 | Spectra FWHM | 3.16 | −1.87E-05 | −0.11 | 0.38 |
| 7 | Image Total Flux | 5.82E-03 | −1.01E-08 | −0.16 | 0.88 |
| 8 | Flux-ratio | 1.68 | −5.12E-04 | −0.32 | 0.56 |
| 9 | Gaussian Peak | 4.55E-04 | −6.08E-11 | 0.27 | 0.77 |
| 10 | Gaussian X | 0.330 | −2.81E-06 | 8.9E-02 | 0.40 |
| 11 | Gaussian Y | 0.480 | 3.20E-05 | −6.3E-02 | 0.46 |
| 12 | Distance to 111 | 0.384 | 3.24E-05 | 0.17 | 0.99 |
| 13 | Gaussian FWHM | 0.422 | −5.55E-05 | −0.27 | 0.84 |
| 14 | Gaussian Ellipticity | 1.59 | 6.51E-04 | 2.3E-03 | 0.85 |
| 15 | Gaussian P.A. | 0.320 | −3.57E-06 | −0.19 | 0.94 |
| | All | | | | 0.17 |
| | 1 +12+13 | | | | 0.79 |

### 3.2 Index definition

We define a photometric index probing the NaD lines as:

$$I = \frac{f_1 - f_2 + f_3}{f_1 + f_2 + f_3}, \tag{1}$$

where $f_1, f_2$, and $f_3$ represent the sky subtracted flux (on a scale of photo-electrons) in the following spectral ranges: 582.9200–588.5643, 588.5643–590.0887, and 590.0887–595.7321 nm, respectively (see Figure 1). As defined in eq. (1), the index $I$ will probe the NaD lines under the assumption that $f_1$ and $f_3$ are stable continua, as it has been demonstrated by Charbonneau et al. (2002).

From Eq. (1) it can be found that a relative change of the flux in the NaD band translates into a relative change $\sim 4$ times smaller for the index, with the opposite sign. So, taking into account that in a typical exposure we collect about $7.15{\cdot}10^6 e^-$ in $f_2$, the standard deviation due to photon noise for $f_2$ would be $3.74{\cdot}10^{-4}$ and, therefore, for the index $I$ should be $\sim 9.5{\cdot}10^{-5}$.

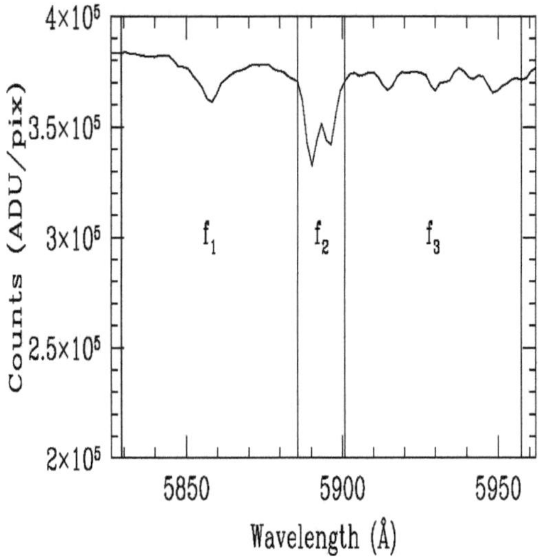


**Fig. 1.** Typical integrated spectrum a single 120 sec exposure. Vertical lines indicate the boundaries of the different bands used for the definition of the photometric index.

### 3.3 Autocalibration

The index as defined above was calculated for the 37 brightest spectra of each exposure. The resulting 5254 (37 spectra $\times$ 142 exposures) values are represented in Figure 2 (top) versus the total flux in the band (i.e. $f_t = f_1 + f_2 + f_3$). Clearly there are some systematic deviations from the behavior expected in the case that noise were only due to photon noise. This is particularly obvious for the largest $f_t$ values, which are clearly non-linear. This can be well calibrated using the data themselves (i.e. *autocalibration*). To this aim, we fit all the data points to a smooth function (splines3 of order 16) of the total flux, $I_{fit}(f_t)$, and we redefine the index as $I_c = I - I_{fit}(f_t)$. The corrected photometric index, $I_c$, for the 5254 spectra are represented as a function of the total flux in Figure 2 (bottom).

The next step was to obtain the photometric index value for each exposure (time),$I'$. This was done averaging the 37 contemporaneous individual values obtained in each exposure. We use an inverse variance weighting of the index (which is proportional to the total flux, $f_t$) in order to take into account the statistical errors of each individual $I_c^i$ value.

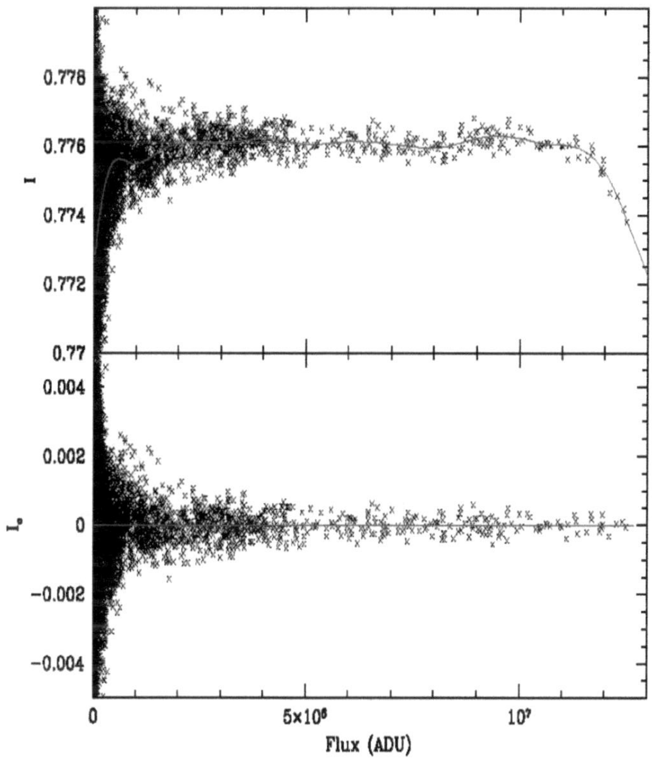


**Fig. 2.** Distribution of the index as a function of the total flux before (top) and after (bottom) autocalibration (see text).

### 3.4 Decorrelation

By *de-correlation* we understand the process of removing trends in the index correlated with the auxiliary variables. This was based on a multiple-regression (least-square) fit to the data with a function which is linear in the coefficients of independent variables. Column 4th of Table 1 shows the correlation parameter for the 15 selected auxiliary variables. In order to have an idea of the effects of the de-correlation procedure reducing noise, we can compare the global standard deviation of the 142 data values before ($1.85 \cdot 10^{-4}$) and after ($1.11 \cdot 10^{-4}$) this procedure was applied. Note that the standard deviation of the de-correlated values is remarkably close to the one expected in a purely photon-noise dominated regime. In fact, considering a typical level of counts per exposure of $3.1 \cdot 10^6$ ADU (counts) for $f_2$, a gain of 2.3 $e^-/ADU$, and eq. (3), a standard deviation of the index of $\sim 0.95 \cdot 10^{-4}$ is expected in a photon-noise dominated regime.

In Figure 3 (left) we show the original data (upper panel), the fit to a function linear to the airmass (center), and the decorrelated data (lower), as a funtion of time. Figures 3 (center) and (right) are similar but fitting the data to a function linear to 14 (all but the airmass), and to all the 15 selected auxiliary variables, respectively. It is interesting to note how the rest of the variables also carry information about the effects of the airmass.

Although this decorrelation method can be applied safely if the functional dependence of the auxiliary variables are different to that of the signal to be detected, it may destroy part of it when that is not the case. To study the degree of destruction of the signal during the transit due to the decorrelation process we proceed as follows. Similarly to Brown et al. (2002) we injected an artificial signal to the data (during the period that the transit took place) before the decorrelation was applied, and we analyzed how much of this signal was recovered after decorrelation. Column 5th in Table 1 indicates the fraction of signal recovered, when the decorrelation is done using only this variable. Table 1 shows that in several cases more than 80 percent of the signal is recovered, while in other cases this is less than 30 percent. Then we studied the combined effects of several variables. We found that when all the 15 variables are used to remove noise (i.e. decorrelation process) most of the signal was actually destroyed: only 17 percent of the signal artificially injected is recovered after decorrelation.

The next step in the analysis was to select a subset of auxiliary variables, which lead to a significant reduction of noise while preserving a substantial fraction of the signal. After several trials based on the individual destruction factors, we found that three variables (airmass, FWHM of the image, distance to 111) give a good compromise. Specifically, the standard deviation of the 142 points was $1.19 \cdot 10^{-4}$, while 79 percent of the signal is preserved after decorrelation.

### 3.5 Time Series

Figure 4 (upper panel) shows the time series data values. Vertical lines indicate the contacts. Following Charbonneau et al. (2002) we define in-transit observations as those that occurred between second and third contact (i.e. $|t - T_c| < 66.1$ m.), and out-of-transit those that occurred before first or after fourth contact (i.e. $|t - T_c| > 92.1$ m.).

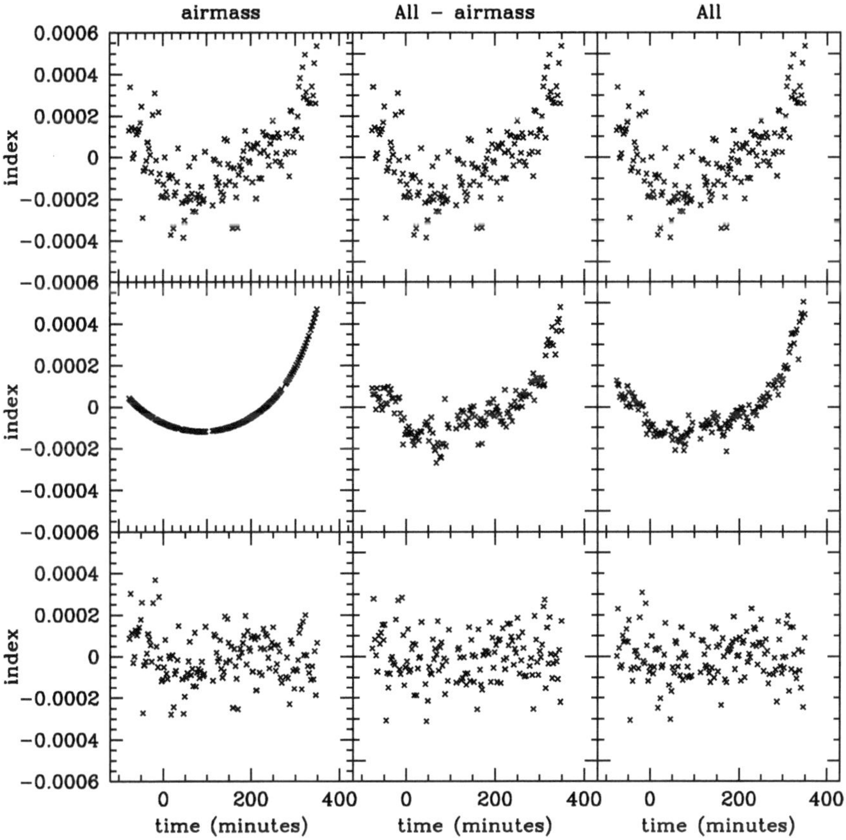


**Fig. 3.** Temporal variation of the index $I_c$. The upper panels indicate the original distribution of values. Intermediate panels represent the fit of the original distribution to a function linear in the variables indicated at top (i.e. left: airmass; center: the 15 variables described in Table 1, excluding the airmass; and right: the 15 variables). The lower panels show the decorrelated distribution, which is obtained subtracting the fit to the original values. The standard deviation for these distributions are: $1.2 \cdot 10^{-4}$(left); $1.15 \cdot 10^{-4}$(middle): $1.1 \cdot 10^{-4}$(right).Units for the horizontal axis are in minutes from the center of the transit.

As mentioned above the standard deviation of all the 142 values represented in Figure 4 is $1.19 \cdot 10^{-4}$. However, the standard deviation in-transit ($1.46 \cdot 10^{-4}$) is significantly larger than out-of-transit ($1.03 \cdot 10^{-4}$). In any case it is remarkable the fact that out-of transit, and considering data that expand for a period of $\sim$ 4 hours, the standard deviation is very similar to that expected from Poison-noise.

The lower panel in Figure 4 shows the mean in-transit and out-of-transit values of the index: $(0.12 \pm 2.17) \cdot 10^{-5}$ and $(0.06 \pm 1.12) \cdot 10^{-5}$, respectively. The quoted 1-$\sigma$ errors correspond to the error in the mean of the statistical distribution (i.e. standard deviation/$\sqrt{n}$). These results are compatible with no variation of the index during the transit (i.e. the relative variation is $[0.06 \pm 2.44] \cdot 10^{-5}$) and, taking into account equation (3), no variation of the relative the flux in the NaD lines (i.e. $f_2$) with respect to the continuum during the transit. (Note that the small value compared with the error bars is an statistical curiosity. If samples of alternative data points [e.g. even / odd position in the time sequence] are considered, values of $[-1.24 \pm 3.7] \cdot 10^{-5}$ and $[1.35 \pm 3.2] \cdot 10^{-5}$ respectively are obtained, instead of $[0.06 \pm 2.44] \cdot 10^{-5}$). We also show in this figure the expected change in the index due to a change in $f_2$ as the one detected by STIS–HST (Charbonneau et al. 2002). Therefore, if a signal with an amplitude as the one reported by Charbonneau et al. were real, and after repeating the current experiment many times, the (gaussian) distribution of results should peak at $4.58 \cdot 10^{-5}$ with a sigma of $2.44 \cdot 10^{-5}$. However, we (and Charbonneau et al.) have performed the experiment only once, and the obtained result is about 2 $\sigma$ away from the expected value (for a signal as the one detected by Charbonneau et al. data). If we take into account the quoted errors, the present data and the HST/STIS results disagree by more than 1 $\sigma$, but they are compatible within 1.5 $\sigma$. This limited statistical significance makes us cautious to over interpret the present results. The recent ground based observations by Narita et al. (2005) could not confirm nor contradict the Charbonneau et al. result. Although these Subaru observations have much higher spectral resolution, its lower S/N and temporal resolution lead to an error in the predicted relative flux in NaD lines (for a 12 Å band) larger (by a factor of about 4 times) than that of the current observations.

## 4 Conclusions

We have shown that Integral Field Spectroscopy is a powerful technique for obtaining accurate relative time series photometry and, therefore, of great potential for studying transits of extrasolar planets. The three main advantages with respect to previously used methods are: i) it allows an increase in the duty cycle of the observations (and, therefore, the S/N) by distributing the light into the two dimensions of the detector; ii) the data can be *autocalibrated* from detector non-linearities and from background effects; and iii) since the photometric index, the image of the star, and several variables tracking the instabilities of the instrument are extracted from the same data-cube, the correlated noise can be well removed.

Using a photometric index that probes the strength of the NaD lines, a standard deviation about a factor 2 larger than that expected due to photon noise was found. However, after removing correlated noise by fitting the data to a function linear in three selected auxiliary variables the standard deviation drops to only ~25 percent larger than that due to photon noise. Our results are compatible with no extra depression of the NaD during the transit. Although the present results are in apparent contradiction with the

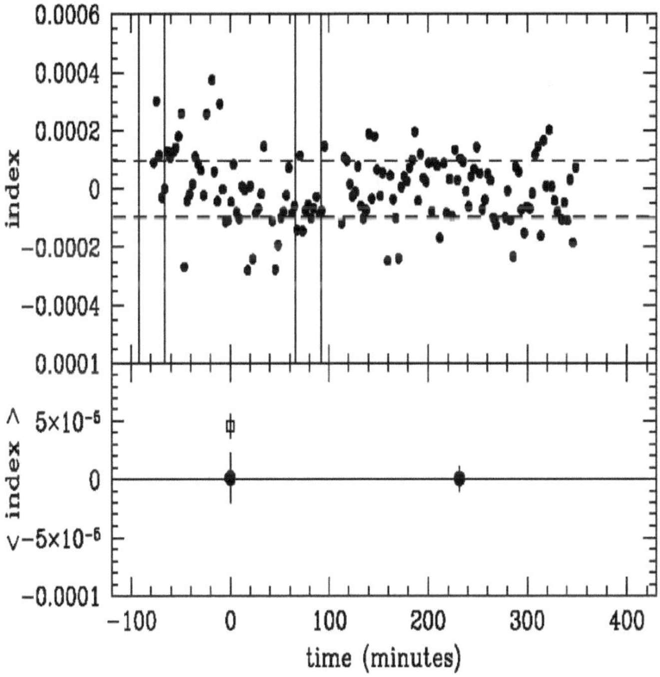


**Fig. 4.** Upper panel: Temporal variation of the index $I_c$ after decorrelation with three selected variables (airmass, image FWHM, distance to 111). Vertical lines indicate (from left to right) first, second, third and fourth contact. Horizontal dashed lines show the expected $\pm 1\text{-}\sigma$ according to photon noise. Lower panel shows the averaged values in-transit (i.e. between second and third contact) and out-of-transit (i.e. after fourth contact). The value corresponding to the HST observations by Charbonneau et al. (2002) is shown with an open square (see text).

HST–STIS results, we stress the limited statistical significance of this disagreement: they disagree at $1\text{-}\sigma$, but they are consistent within $1.5\text{-}\sigma$.

The present results can be improved with multi-transit observations using the recently developed new integral field spectrographs for 8m. class telescopes (e.g. GMOS: Allington-Smith et al. 2002, SINFONI/SPIFFI: Eisenhauer et al. 2003, VIMOS: Le Fevre et al. 2003).

## References

[1] Allington-Smith, J.R. et al. 2002, PASP, 114, 892
[2] Arribas, S., et al. 1998, SPIE, 3355, 821
[3] Arribas, S., Mediavilla, E., Fuensalida, J.J., 1998, ApJL, 505, 43.
[4] Arribas, S., Gilliland, L.R., Sparks, W., Martín-López, L., Mediavilla, E., and Gómez-Alvarez, P. (astro-ph/0509407 )

[5] Brown, T.M., Charbonneau, D., Gilliland, R.L., Noyes, R.W., and Burrows, A. 2001, ApJ, 552, 699
[6] Brown, T.M. 2001, ApJ, 553, 1006
[7] Brown, T.M., Libbrecht, K.G., and Charbonneau, D. 2002, PASP, 114, 826
[8] Bundy, K.A., and Marcy, G.W. 2000, PASP, 112, 1421
[9] Charbonneau, D. et al. 2000, ApJL, 529, L45
[10] Charbonneau, D., Brown, T.M., Noyes, R.W., and Gilliland, R.L. 2002, ApJ, 568, 377
[11] Coustenis,A. et al. in 'Brown Dwarfs and Extrasolar Planets', Ed. R. Rebolo, E. L. Martin, M. R. Zapatero Osorio, ASP Conference Series, 134 , 296
[12] Deeg, H. 2002, 36th ESLAB Symposium, Eds.: B. Foing, B. Battrick. ESA SP-514, 237
[13] Deming, D., Seager,S., Richardson,L.J., Harrington,J., 2005a, Nature, 434, 740
[14] Deming, D., Brown, T.M., Charbonneau, D., Harrington, J., Richardson, L.J. 2005b, ApJ, 622, 1149
[15] Eisenhauer, F. et al. 2003, SPIE, 4844, 1548
[16] Gilliland, L.R., Goudfrooij, P., and Kimble, R.A 1999, PASP, 111, 1009
[17] Gilliland, L.R., and Arribas, S. 2003, ISR-NICMOS-03-01, STScI
[18] Le Fevre, O. et al. 2003, SPIE, 4841, 1670
[19] Moutou,C., Coustenis,A., Schneider, J., StGilles,R., Mayor,M., Queloz,D., Kaufer,A. 2001, A&A, 371, 260
[20] Narita, N., Suto, Y., Winn, J. N., Turner, E. L., Aoki, W., Leigh, C. J., Sato, B., Tamura, M., Yamada, T., 2005, PASJ, 57, 471
[21] Rauer,H., Bockele-Morvan,D., Coustenis,A., Guillot,T., Schneider,J. 2000, A&A, 355, 573
[22] Sparks, W. B., Ford, H.C., 2002, ApJ, 578, 543.
[23] Vidal-Madjar, A. et al. 2003, Nature 422, 143
[24] Vidal-Madjar, A. et al. 2004, ApJ, 604, L69

# Solar System Objects with the NACO Fabry-Perot and SINFONI

M. Hartung[1], C. Dumas[1], T. M. Herbst[2], A. Coustenis[3], M. Hirtzig[3,4], M. Ádámkovics[5], F. Eisenhauer[6], C. deBergh[3] and A. Barucci[3]

[1] European Southern Observatory, Alonso de Cordova 3107, Santiago 19, Chile
mhartung@eso.org, cdumas@eso.org
[2] Max-Planck-Institut für Astronomie, Königstuhl 17, 69117 Heidelberg, Germany
[3] LESIA, Observatoire Paris-Meudon, 92195 Meudon Cedex, France
[4] Laboratoire de Planétologie et Géodynamique, 44322 Nantes, France
[5] 601 Campbell Hall, Department of Astronomy, University of California, Berkeley, CA 94720, USA
[6] Max-Planck-Institut für extraterrestrische Physik, Giessenbachstraße, 85741 Garching, Germany

**Summary.** 3-D spectroscopy assisted by adaptive optics (AO), provides an extremely efficient way to study the surface or atmospheric composition of planetary bodies. AO supported near-infrared instruments such as NACO and SINFONI – which provide both the high-angular resolution required for spatially resolving the disks of small solar system bodies and the spectral resolution adequate for compositional studies – open new windows of investigation to planetary scientists. Since their installation in UT4 (Yepun) at ESO/VLT, these instruments have been used to map the surface and atmospheric composition of small angular-size bodies such as Vesta and Titan, to study the spatial compositional variation of Pluto's moon Charon and to obtain spectra of faint bodies such as the trans-neptunian objects. Focusing on these targets, we demonstrate and compare the capabilities of the NACO Fabry-Perot (FP) imager (12) and the new integral-field spectrograph SINFONI (6; 10). The prospects for 3-D spectroscopy applied to the understanding of our solar system are discussed.

## 1 Introduction

Titan is an exciting example of the potential of 3-D spectroscopy in combination with AO. In particular, in the context of the Huygens-Cassini mission and the ground-based associated observing campaign (11), the synergy between ground-based observations and space missions becomes obvious. At the epoch of the descent several spectroscopic 3-D data sets were acquired with the VLT. A week before the Huygens probe was released from Cassini, NACO FP data were taken to trace $CO_2$ ice on Titan's surface (7), and one night after the Huygens descent that took place on 14 January 2005, a NACO FP scan (8) of Titan was performed at wavelengths sampling different atmospheric depths and the surface. Six weeks later (28 February 2005) Titan was observed during a commissioning run of the SINFONI integral field spectrometer. Using these data, (18) analyzed the aerosol distribution in Titan's atmosphere and derived H and K-band surface albedos. Both, the NACO and SINFONI data cubes have similar scientific potential (study of surface and atmospheric characteristics, tracing of chemical compounds). The SINFONI data have the advantage of a slightly higher spectral resolution (4000) than the NACO FP data (1100). In addition, the static PSF ensures a usable dataset even at variable seeing condition, whereas the FP scan relies on a stable seeing and AO correction during the scan. On the other hand, the imaging quality of the individual NACO FP frames is superior to the SINFONI ones. The latter should benefit in the near future of better image reconstruction software tools. Nevertheless, image quality of integral field units might always be slightly degraded with respect to "real" imaging devices.

Hereafter, we give a short insight into the potential of these data sets to study Titan and other (small angular size) solar system objects as Vesta and Pluto's moon Charon.

## 2 Tracing different layers of Titan's atmosphere

Spectral sampling in and outside the methane absorption bands allows the probing of different altitude layers in Titan's troposphere and stratosphere down to the surface. The width of the traced layer depends on the bandpass of the registered wavelength. Fig. 1 shows the images of a FP scan sampling the wing of a methane window from 2000 nm to 2180 nm. In this application, the FP can be considered as a tunable filter with a bandpass of 2 nm. Using a radiative transfer model of Titan's atmosphere each wavelength can be allocated to a specific altitude. The derived altitude can differ significantly for adjacent wavelength settings of the FP because of the varying methane absorption. Fig. 2 shows the nominal, and the minimum and maximum altitudes derived by the radiative transfer code of (17) according to the central wavelength and bandpass. The values are extracted from Tab. 2 from (8). These data monitor the long (seasonal) and short term dynamics of the aerosol distribution (haze) and clouds to constrain models for Titan's atmosphere. Ground-based observations allow to cover the periods between the Cassini flybys and to get a global snapshot during the flybys to cross-check and counter-calibrate particularly the highly model-depending results. The scientific outcome will be enhanced through 3-D spectroscopic monitoring during the on-going Cassini mission (until 2008 in total 44 planned Titan flybys).

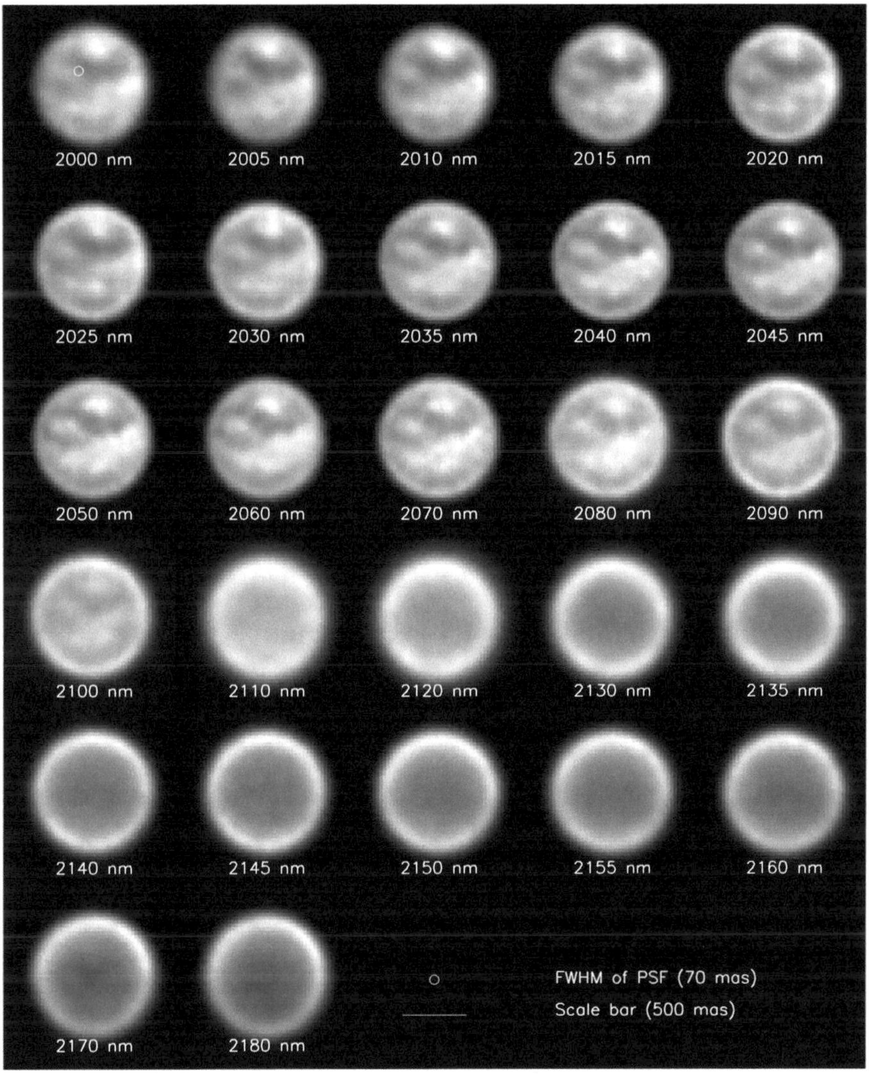


**Fig. 1.** Images of Titan taken at sub-Earth longitude 176W on 15 January 2005, the night after the Huygens-descent. The landing site is marked on the first image. The FP is tuned onto the surface (2000 - 2030 nm) and through different layers of the troposphere (2030-2150 n) and stratosphere (2150-2180 nm). The images are slightly deconvolved with a corresponding PSF star to enhance the contrast. (Lucy-Richardson with IRAF. The number of iterations ranges between 4 for the lowest signal-to-noise frames sampling the stratosphere and 26 for the highest signal-to-noise frames sampling the surface.) Linear intensity stretch scaling with maximum intensity for the individual frames.

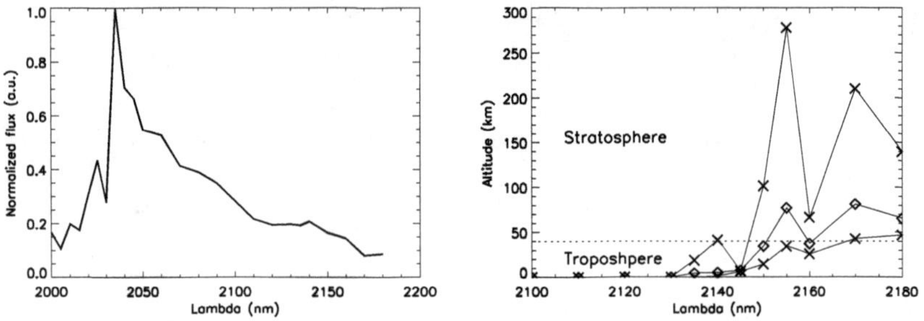


**Fig. 2. Left:** Total flux integrated signal over Titan's disk for the different wavelength settings in arbitrary units. The shape is dominated by the methane absorption in Titan's atmosphere. No telluric flux correction is applied. The highest flux levels occur within the methane absorption window where the surface is accessible. **Right:** The diamonds display the nominal altitude of the atmospheric layer sampled for a FP wavelength setting. The upper and lower crosses define the width of the sampled layer corresponding to a bandwidth of the FP (2 nm). The values are determined by radiative transfer calculations.

## 3 The Titan surface $CO_2$ ice experiment

It is known that planetesimals can be expected to contain large amounts of $CO_2$ ice. Large surface deposits of $CO_2$ ice have been found on Uranus' moon Ariel (13), as well as on Neptune's moon Triton (14). (16) reported laboratory results on near-infrared spectroscopy of simple hydro-carbons and carbon-oxides. In the case of Triton and Ariel, some narrow K-band absorption lines of these laboratory spectra allowed for an unambiguous detection of $CO_2$ ice, thereby setting a new challenge for models of the chemical pathways active within the surface ices of these satellites.

In principle, ground-based AO observations in combination with 3-D spectroscopy allow to map $CO_2$ ice on Titan's surface via spectroscopic features outside the atmospheric methane absorption bands. Two solid $CO_2$ ice vibration modes correspond to narrow absorption lines at 2011.6 nm and 2070.2 nm, and lie in the 2.05 $\mu$m methane window, which make this experiment possible. With a bandwidth of 2 nm the NACO FP matches well the width of these $CO_2$ ice absorption lines. These lines were scanned in steps of 1 nm for 11 nm spanning the central wavelength. Fig. 3 shows the collapsed, continuum, absorption line and difference images that are produced from the data cube. No differential $CO_2$ ice signal is detected at sub-Earth longitudes 284W and 307W (full disk coverage). Fig. 4 demonstrates an Ariel like $CO_2$ ice spot with a size of our PSF that would show up as 6$\sigma$ detection.

A detailed description of this experiment can be found in (7). No $CO_2$ ice has been found at sub-Earth longitudes 284W and 307W, neither on a NACO spectrum taken in K-band on 15 January 2005 at 199W (9). For the FP data we can conclude that the $CO_2$ ice coverage is less then 14% or 28%

inside the PSF region ($12.6 \cdot 10^5\,km^2$) for bright or dark regions, respectively. There may be a chance that solid $CO_2$ ice spots could be detected at other longitudes with this technique (1).

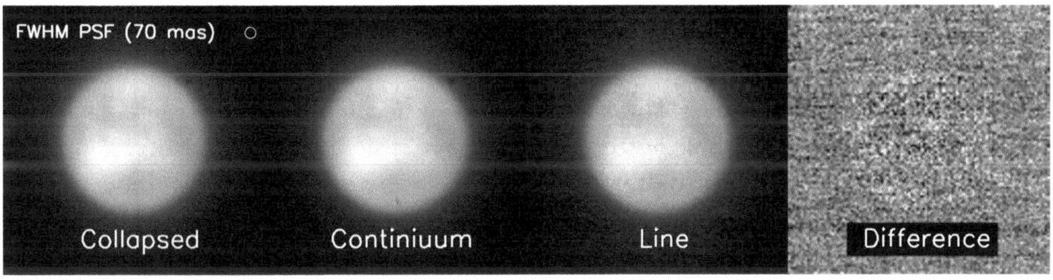


**Fig. 3.** From left to right a collapsed, continuum, absorption line and difference image is show from Titan at sub-Earth 307W and 2070 nm. There is no $CO_2$ ice region revealed at either of the scanned absorption lines at this rotation phase. (The corresponding images for the 2012 nm absorption line is not displayed.) The grey scale of the difference images stretch from -2 to 2 ADU (Analog Digital Unit), adopted to an average signal of 0 ADU with a noise of $1\sigma$.

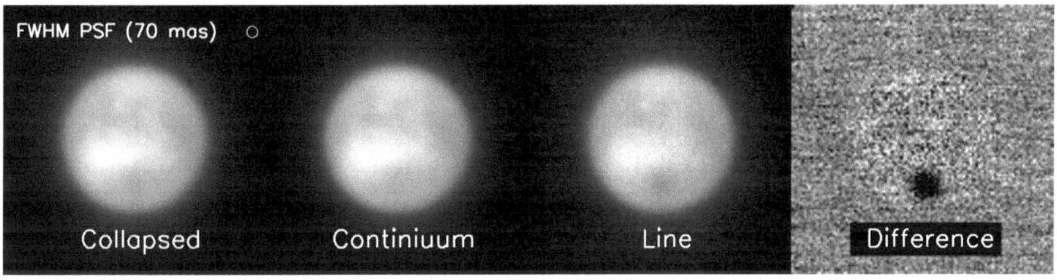


**Fig. 4.** *Simulation of an Ariel-like detection.* From left to right a collapsed, continuum, absorption line and differenced image is shown. An Ariel like $CO_2$ ice spot with a size of our PSF (65 mas corresponding to 400 km) is simulated on a dark region (harder to detect due to lower contrast) of Titan. The absorption signal already can be seen on the line image, but is clearly revealed ($6\sigma$) on the difference image as a dark spot. The grey scale of the difference image stretches from -2 to 2 ADU.

## 4 3-D reflectance spectroscopy of Charon and Vesta

The study of the small bodies of our solar system, in particular the populations of asteroids and trans-neptunian objects, can provide insight into early stages of the solar system evolution. Here we describe preliminary results obtained from using the combination of 3-D spectroscopy and high-angular resolution techniques provided by SINFONI on Pluto's moon Charon and the main-belt asteroid Vesta.

### 4.1 Vesta

SINFONI was used to carry out a compositional study of the surface of the main-belt asteroid Vesta. In comparison to other small solar system bodies, Vesta is known to display the second largest hemispheric contrast in albedo (after Saturn's satellite Iapetus). Also, Vesta is considered as being unique among the asteroid main-belt, as the sole intact asteroid that may have undergone complete planetary-style differentiation (i.e. the heaviest material has sunk towards the body's center). The similarity between its disk integrated spectrum and the spectrum of the HED meteorites (Howardites, Eucrites, Diogenites) collected on Earth support the idea that Vesta could be their parent body. Eucrites and diogenites have formed from magmas produced inside the asteroid and migrated toward the surface. Studying their distribution and characteristics, including the material between them, the howardites, will help understand the igneous processes that occurred at an early stage of formation of what can be considered as the smallest terrestrial planet in the solar system. HST images of Vesta obtained in 1997 (15) revealed that its southern regions display a large impact crater which has excavated the surface material to reveal the internal composition of the asteroid. The low latitude regions of the asteroid provide a unique terrain to explore the composition of Vesta's mantle.

The asteroid Vesta (rotation period 5.3h) was observed at three occasions as part of the SINFONI science verification program. Vesta, which was transiting near zenith (DEC -15deg) for Paranal, displayed a 0.5" diameter during the August and October 2004 runs, and its brightness (V 6) provided a case for high-Strehl performances by MACAO. Furthermore, with an aspect angle of $\sim$90 degrees at that time, the geometry of this opposition was particularly favorable to carry out a detailed mineralogical study of Vesta across the 1-2.4 $\mu$m region of the spectrum (Fig. 5). We used the J and H+K grisms of SINFONI to spectroscopically map, at several rotational phases, the surface of Vesta. At these wavelengths, its reflectance spectrum is modulated by the absorption bands of pyroxene, feldspar and olivine. The detailed analysis and spectroscopic modelling of the SINFONI's data is still underway, but by quantifying the spatial distribution of eucrites, and diogenites at an unequalled spatial resolution, and measuring the abundance of olivine-rich regions excavated from the below the crust, we expect to better understand how, and how much, Vesta's interior melted soon after its formation within the proto-planetary nebula. Fig. 6 shows the relative variation in the depth of the pyroxene band across both hemispheres of the asteroid. Variations in the downward slope of this band can notably be seen and are consistent with older eucritic rock in the northern hemisphere, while younger diogenitic rock is found near the South pole region. The J-band analysis is undergoing and will provide additional insights on the location of olivine-rich regions on the surface of Vesta, which could be associated to regions where deep cratering and removal of the mantle material occurred in the early stages of Vesta's history.

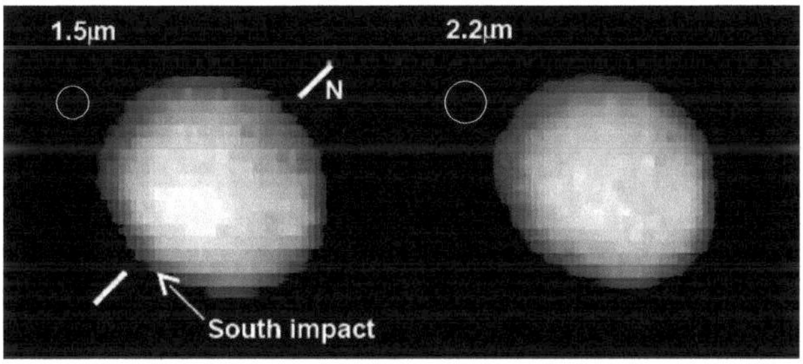


**Fig. 5.** H (left) and K (right) band images of Vesta extracted from the reconstructed SINFONI data-cube. Note the Southern pole peak, which is better visible at shorter wavelengths due to the improvement in spatial resolution. The direction of the asteroid spin axis is also noted on the figure. Vesta's angular diameter was near 0.5" at the time of these observations. The FWHM of the PSF is approximately 55 mas and 70 mas for 1.5 $\mu$m and 2.2 $\mu$m, respectively (indicated by the circles).

## 5 Charon

Until recently (3), a separate study of Pluto and its satellite Charon has long been hampered by the small angular separation between the two bodies (0.9" max elongation). The mid-1980's occultation events revealed that the surface of Charon was mainly covered with water ice, but a precise compositional diagnostic could only be carried out recently from high-contrast imaging. (4) and (5) obtained spatially separated spectra of Charon and Pluto with Keck and HST/NICMOS which show that water ice is in its crystalline state over Charon, supporting the hypothesis of a young, continuously "refreshed surface". In addition, these data revealed the presence of an absorption feature at 2.21 $\mu$m that was attributed to ammonia hydrates on Charon. Although this compound is expected to play an important role into the geological activity of the small bodies of outer solar system, it has only been scarcely reported in its solid form (e.g. (2) has observed a similar feature in the spectrum of Uranus moon Miranda). Also, ammonia ice is expected to sublimate rapidly on Charon's surface and competing mechanisms must be invoked to explain its presence over the age of the solar system. Its production could be for instance of endogenic origin, ammonia ice slowing migrating toward the surface from a reservoir within Charon, or exogenic through bombardment of its icy surface by $N^+$ ions released from Pluto's atmosphere. In addition, the HST

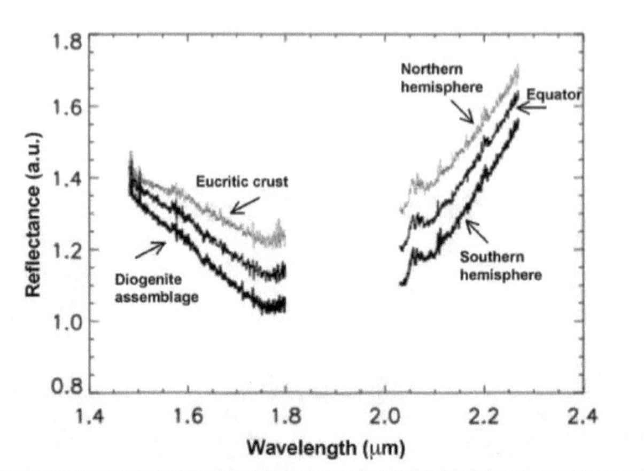


**Fig. 6.** Reflectance spectra of Vesta extracted from the SINFONI data-cube for three different latitudes of the asteroid. At these wavelengths, the spectrum of Vesta is shaped by the strong absorption band of pyroxene. The spectra have been displayed with a vertical offset to improve clarity. Note the variation of the downward slope of the pyroxene band, which reflects the presence of various types of igneous assemblages on the surface. The Northern hemisphere is dominated by an older eucritic crust (shallower band) while the Southern regions (where a large impact occurred) is dominated by diogenetic assemblage, displaying a deeper pyroxene band. Similar analysis of the J-band spectra will lead to the identification of olivine-rich regions, resulting in deeper mantle material to be exposed to the surface after impact resurfacing processes.

data obtained on Charon revealed a possible hemispheric variation in the band strength of ammonia, which could be interpreted as some supporting evidence for exogenic production of this compound.

To address these questions SINFONI observations of the Pluto-Charon system were performed at several rotational phases (Charon's orbit is tidally locked). The main goal was to measure the variation in the spatial distribution of the ices and test particularly whether ammonia ice was present over the entire surface or concentrated on one hemisphere. We used the H+K grism of SINFONI and a pixel scale of 0.025mas/pix, (FOV 0.8" x 0.8") to separate both components and limit the amount of scattered light contamination from Pluto. The AO system was locked on Pluto (V 13) while we offset the field of view (FOV) of SINFONI to its satellite Charon. Fig. 7 shows two spectra obtained for different dates and intermediate separations (0.6"). The 2.21 $\mu$m feature is clearly visible with similar strength in both spectra. This result supports a uniform distribution at the surface of the satellite and weakens the scenario of an exogenic production of ammonia from implantation of N+ ions on its surface. A reservoir of ammonia, relic of the early age of the solar system, and buried under the icy crust, could still be present on Charon.

Additionally, Pluto's disk being resolved in our image-cube, we will attempt to measure a possible hemispheric variation of ices such as CO, $N_2$, and $CH_4$ on Pluto.

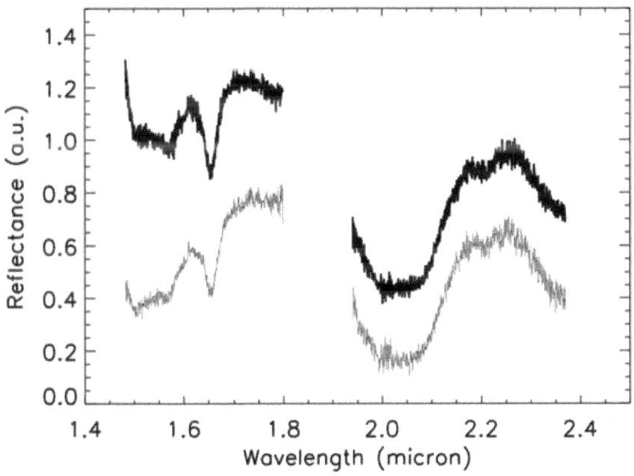


**Fig. 7.** H/K band reflectance spectra of Charon at two different orbital positions of the satellite. The 1.65 $\mu$m feature due to the crystalline state of water ice over Charon is clearly visible, as well as the 2.21 $\mu$m feature attributed to the presence of ammonia hydrate on the surface. These results tend to support a uniform abundance of ammonia within the water ice coating the surface of the satellite.

## References

[1] Barnes, J. W., Brown, R. H. Turtle, E. P. et al: Science, **310**, 92 (2005)
[2] Bauer, J. M., Roush, T. L., Geballe, T. R. et al: Icarus, **158**, 178 (2002)
[3] Brown, M. E., van Dam, M. A., Bouchez, A. H. et al: ApJ, **639**, L43 (2006)
[4] Brown, M. E., Calvin, W. M., Science, **287**, 107 (2000)
[5] Dumas, C., Terrile, R. J., Brown, R. H. et al: AJ, **121**, 1163 (2001)
[6] Eisenhauer, F., Abuter, R., Bickert, K., et al: SPIE **4841**, 1548 (2003)
[7] Hartung, M., Herbst, T., Dumas, C., et al: Journal of Geophysical Research, (2006)
[8] Hirtzig, M., Coustenis, A., Gendron, E., et al: Journal of Geophysical Research, in press (2006)
[9] Negrão, A., Hirtzig, M., Coustenis, A., et al: Journal of Geophysical Research, in press (2006)
[10] Bonnet, H., Conzelmann, R., Delabre, B. et al: SPIE, **5490**, 130 (2004)
[11] Wittasse, O., Lebreton, J.-P., Bird, M. K., et al: Journal of Geophysical Research, in press (2006)

[12] Hartung, M., Lidman, C., Ageorges¡ N., et al: SPIE, **5492**, 1531 (2004)
[13] Grundy, W. M., Young, L. A., Young, E. F., Icarus, **162**, 222 (2003)
[14] Cruikshank, D. P., Roush, T. L., Owen, T. C. et al: Science, **261**, 742 (1993)
[15] Thomas, P. C., Binzel, R. P., Gaffey, M. J. et al: Science, **277**, 1492 (1997)
[16] Quirico, E., Schmitt, B., Icarus, 128, **181** (1997)
[17] Rannou, P., McKay, C. P., Lorenz, R. D., Planet. Space Sci., **51**, 963 (2003)
[18] Ádámkovics, M., de Pater, I., Hartung, M., Journal of Geophysical Research, in press, (2006)

# 3D Spectroscopy with an Imaging FTS

J.-P. Maillard

Institut d'Astrophysique de Paris, 98b Blvd Arago, F-75014 Paris, France
maillard@iap.fr

**Summary.** Among the solutions for 3D spectroscopy, an Imaging Fourrier Transform Spectrometer (IFTS) has the unique capability of making possible flexible high spectral resolution on a wide field. After a brief review of scientific programmes which need this combination, illustrated with results from BEAR, a prototype IFTS, the perspectives for IFTS at Dome C and on an ELT are presented.

## 1 Introduction

There are two separate classes of application for 3D spectroscopy, which imply different instrumental solutions: i) compacts objects at very high spatial resolution for which fields of few arcsecs are enough; ii) wide objects which extend over several arcmins. In the first class are, the disks and jets of young stars, the envelopes around AGB stars, the nucleus and disk of distant galaxies. The science cases for wide-field 3D spectroscopy include galactic star-forming regions, nearby planetary nebulae, super-novae remnants, comets and particular crowded regions as the Galactic Centre. Quite often, the active parts of these regions are deeply embedded, attenuated by high dust extinction. Therefore, the spectral domain of analysis is the infrared or the submillimetric range. We focus here on the wide-field case in the infrared domain which makes possible to reach the cold to warm parts of these regions. But, most 3D spectrometers are only capable of small fields. They are built around a medium resolution, long-slit grating spectrometer. An integral field unit (IFU) is placed before the spectrometer slit, which provides the spatial sampling of the field and feeds the entrance slit (e.g. fiber bundle, microlens array, image slicer). In the infrared, the system of choice is the image slicer, an all-mirror solution (see review by J. Allington-Smith in these proceedings). To approach the wide-field problems, almost absent in this 3D Conference, another type of instrumentation is required. This goes for an interferometric solution, Fabry-Perot or imaging Fourier transform spectrometer (IFTS). If in addition, a high spectral resolution ($\geq 5\times10^4$) is needed, particularly for the study of the gas content, the only possible solution is an IFTS.

## 2 High Spectral Resolution with a Wide-Field IFTS

Details on the principle and the properties of this type of instrument have already been presented (1), (2). The core instrument is a step-by-step FTS. The optical path difference (OPD) of the Michelson interferometer is stopped to integrate the interferometric signal from the source. The signal is stored and the moving arm displaced rapidly by a precise step, whose size is related to the spectral bandpass of the signal, for a new integration. The constant elementary integration time (at each step) is chosen to record data with the appropriate signal-to-noise ratio in a single scan.

### 2.1 Wide-Field Imaging

To make an imaging FTS, the single pixel detector of a standard FTS is replaced by a 2D-array. The entrance field is directly imaged onto the detector array. As a dual-output FTS is preferred for not loosing half of the energy, the two detectors must be replaced by two detector arrays. For economy reasons the adaptation of an interface to bring the two output beams on a single camera was chosen for BEAR, a prototype IFTS which was in operation on the Canada-France-Hawaii Telescope until 2001. An image is recorded at each step of the OPD. Each pixel acts as the detector of the corresponding point of the entrance field. The number of spatial elements analysed in a single scan is independent of the number of spectral elements. The detector array defines the usable field. With a large array, a wide-field IFTS can be obtained. This property is well illustrated with SpIOMM (3) developed for the 1.6-m Mt Megantic telescope (Quebec), equipped with two CCD cameras. The imaged field has a size of 12'×12'. With 1K×1K CCDs the plate scale is equal to 0.55"/pixel, which is well adapted to the seeing quality of the site. The optical design is based on a slightly off-axis Michelson interferometer to have access to the two complementary beams. With this design, the maximum OPD is limited to $\pm 0.5\,\text{cm}$.

### 2.2 High Spectral Resolution

By adopting the cat's eye design (1) as on BEAR, a high spectral resolution is compatible with a large field. The total number of steps defines the spectral resolution, which translates into the number of planes in the data cube. However, there are interferometric limitations, which are not so much of a constraint thanks to the throughput advantage of an interferometer. At OPD $\delta$, a ring of the fringe pattern of order $k$ at the wavelength $\lambda$ has an angular radius $i_k$ such as:

$$i_k = (2\lambda/\delta \times k)^{1/2} \tag{1}$$

From the derivative of Eq. 1 it comes:

$$i\mathrm{d}i = \lambda/\delta \times \mathrm{d}k \tag{2}$$

The angular spacing from one ring to the next ($\mathrm{d}k = 1$) corresponds to $\mathrm{d}i$. When $\mathrm{d}i$ becomes equal to the projected pixel size, the fringe modulation is null. For a square array of $N \times N$ pixels adapted to a field-of-view of angular diameter $\theta$, at the edge of the field:

$$\mathrm{d}i = 2i_{max}/N \quad \text{with} \quad i_{max} = \theta \times (D_T/2D_I) \tag{3}$$

where $D_T$ is the telescope diameter and $D_I$ the parallel beam diameter in the interferometer. By combining Eq. 2 and 3 it comes:

$$i\mathrm{d}i = \lambda/\delta_{max} = 2/R_{max} \tag{4}$$

with $R_{max}$ the maximum resolution reached at null modulation. Finally, from Eq. 3 and 4, there corresponds a maximum field $\theta_{lim}$ defined by:

$$\theta_{lim} = 2\frac{D_I}{D_T} \times \sqrt{\frac{N}{R_{max}}} \tag{5}$$

## 3 Science Highlights of BEAR at CFHT

The BEAR instrument (1) (2) resulted from the coupling of the facility high-resolution, Cassegrain, infrared FTS on the 3.6-m CFH Telescope with a HgCdTe NICMOS3-type camera (258×258) as imaging detector. The plate scale on the detector was chosen equal to 0.35"/pixel for a circular field of 24" (70×70 pixels size). The camera was equipped with a set of narrow-band filters ($\Delta\lambda/\lambda \simeq 1\,\%$) centered on the prominent lines of the K band, like Br$\gamma$ (2.16 $\mu$m), $H_2$ 1-0 S(1) (2.12 $\mu$m), HeI (2.06 $\mu$m), making possible high resolution data cubes up to R $\simeq$ 30,000 with $\sim$ 1200 planes in the spectral dimension. The programmes, which illustrate scientific cases of 3D spectroscopy at high resolution, were devoted to various types of emission line objects. In many cases, lines other than the main line were detectable within the filter bandpass, which is a major interest of the method. The filter width, corresponding to at least a velocity range of 3000 km s$^{-1}$, allows a large velocity field (for example in the Galactic Centre) to be covered.

*Planetary atmospheres:*

The north and south Jupiter auroral zones (5) were observed, with the most complete longitude coverage, in order to map the $H_2$ and $H_3^+$ lines. For the first time a mapping of the $H_2$ aurora was obtained, simultaneously with the $H_3^+$ aurora, showing evidence of a noticeable difference of morphology. A temperature map of the $H_3^+$ emission showing a hot spot, not present in $H_2$, was deduced.

*Star-forming regions:*

The environment of S106-IR (6), a young massive star ($> 15\,M_\odot$), source of an important bipolar HII region, was observed in $H_2$ 1-0 S(1) and 2-1 S(1) and in Br$\gamma$, probing respectively the molecular cloud and the ionized region in which localised emissions of HeI and [FeIII] were also detected.

*Young planetary nebulae:*

The same filters as above were used to observed pre-planetary nebulae (PN), representing steps in the evolution from AGB to PN. For NGC 7027, from the high resolution velocity information, it was possible to construct a 3D-map of the molecular envelope (Fig. 1) demonstrating the fundamental role of bipolar jets from the central star in dispersing the molecular envelope in pre-PNs.

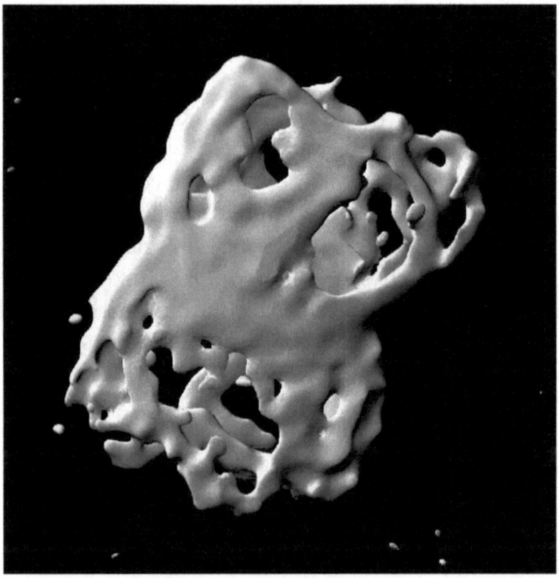

**Fig. 1.** 3D View of the $H_2$ molecular envelope of the young PN NGC 7027 (from Cox *et al.* (7)). The images of a sub-cube extracted from the original data and covering the full velocity width of the $H_2$ line have been deconvolved with a stellar PSF (from Mekarnia (8)). The velocity is used as the third dimension in the 3D representation showing clearly the holes in the molecular envelope made by high-velocity jets from the central white dwarf.

*Galactic Centre:*

The central parsecs of the Galaxy centered on the massive black hole SgrA$\star$ harbours a cluster of massive stars and an HII region. The members of the cluster were identified with their HeI 2.06 µm emission line and the ionized gas with the same line and Br$\gamma$. The hot stars can be grouped in two classes: Luminous Blue Variables and Wolf-Rayets, two main stages of massive star evolution. The HII region was decomposed in 9 separate velocity structures (9), which appear as the excited surfaces of molecular clouds gravitationally extracted from the circumnuclear disk surrounding the central parsecs.

# 4 Science Perspectives of Imaging FTS

Based on the BEAR results, several proposals were made for building an Imaging FTS to do astronomical 3D spectroscopy, which could make use of the unique properties of such an instrument. They were first presented at the call for ideas for the Next Generation Space Telescope (currently JWST) (10). Space can be considered as the ideal solution for an infrared IFTS, with extremely low thermal background, stable transparency and diffraction-limited images over a wide field. The proposals were not successful. However, from this initiative resulted a proposal for a ground-based instrument - SpIOMM (3). Then, a proposal was made for the VLT second generation instrumentation (4), specifically to equip one VLT with a capability of 3D spectroscopy at high spectral resolution, in the near infrared. The interest of integral field spectroscopy was recognized, but other choices were made. The future developments of two new observing capabilities justify the renewal of this proposition.

## 4.1 IFTS at Dome C

The rationale for proposing an IFTS in Antarctica (11) is based on the combination of the exceptional qualities of the site relevant to IR astronomy and the properties of the instrument itself. Although not as low as in space, the sky brightness in the 2.3 to 25 $\mu$m range is nevertheless the lowest that can be obtained from the ground. The low water vapor content ($\sim$ 200 $\mu$m p.w.v.) ensures optimum transparency in the atmospheric windows. Seeing is an issue but which can be probably corrected with low-order adaptive optics. In addition, being a ground-based site, a high resolution instrument can be seriously considered without the limitations of weight and size of a spatial instrument. A cryogenic instrument covering all the infrared spectral range, on a 7'×7' field, and with capabilities of high spectral resolution (limit of resolution of 0.01 cm$^{-1}$) through narrow-band filters would be a unique facility.

## 4.2 IFTS on an ELT

The next generation of telescopes, called "Extremely Large Telescopes" (ELTs), pose a serious challenge to the instrument builders to make spectrometers able to accept the huge étendue delivered by a 30-meter, or larger, telescope. With an increase in sensitivity of 3 mag for a 30 m compared to a 8-m telescope, all the astronomical fields will become crowded by the underlying components on the same line of sight. 3D spectroscopy will be more mandatory to properly separate the background contribution from the sources of interest. Again, the throughput advantage shown by Eq. 5 of an IFTS could be used to get uncontaminated spectra of particular objects. For an IFTS matched to a 30-m ELT, with a field of 3' ($\theta_{lim}$) , a 1K×1K detector array, a plate scale of 0.175"/pxl, and beams in the interferometer ($D_I$) of 50.0 mm, one obtains:

$$R_{max} \simeq 1.4 \times 10^4 \quad \text{or} \quad dv \simeq 22\,km\ s^{-1} \tag{6}$$

in the 0.6 to 5 $\mu$m region, chosen to minimize the thermal background. This performance could be obtained with an instrument of modest size compared to the huge grating spectrometers which are being proposed.

## 5 Conclusions

As illustrated in this conference, the 3D spectrometers currently in operation behind the large telescopes are based on IFU's, the most advanced with multiple IFUs. These instruments are well adapted to deep extra-galactic projects requiring small fields and low spectral resolution. To undertake the study of galaxy clusters, a larger field is needed. That is the rationale of the MUSE project (12). Using the IFU technique 24 image slicers and 24 grating spectrometers in parallel are needed to cover a 1'×1' field. It should be accepted to explore some other avenues for wider fields and the capability of high spectral resolution.

## References

[1] J.P. Maillard: 3-D Spectroscopy with a Fourier Transform Spectrometer. In: *Tridimensional Optical Spectroscopic Methods in Astrophysics*, ed. by G. Comte, M. Marcelin, IAU Coll. 149, ASP Conf. Ser. vol 71, pp 316–327, (1995)

[2] J.P. Maillard: BEAR imaging FTS: High Resolution Spectroscopy in Infrared Emission Lines. In: *Imaging the Universe in Three Dimensions: Astrophysics with Advanced Multi-Wavelength Imaging Devices*, ed. by E. van Breughel, J. Bland-Hawthorn, ASP Conf. Ser. vol 195, pp 185–197, (2000)

[3] F. Grandmont, L. Drissen, G. Joncas: Development of an Imaging Fourier Transform Spectrometer for Astronomy. In: *Specialized Optical Developments in Astronomy*, ed. by E. Atad-Ettedgui, S. D'Odorico, (SPIE 2003), vol 4842, pp 392–401

[4] J.P. Maillard, R. Bacon: A Super-Imaging Fourier Transform Spectrometer for the VLT. In: *Scientific Drivers for ESO future VLT/VLTI Instrumentation*, ed. by J. Bergeron, G. Monnet, ESO Astroph. Symp. (Springer 2002), pp 193–198
[5] E. Raynaud, E. Lellouch, J.P. Maillard, et al.: Icarus **171**, 133 (2004)
[6] B. Noel, C. Joblin, J.P. Maillard, T. Paumard: A&A **436**, 569 (2005)
[7] P. Cox, P. Huggins, J.P. Maillard, et al.: A&A **384**, 603 (2002)
[8] D. Mekarnia, private communication (2005)
[9] T. Paumard, J.P. Maillard & M. Morris, A&A **426**, 81 (2004)
[10] J.P. Maillard: Comparison of two concepts of imaging FTS. In: *NGST Science and Technology Exposition*, ed by E.P. Smith, K.S. Long, (2000) ASP Conf. Ser. vol. 207, pp 479–483
[11] J.P. Maillard, C. Joblin: Wide Field, Integral Field Spectroscopy at Dome C with an Infrared Imaging FTS. In *Dome C Astronomy and Astrophysics Meeting*, ed by M. Giard, F. Casoli, F. Paletou, (EDP Sciences 2005), EAS Pub. Series, vol 14, pp 187–192
[12] M.M. Roth, R. Bacon: MUSE - The Multi Unit Spectroscopic Explorer. In: *Science Perspectives for 3D Spectroscopy*, this conference (2006)

# Part III

# Nearby Galaxies

# Nearby Galaxies: 3D Spectroscopy as a Tool, not as a Goal

E. Emsellem[1]

Centre de Recherche Astrophysique de Lyon emsellem@obs.univ-lyon1.fr

**Summary.** We briefly review here the progress made using integral-field spectroscopy in the field of nearby galaxies, via the illustration of a few recent results.

## 1 A turn we need to take

There is no need for a long diatribe to demonstrate that 3D spectroscopy has now reached a certain maturity, at least in the sense of not being anymore a private tool for a restricted number of aficionados. It is however not so clear if we, astronomers, have yet attained that level in the way we make use of these facilities. There were times when long-slit spectroscopy by itself would solely motivate the writing of a paper: this naturally changed with the democratization of such instruments, that we today call "standard" (even sometimes with a slight distaste). The fact that we still need a scientific Workshop devoted to 3D spectroscopy seems to imply that the final step is still to be taken for integral-field units (IFUs): the one which will make them one tool among many to probe our pet astronomical targets, to be used appropriately in combination with other instrumentation. This conference nevertheless marks the turn of such an era, and I wish to illustrate this by mentioning some of the progress made via 3D spectroscopy in the context of nearby galaxy studies.

## 2 Potential for discovery

The ability to produce two-dimensional maps of our favorite physical quantity (e.g., velocity, stellar population age, molecular gas distribution) is probably the most easily advertised asset of 3D spectroscopy. Long-slit radial profiles now appear like skimpy versions of these maps, which at last allow to reveal the full extent of a presumed structure, or even serve as a source for discovery. This is illustrated in Fig. 1, where counter-rotating structures are unambiguously exposed to view in the stellar velocity maps of e.g. NGC 4550 (left panel, (15)) or NGC 770 (right panel, (21)).

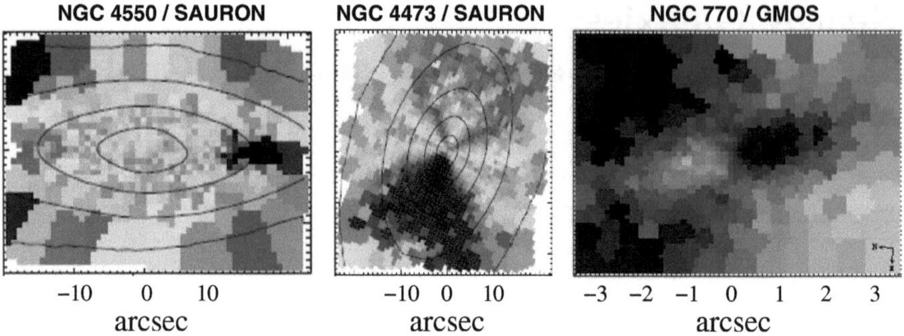


**Fig. 1.** Counter-rotating structures in nearby galaxies viewed by IFUs. From left to right: stellar velocity fields of NGC 4550 and NGC 4473 (SAURON; extracted from (15)) and NGC 770 (GMOS; extracted from (21)). Note the different scales in arcseconds in the panels. In the case of NGC 4473, a detailed dynamical modeling is required to reveal the counter-rotating systems (5).

However, the detection of such structures is not always straightforward. This is illustrated in the case of NGC 4473 (15), where only a detailed modeling of the full photometry and kinematics unequivocally identifies the presence of two large-scale counter-rotating disk-like systems (5).

## 3 Probing complex regions

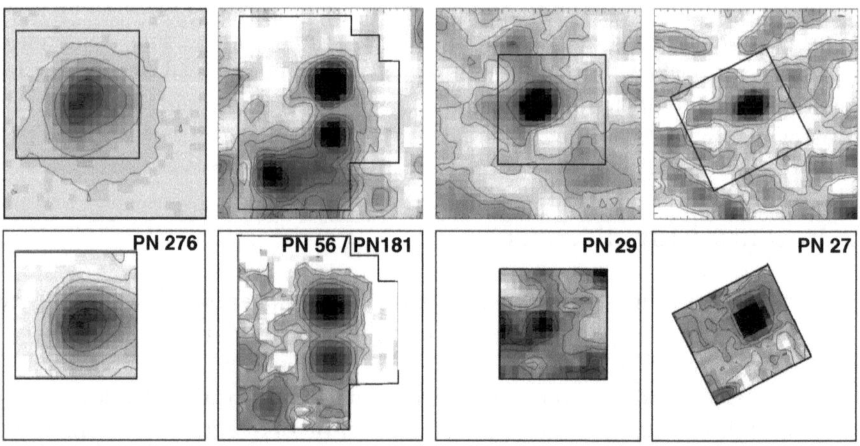


**Fig. 2.** Images of planetary nebulae in M 31 (top panels), and as observed by (30) with the PMAS IFU (bottom panels).

IFUs are obviously ideal tools to extract sources in crowded regions (Fig. 2), or to cover regions were multiple components spatially overlap. A small survey of 20 interacting systems was performed by Sanchez and collaborators (31) with VIMOS and PMAS, to analyze the ionization processes in different parts of the galaxies via their gas emission lines. IFUs are then very useful to disentangle regions with different physical states without too strong a priori assumptions on their relative spatial distribution. A more central view at such objects (1) hold keys to the determination of the dynamical phase of the merger, and its possible link with efficient central gas accumulation. In such extreme environments, super stellar clusters can sometimes be formed at the interface region, the study of which can certainly help us constrain relevant mechanisms for the triggering of star formation (3).

## 4 Dynamics in the central kiloparsec

The wealth of structures revealed by 3D spectroscopy in nearby galaxies has led to a rather impressive number of papers on individual targets, specifically focusing on the inner regions, e.g. probing the kinematics and intrinsic shapes of early-type galaxies (34; 25), searching for massive black holes (13; 6; 24), studying the Narrow Line Regions of Seyfert galaxies (23; 19; 28). More recently, Davies et al. (11) had a closer view on a few active nuclei with the advent of adaptive optics assisted near-infrared IFUs such as SINFONI on the VLT, probing the radius of influence of the presumed nuclear black holes at a scale of a few parsecs.

Asymmetries and offsets are often observed in the kinematics of Seyferts, but should be interpreted with caution, considering the eventual role of dust extinction, and the fact that the observed gaseous component is generally the result of overlapping (unresolved) systems with rather different dynamical and ionization status (shocks, photoionization, AGN related). We should also not be surprised to observed large differences between the stellar and gaseous kinematics, as stars do not share the dissipative nature of the gas clouds. Dissipation is one of the key to explain the accumulation of gas in the central few tens of parsecs. It is not clear yet if inner bars generally have a significant role at these scales (see work by Dumas, these Proceedings; (29)), although a few cases, such as the prototypical Seyfert 2 NGC 1068 (Fig. 3, (16; 22)), have clear signatures of a bar driven inflow. This sometimes leads to the formation of a cold inner disk-like structure, the signature of which is revealed in stellar velocity dispersion map as a central dip, a so-called $\sigma$-drop (14; 35). These $\sigma$-drops seem rather common in spiral galaxies and could be linked with recurrent star formation episodes. Line indices, as provided by instruments such as MPFS (33), SAURON (26) or the ARGUS mode of FLAMES (17), would set constraints on the underlying stellar populations, and help to quantify the age and extent of these central starbursts.

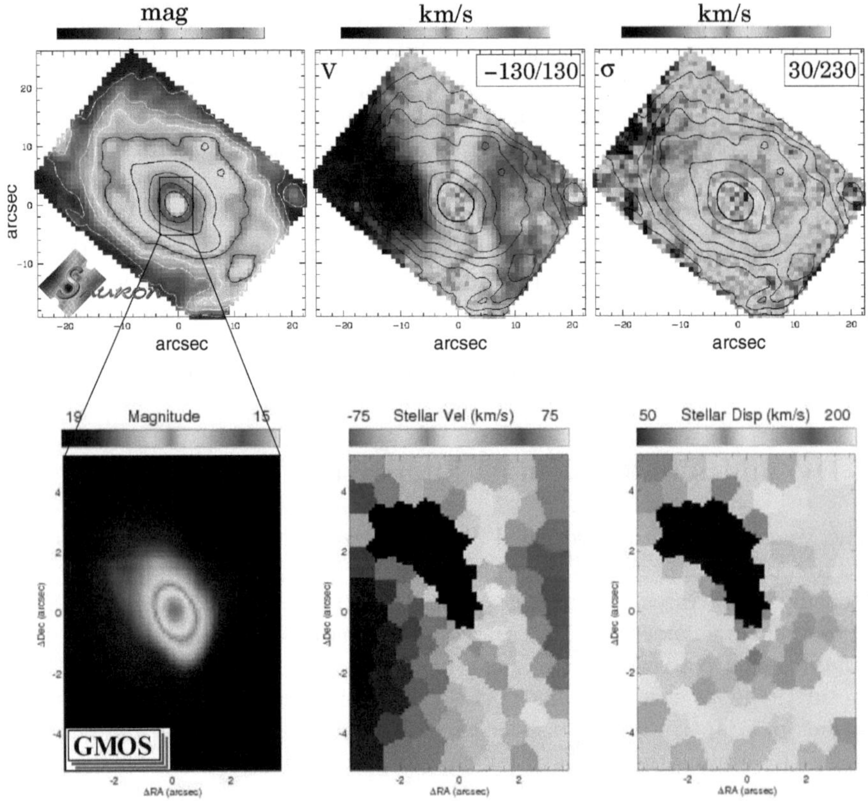


**Fig. 3.** IFU observations of NGC 1068: the central 2 kpc from SAURON (top panels, extracted from (16)), and a zoomed version with GMOS (bottom panels, extracted from (22)). From left to right: reconstructed images, velocity and velocity dispersion maps. The velocity twist due to the presence of the bar is clearly seen in the SAURON data. The $\sigma$-drop hinted in the SAURON dispersion map is emphasized in the GMOS data (see (22) for details).

## 5 Surveys at last

A truly remarkable change in our usage of IFUs is the advent of the first "surveys" of nearby galaxies. We are obviously not (yet) considering hundreds of targets, but surveys of a few tens of galaxies are still a remarkable achievement in this context. Heterogeneous samples were already obtained during the last decade with IFUs such as e.g., TIGER and OASIS (CFHT) by our group in Lyon, INTEGRAL (WHT, work by Arribas, Mediavilla, et al.) and with MPFS (Russian 6m, Sil'Chenko and collaborators). The SAURON project (Lyon/Leiden/Oxford) is the first to address a representative sample of nearby objects (2; 12), with 72 early-type (E/S0/Sa) galaxies with total luminosity $M_B$ ranging from -18 to -22. Besides the wealth of structures

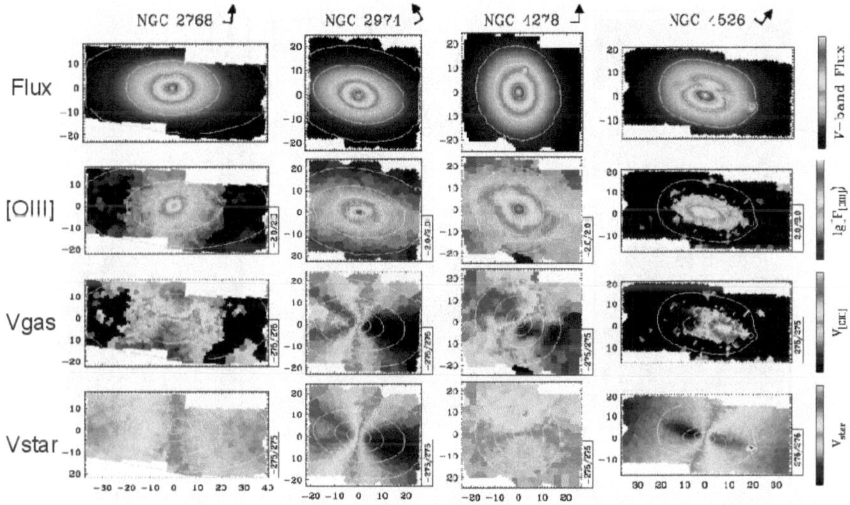


**Fig. 4.** SAURON maps of 4 early-type galaxies (32). From top to bottom: reconstructed image, [OIII] emission line flux, gas velocity and stellar velocity maps. See (32) for details.

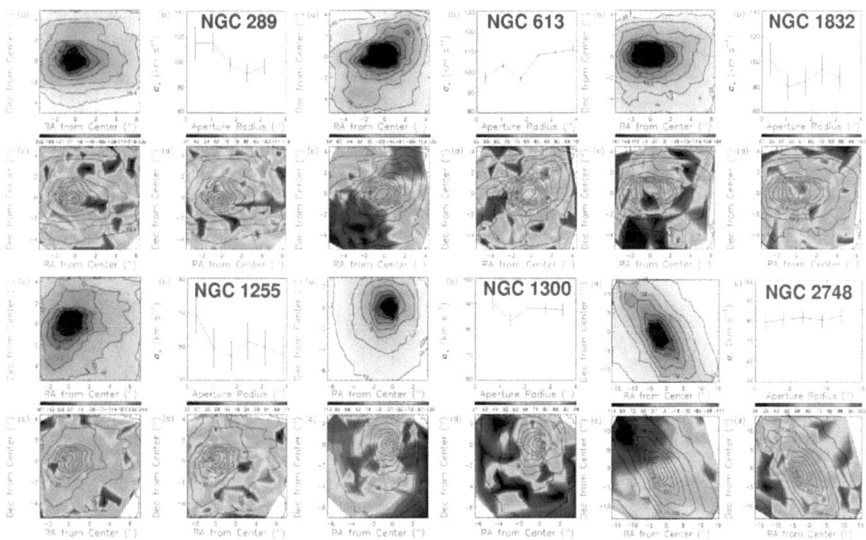


**Fig. 5.** 6 bulges of spiral galaxies observed with INTEGRAL (see (4) for details), with the reconstructed image and growth curve (top), and the velocity and dispersion maps (bottom).

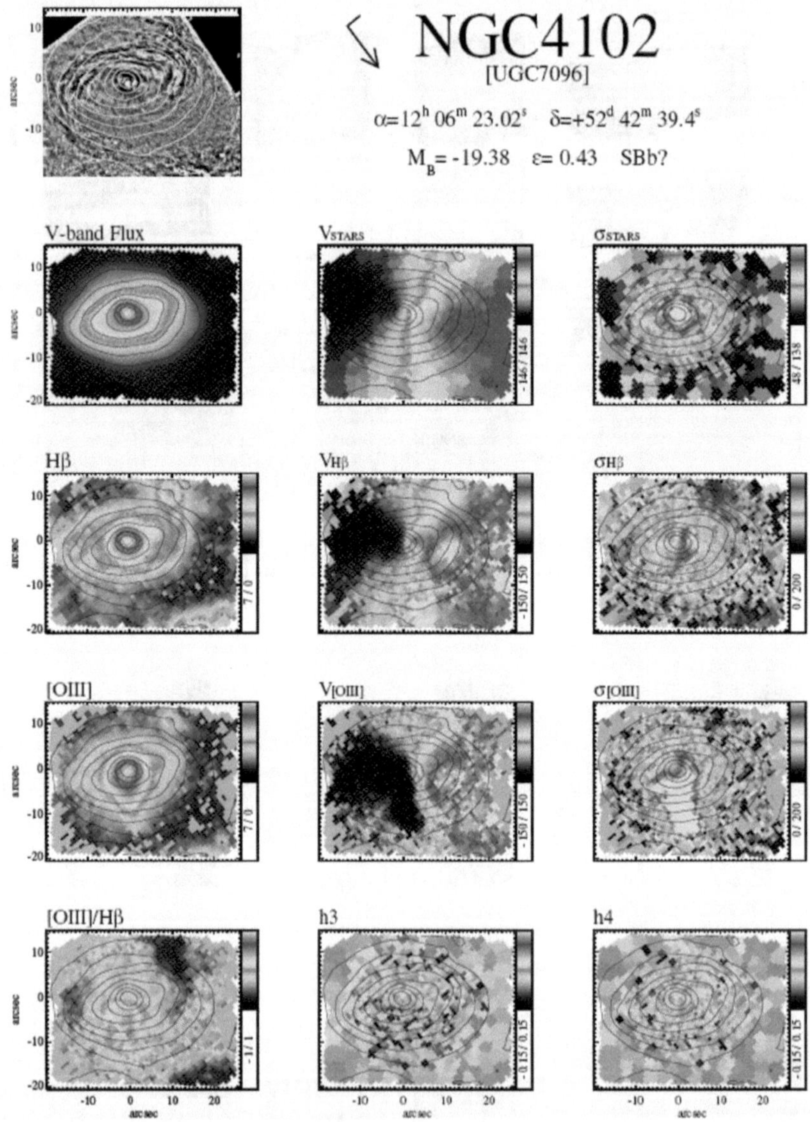


**Fig. 6.** SAURON maps of NGC 4102 obtained with SAURON (20). The top panel shows an unsharp masked HST image of this galaxy within the SAURON field of view. Other maps present the stellar + gas distributions and kinematics as well as an emission line ratio map. Galaxy extracted from the sample of (20).

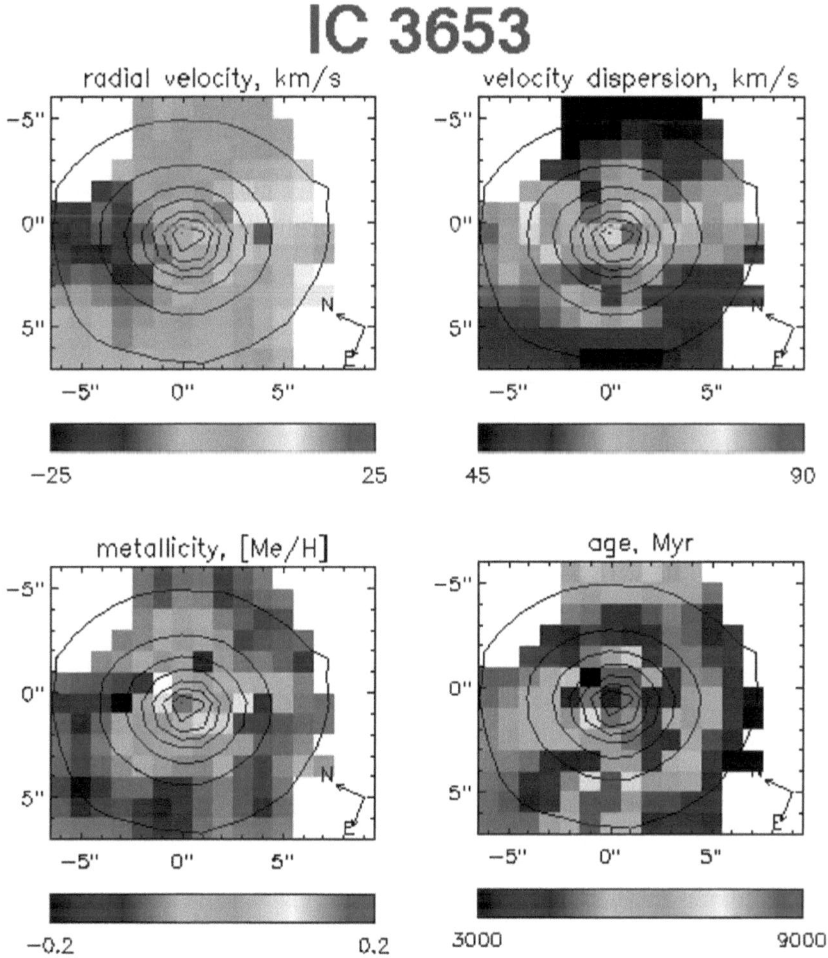


**Fig. 7.** MPFS maps of the dwarf elliptical IC 3653 as observed by (10). From top to bottom and left to right: stellar velocity, velocity dispersion, metallicity and age maps.

revealed by this unique dataset (Fig. 4), supplemented by higher resolution OASIS (CFHT and WHT) data (27), this successful observational campaign emphasizes the need to revise our view of these objects (15; 32; 26; 18), and bring new constraints on their formation and evolution (7).

Bulges of spirals have also been recently examined by (4) (Fig. 5) with INTEGRAL (WHT) and SPIRAL (AAT), and later-type spirals were finally scrutinized with SAURON (20) (Fig. 6), providing simultaneous gas, stellar kinematics and line-strengths maps of the central regions of galaxies for a lower range of stellar velocity dispersions. A SAURON campaign has also been devoted to a sample of active and non-active nearby galaxies to test a presumed link between the inner gravitational potential and the nucleus activity (Mundell et al.). Chilingarian and collaborators (10) obtained MPFS data on 5 dwarf ellipticals (Fig.7), in an attempt to probe the link between the stellar population and kinematical substructures. It is finally worth emphasizing a scanning Fabry-Perot H$\alpha$ study (even though not a true IFU) of the effect of environment on the gas content and kinematics (9) with a survey of 30 galaxies of the Virgo cluster.

## 6 New ways

We are of course waiting for better and bigger: higher spatial resolution IFUs using adaptive optics on large telescopes (VLT, Keck), higher spectral resolution pioneered by instruments such as FLAMES (Argus mode), larger fields of view following the trend shown by PMAS, higher number of spatial elements after the VIMOS IFU. In this context, the second generation VLT instrument MUSE, with its 90 000 spatial element, its field of view of 1 square arcminute with a sampling of 0.2 arcsecond, and its simultaneous coverage of most of the optical domain will probably remain the only specimen of its kind for some time. Such an ambitious spectrograph will benefit from parallel observations with ALMA and the NGST to probe the different ingredients of nearby galaxies (e.g., stars, ionized, molecular and neutral gas), their internal dynamics and star formation history, and with the help of state-of-the-art models, constrain the origin and evolution mechanisms of the structures we observe today.

But, as mentioned above, we already witness a shift in our approach: IFUs are now present on most 8m class telescopes, so that the ones mounted on smaller apertures can at last be used to conduct reasonably large surveys. We can also devote more time to individual targets, by e.g., mosaicing on nearby objects to maximize the field of view and cover structures of interest (the bar of NGC 936 with SAURON), or integrate for hours to explore the outer regions of nearby galaxies where dark matter is presumed to dominate (8). 3D spectroscopy should thus soon serve our scientific goals, and not just our frantic need to publish papers.

I would like to thank Nate Bastian, Ralf Bender, Joss Bland-Hawthorn, Laurent Chemin, Ric Davies, Igor Chilingarian, Frank Eisenhauer, Joris Gerssen, Alexei Moiseev, Francisco Mueller-Sanchez, Nina Nowak, Olga Sil'Chenko, and the Sauron team, for letting me present some of their results, and for useful discussions, as well as Jeremy Walsh and Markus Kissler-Patig for their invitation to this workshop.

## References


[1] Arribas, S., & Colina, L.: ApJ **591**, 791 (2003)
[2] Bacon R., et al.: MNRAS **326**, 23 (2001)
[3] Bastian, N., Emsellem, E., Kissler-Patig, M., Maraston, C.: A&A 445, 471 (2006)
[4] Batcheldor, D., et al.: ApJS **160**, 76 (2005)
[5] Cappellari, M., et al.: Carnegie Observatories Astrophysics Series, Vol. 1: Coevolution of Black Holes and Galaxies, ed. L. C. Ho (Pasedena: Carnegie Observatories), 1 (2003)
[6] Cappellari, M., & McDermid, R. M.: Classical and Quantum Gravity **22**, 347 (2005)
[7] Cappellari, M. et al.: MNRAS 366, 1126 (2006)
[8] Cappellari, M. et al.: XXIst IAP Colloquium "Mass Profiles and Shapes of Cosmological Structures", Eds G. Mamon, F. Combes, C. Deffayet, B. Fort, EDP Sciences (astro-ph/0511530) (2006)
[9] Chemin, L. et al.: MNRAS 366, 812 (2006)
[10] Chilingarian, I., Prugniel, P., Sil'Chenko, O., & Afanasiev, V.: IAU Colloq. 198: Near-fields cosmology with dwarf elliptical galaxies, 105 (2005)
[11] Davies, R. I., et al.: ApJ, submitted
[12] de Zeeuw, P. T. et al.: **329**, 513 (2002)
[13] Emsellem, E., Dejonghe, H., & Bacon, R.: MNRAS **303**, 495 (1999)
[14] Emsellem, E., Greusard, D., Combes, F., Friedli, D., Leon, S., Pécontal, E., & Wozniak, H.: A&A **368**, 52 (2001)
[15] Emsellem, E., et al.: MNRAS **352**, 721 (2004)
[16] Emsellem, E., Fathi, K., Wozniak, H., Ferruit, P., Mundell, C. G., & Schinnerer, E.: MNRAS **365**, 367 (2006)
[17] Emsellem, E., et al., in prep.
[18] Falcon-Barroso, J. et al.: MNRAS 369, 529 (2006)
[19] Ferruit, P., Mundell, C. G., Nagar, N. M., Emsellem, E., Pécontal, E., Wilson, A. S., & Schinnerer, E.: MNRAS **352**, 1180 (2004)
[20] Ganda, K., et al.: MNRAS 367, 46 (2006)
[21] Geha, M., Guhathakurta, P., & van der Marel, R. P.: AJ **129**, 2617 (2005)
[22] Gerssen, J., Allington-Smith, J., Miller, B. W., Turner, J. E. H., & Walker, A.: MNRAS **365**, 29 (2006)
[23] González Delgado, R. M., Arribas, S., Pérez, E., & Heckman, T.: ApJ **579**, 188 (2002)
[24] Kowak, Bender, Saglia, Thomas, et al., in preparation
[25] Krajnović, D., Cappellari, M., Emsellem, E., McDermid, R. M., & de Zeeuw, P. T.: MNRAS **357**, 1113 (2005)
[26] Kuntschner, H., et al.: MNRAS 369, 497 (2006)
[27] McDermid, R. et al.: MNRAS submitted

[28] Mediavilla, E., Guijarro, A., Castillo-Morales, A., Jiménez-Vicente, J., Florido, E., Arribas, S., García-Lorenzo, B., & Battaner, E.: A&A **433**, 79 (2005)
[29] Moiseev, A. V., Valdés, J. R., & Chavushyan, V. H.: 2004, A&A **421**, 433 (2004)
[30] Roth, M. M., Becker, T., Kelz, A., & Schmoll, J.: ApJ **603**, 531 (2004)
[31] Sánchez, S. F., Becker, T., Garcia-Lorenzo, B., Benn, C. R., Christensen, L., Kelz, A., Jahnke, K., & Roth, M. M.: A&A **429**, L21 (2005)
[32] Sarzi, M. et al.: MNRAS 366, 1151 (2006)
[33] Sil'chenko, O. K., Afanasiev, V. L., Chavushyan, V. H., & Valdes, J. R.: ApJ **577**, 668 (2005)
[34] Statler, T. S., Emsellem, E., Peletier, R. F., & Bacon, R.: MNRAS **353**, 1 (2004)
[35] Wozniak, H., Combes, F., Emsellem, E., & Friedli, D.: A&A **409**, 469 (2003)

# Nearby Galaxies Observed with the Kyoto Tridimensional Spectrograph II

H. Sugai[1], T. Ishigaki[2], T. Hattori[3], A. Kawai[1], S. Ozaki[4] and G. Kosugi[3]

[1] Department of Astronomy, Kyoto University, Sakyo-ku, Kyoto 606-8502, Japan. sugai@kusastro.kyoto-u.ac.jp,kawai@kusastro.kyoto-u.ac.jp
[2] Asahikawa National College of Technology, Asahikawa-shi, Hokkaido 071-8142, Japan. ishigaki@asahikawa-nct.ac.jp
[3] Subaru Telescope, National Astronomical Observatory of Japan, 650 North A'ohoku Place, Hilo, Hawai'i 96720, U.S.A. hattori@subaru.naoj.org,george@subaru.naoj.org
[4] Nishi-Harima Astronomical Observatory, Sayo-cho, Hyogo 679-5313, Japan. ozaki@nhao.go.jp

**Summary.** The integral field spectrograph mode of the Kyoto tridimensional spectrograph II is optimized for a high spatial resolution with the $\sim 0''.1$ sampling when mounted on the Subaru Telescope, while a wider field of view is emphasized on the University of Hawaii 88-inch Telescope (UH88). One of our main targets is to resolve activities in nearby galaxies. Examples are shown: one is the low ionization emission line region in the galaxy NGC 1052 observed with the Subaru. The spatial resolution of $\sim 0''.4$ has revealed the structures, both in space and velocity, of the AGN outflow. An example observed with the UH88 is an interacting galaxy system NGC 6090. Through analyses of the spatial variations of emission line ratios, we have found rapid large-scale metal/abundance enhancement in starbursts.

## 1 The Kyoto Tridimensional Spectrograph II

The Kyoto tridimensional spectrograph II (Kyoto 3DII) is a multimode optical spectrograph with four observational modes: Fabry-Perot imager; integral field spectrograph (IFS) with a lenslet array; long-slit spectrograph; and filter-imaging modes (14; 15; 17; 19). The pixel/lenslet of $\sim 0''.1$ well samples image sizes obtained by the Subaru 8.2-m Telescope so that we can take advantage of its good image quality. This compact spectrograph is used also at the UH 88-inch Telescope (UH88), which provides us a wider field of view: e.g., $\sim 15''$ in the IFS mode. We have succeeded in test observations in all the instrument's modes on the UH88 as well as on the Subaru. In this paper we show some of scientific results obtained in the IFS mode.

## 2 Young AGN-Driven Outflow Observed with Subaru

Low Ionization Nuclear Emission-line Regions (LINERs) are characterized by their relatively strong, low ionization emission lines and are found in over 30%

of all galaxies and in 60% of Sa-Sab spirals with $B \leq 12.5$ mag (7; 6). Narrow-band filter imaging with *Hubble Space Telescope* obtained by (12) showed that the emission line regions have complex and disordered morphologies. In order to understand phenomena that occur in such complex systems, it is essential to obtain the velocity field of the gas. The Kyoto 3DII IFS on the Subaru telescope enables us to **resolve these faint and small objects in data cubes, with Subaru's high image quality and large aperture, as well as with a fine sampling of the instrument.**

We show an example of NGC 1052 ((18)), which is often considered as a prototypical LINER. It is an important object also in the following two aspects. **(1) Pure AGN outflow:** AGN outflows coexist with starburst winds in many cases. This complicates the interpretation of AGN-driven outflows (21). We find no evidence of young starbursts in NGC 1052 (18). This suggests that the observed outflow is a pure AGN-driven one. **(2) Young activities:** This object is similar to Gigahertz-Peaked-Spectrum or Compact-Steep-Spectrum sources (11), based on the fact that the entire radio source is contained within the host galaxy and that it shows a convex radio spectrum peaked at 10 GHz (3). These sources are generally thought to be young jets, propagating through, and perhaps interacting with, a rich inner galactic medium (11; 22).

Figure 1 represents our data cube in a "pseudo-slit" scanning form. In this resampled form, spatial cuts in the data cube mimic slit scanning observations. The emission line regions consist mainly of three components: a high velocity bipolar outflow, low velocity disk rotation, and a spatially unresolved nuclear component. In order to understand the spatial distribution of the above components, particularly of the high velocity component, we have constructed velocity channel maps from the same data cube (Figure 2). This figure highlights the bipolar nature of the outflow. We have found the following two aspects: i) the opening angle of the outflow decreases with velocity shift from the systemic velocity *both* in bluer and redder velocity channels. This is explained only if the outflow has intrinsically higher-velocity components inside, i.e., in regions closer to the outflow axis; ii) at both sides of the bipolar outflow, we find that the highest velocity components are detached from the nucleus. This gap can be explained by an acceleration of at least a part of the flow or the surrounding matter, or by bow shocks that may be produced by even higher velocity outflow components that are not detected so far.

Along the edges of the outflow and extending ENE and WSW, there exist strong [O III] emission ridges. These are closely related with the radio jet-counterjet structure found in the 1.4 GHz image (8). The abrupt change in the velocity field of the ionized gas and a large [O III]/H$\beta$ line flux ratio in this region suggest a strong interaction of the jets and possibly also of some ridge components of the line emitting gas with the interstellar matter (18).

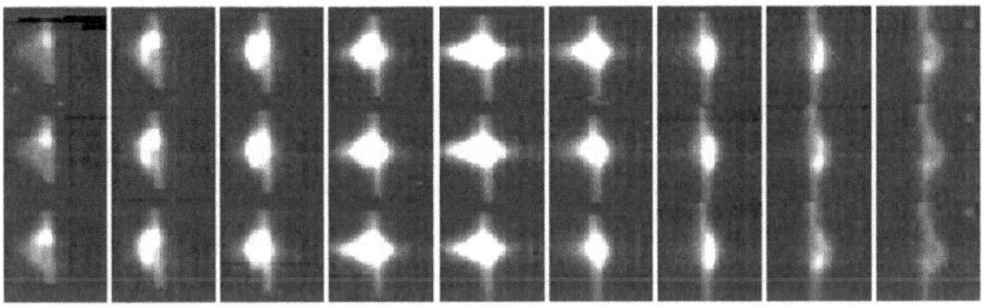

**Fig. 1.** A "pseudo-slit" representation scan through the continuum-subtracted data cube. Only the [O III]$\lambda$5007 line is shown. The twenty seven panels are for each $0''.096$-width pseudo slit, stepped every $0''.096$. The central panel includes the location of the line-free continuum peak. For each panel, the horizontal axis runs over the wavelength range of 5001.9 Å(left) to 5068.2 Å(right). The vertical axis is spatial: 37 lenses with $0''.096$ sampling.

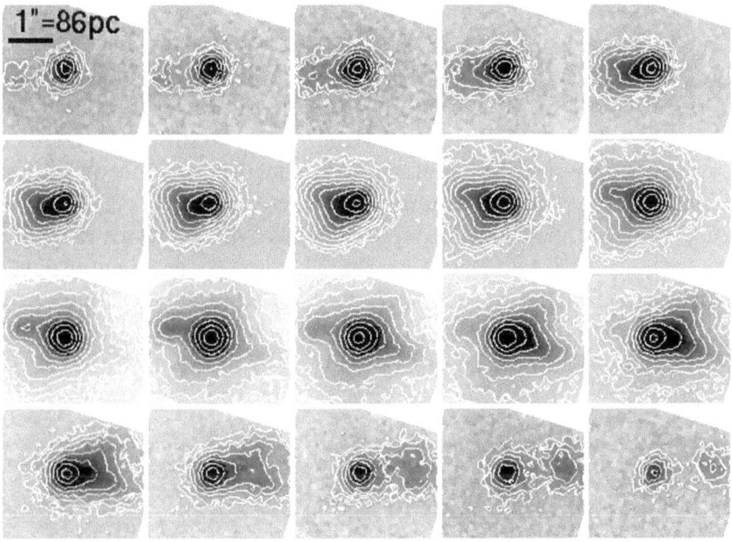


**Fig. 2.** Velocity channel maps of the [O III]$\lambda$5007 line for every 1.7 Åstep, which corresponds to 102 km s$^{-1}$. The wavelength ranges from 5012.1 Å(top left) to 5044.4 Å(bottom right). Greyscale shows the [O III] surface brightness in linear scaling and is arbitrary among the channels. Contours are drawn every 0.5 mag.

## 3 Rapid Large-Scale Metallicity/Abundance Enhancement Observed with UH88

Galaxies form stars, and only stars can produce heavy elements such as carbon, nitrogen, and oxygen. In order to understand the chemical evolution of the universe, therefore, it is necessary to understand how stars produce and eject those elements in galaxies. Stars form from interstellar gas while stellar winds/mass losses and supernova explosions provide feedback of gas, increasing a fraction of metals. It is not yet known, however, what is the timescale for heavy elements ejection and distribution into interstellar space and how their origin differs between elements. From an observational viewpoint, it has been difficult to provide an image of *on-going* metallicity enrichments on a galactic scale. Bursts of star formation are often found in galactic centers, where the underlying metallicity is already high. This makes it difficult to distinguish on-going enrichments from the underlying radial metallicity gradients.

In order to solve these problems, we have observed NGC 6090 (Figure 3; (16)). This interacting galaxy system provides the unique opportunity in the following two aspects: (1) bursts of star formation, which are expected to increase heavy elements, do *not* occur in the galactic centers of these systems but occur in regions *offset* from the centers (Figure 4 *left*). **This axially asymmetric configuration highlights the increase of heavy elements caused by the starbursts.**; (2) in these systems, starbursts occur at interfaces between regions in the galaxies themselves and those located adjacent to the galactic nuclei ((13)). Newly-formed stars move away from the interface when the galaxies rotate. The rotation of a galaxy therefore works just like a "clock". **The age of starbursts estimated by this clock, $\sim 10^7$ years, has provided a strong constraint on the origins of the metal/abundance enhancements.**

Figure 4 (*right*) shows an example of spatial distributions of various line flux ratios: the ([O III]+[O II])/H$\beta$ ratio. This is a good indicator of the O/H abundance ratio. Smaller line ratios (higher abundances) are found associated with starbursts, which suggests that *the starbursts* have enhanced the metallicity of the ionized gas. Based on the spatial distributions of the [N II]/[S II]

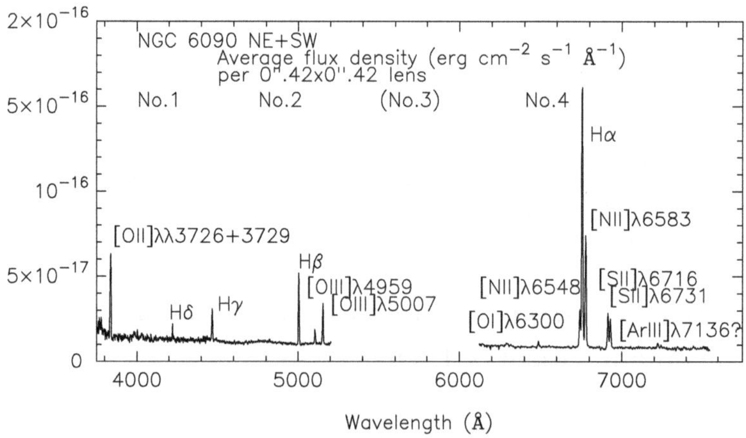


**Fig. 3.** Average spectra of the whole starburst regions in NGC 6090. The absolute flux calibration in each of three observed wavelength regions (No.1,2,&4) was carried out independently by using the same standard star.

and [N II]/[O II] ratios as well as on comparisons with theoretical models of starbursts ((10)), we have also found an excess of nitrogen in the starburst regions ((16)). Nitrogen is a key element in chemical evolution because of its "secondary" nature (e.g., (9; 1; 2)). Because the estimated starburst age is too short for intermediate mass stars to have reached the AGB phase, where they lose most of their mass to the interstellar medium (20), the observed nitrogen enhancement is likely to be caused by winds/mass loss from massive stars.

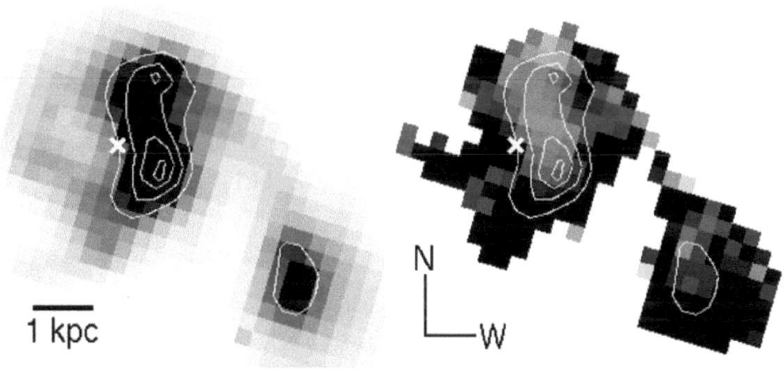


**Fig. 4.** *(left):* Spatial distribution of the observed H$\alpha$ emission in NGC 6090. The X denotes the location of the 2$\mu$m continuum peak in the northeastern galaxy ((4)). The offset of the H$\alpha$ emission peak from the galactic center is also seen in the H$\alpha$ equivalent width map ((16)). *(right):* Greyscale shows the spatial variation of the extinction-corrected ([O III]$\lambda\lambda$4959, 5007+[O II]$\lambda$3227)/H$\beta$ ratio (darker gray means a larger ratio), with the observed H$\alpha$ intensity shown in contours.

## References


[1] Considère, S. et al. 2000, A&Ap, 356, 89
[2] Coziol, R. et al. 1999, A&Ap, 345, 733
[3] de Vries, W. H. et al. 1997, A&Ap, 321, 105
[4] Dinshaw, N. et al. 1999, ApJ, 525, 702
[5] Heckman, T. M. 1980, A&Ap, 87, 152
[6] Ho, L. C. 1996, in Astron. Soc. Pacif. Conf. Ser., Vol. 103, The Physics of LINERs in View of Recent Observations, ed. M. Eracleous, A. Koratkar, C. Leitherer, & L. Ho (San Francisco: Astronomical Society of the Pacific), 103
[7] Ho, L. C. et al. 1997, ApJ, 487, 568
[8] Kadler, M. et al. 2004, A&Ap, 420, 467
[9] Kennicutt, R. C. et al. 2003, ApJ, 591, 801

[10] Kewley, L. J. et al. 2001, ApJ, 556, 121
[11] O'Dea, C. P. 1998, PASP, 110, 493
[12] Pogge, R. W. et al. 2000, ApJ, 532, 323
[13] Sugai, H. et al. 2000a, MNRAS, 317, 447
[14] Sugai, H. et al. 2000b, SPIE, 4008, 558
[15] Sugai, H. et al. 2002, in Astron. Soc. Pacif. Conf. Ser., Vol. 282, Galaxies: The Third Dimension, ed. M. Rosado, L. Binette, & L. Arias (San Francisco: Astronomical Society of the Pacific), 433
[16] Sugai, H. et al. 2004a, ApJ, 615, L89
[17] Sugai, H. et al. 2004b, SPIE, 5492, 651
[18] Sugai, H. et al. 2005a, ApJ, 629, 131
[19] Sugai, H. et al. 2005b, New Astronomy Reviews, in press
[20] van den Hoek, L. B. & Groenewegen, M. A. T. 1997, A&ApS, 123, 305
[21] Veilleux, S. et al. 2005, ARAA, 43, 769
[22] Vermeulen, R. C. et al. 2003, A&Ap, 401, 113

# Scalable N-body Code for the Modeling of Early-type Galaxies

E. Jourdeuil and E. Emsellem

Centre de Recherche Astronomique de Lyon (CRAL)
9 avenue Charles André
69561 Saint Genis Laval Cedex
emilie@obs.univ-lyon1.fr

**Summary.** Early-type galaxies exhibit a wealth of photometric and dynamical structures. These signatures are fossil records of their formation and evolution processes. In order to examine these structures in detail, we build models aimed at reproducing the observed photometry and kinematics. The developed method is a generalization of the one introduced by (12), consisting in an N-body representation, in which the weights of the particles are changing with time. Our code is adapted for integral-field spectroscopic data, and is able to reproduce the photometric as well as stellar kinematic data of observed galaxies. We apply this technique on SAURON data of early-type galaxies, and present preliminary results on NGC 3377.

## 1 Introduction

**Goal** Elliptical galaxies are presumed to be the product of galaxy merging, where different mass ratios result in different classes of elliptical galaxies (disky, boxy...) (8). This seems consistent with the hierarchical scenario of galaxy formation (9), although it is clear that more standard collapse processes and secular evolution contribute to the shaping of galaxies. There are many different and often complementary approaches to constrain these scenarios. The one we adopted is to study the structures of nearby elliptical galaxies in detail, in order to probe the signatures of their evolution and to be able to trace their formation history. This fossil search requires data both on the chemical and dynamical status of the stellar components, and if present of the gaseous system, both being provided by spectroscopy.

**3D spectroscopy** Spectroscopy can indeed deliver information on the stars' dynamics and chemistry via the analysis of their absorption lines. Two-dimensional spatial coverage provided by integral-field spectroscopy is critical to homogeneously sample an extended target, while more traditional long-slit spectroscopy generally restricts our view to a few a priori fixed axes. A qualitative assessment of the obtained maps is clearly not sufficient to fully address the issues mentioned above. We need to go further by modeling the observed galaxies using constraints retrieved from 3D spectroscopy, such as luminosity, stellar velocity distribution and stellar populations. As emphasized by (1), 3D spectroscopy seems to be essential to properly constrain the dynamics of the galaxy under scrutiny.

**Existing models** One of the key goal of galaxy modeling is to retrieve the full distribution function (DF), *i.e.* the density of stars in phase-space. There are already several known methods to address this problem, each of them having both advantages and drawbacks. We can distinguish different classes of techniques :

- DF-based method, where models are often restricted to simple geometries.
- Moment-based method, which consists in solving Boltzmann and Poisson equations via the use of a closed system of relations (Jeans equations). The main issue here is that the final DF may not be positive everywhere.
- Orbit-based method, or Schwarzschild modeling (10). Libraries of orbits are built within a fixed potential, and each orbit is weighted as to reproduce the observed galaxy (photometry and kinematics).
- Particle-based method, where the particles represent groups of stars. One has to guess the right initial conditions which will evolve in a configuration resembling the chosen galaxy.

## 2 Method

The method we wish to present here is an hybrid scheme between the Schwarzschild and N-body techniques. It has been first proposed and developed by (12) and consists in an N-body realization where the weight of each particle is gradually changed in order to fit the observables.

**Previous applications** This method has been tested by (12) on fake galaxy models, adjusting the photometry alone. It has been recently applied on the Milky Way by (2), fitting the photometry of a frozen snapshot extracted from a full N-body simulation.

**Algorithm** The algorithm we use is illustrated in Fig. 1. Particles start from (N-body) initial conditions and are integrated along their orbits. Observables are used as reference input to the modeling, particles being projected as to mimic these observables. A weight prescription is then derived from the difference between the model and the data, this prescription being then applied on each particle accordingly. This procedure (integration, projection, comparison, weight changing) is repeated until the weights converge.

**Initial conditions** The initial conditions have to properly sample the galaxy, *ie* in position/velocity space or using integrals of motions. This is not an easy task when we do not *a priori* know the internal structure of the galaxy we wish to model. If we want to model real galaxies, we have to find appropriate initial conditions, without a priori knowing its DF. We thus first build

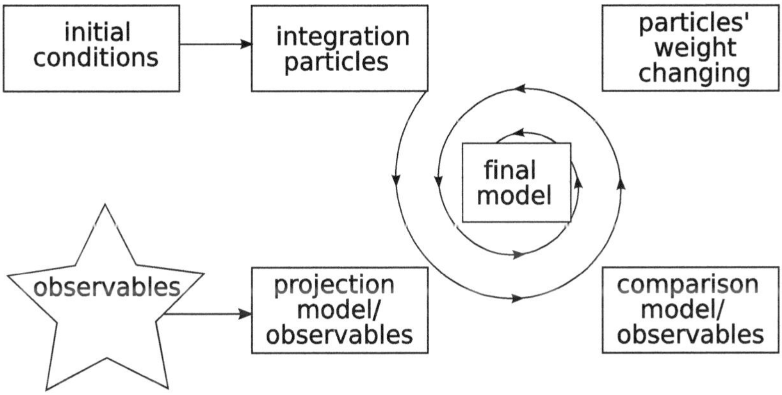


**Fig. 1.** Algorithm of the Syer & Tremaine method.

an initial mass distribution, by using the known photometry and assuming a constant mass-to-light ratio. This is achieved here via the Multi Gaussian Expansion (MGE) proposed by (7), which allows us to decompose the mass distribution into a sum of tri-dimensional Gaussians, and to derive the corresponding gravitational potential analytically (5). Jeans equations are then used to obtain a first guess of the dynamics and set up initial conditions.

**Integration** In a first version of the code, we keep the potential steady. The code has been developed in a very modular and flexible fashion, such as to allow the easy implementation of a self-consistency module : this will be described in a forthcoming paper. As we need to evaluate the observables of all particles at the same time, we chose a second-order synchronized leapfrog scheme, where positions and velocities can be evaluated simultaneously. The leapfrog technique is perhaps not the best choice in terms of accuracy, but was found sufficient for this first development step. Other schemes (e.g. Runge-Kutta) can be easily implemented. Adaptive time-steps have been included, in order to optimally sample all the orbits. An independent integration of each particle would result in a very inefficient algorithm. We thus chose to follow the algorithm described in the N-body code GADGET (11), in which particles advance in bunches so that they are always distributed in a tight time range around the current time.

**Observables** Our code has been designed around two new main items. First, we do not restrict ourselves to fit the photometry : the code also allows the fitting of the kinematics, via the use of Line of Sight Velocity Distributions (LOSVDs). Second, the code has been developed as to permit the use of 3D spectroscopic data, including spatial adaptive binning (Voronoi bins, see (3)). We have thus used the data obtained with the SAURON spectrograph (WHT, Canary Islands) for its survey of early-type galaxies (4; 6).

**Prescription** The heart of the code is the prescription developed by (12), which rules the weight changing of each particle so that the cumulated observables reproduce the observations :

$$\frac{dw_i(t)}{dt} = -\epsilon w_i(t) \sum_{j=1}^{J} \frac{K_{ij}}{Z_j} \Delta_j \tag{1}$$

where $i$ describes the particles, $j$ the observables, $w$ the weights, $K_{ij}$ the contribution of the particle $i$ to the observable $j$, and $\Delta_j \equiv y_j(t)/Y_j - 1$ the contribution of all the particles to this same observable. $\epsilon$ is the strength of the weight changing, $w_i$ avoids the weights to be negative when $w_i$ approaches 0. The input parameters are the following :

- $N$ is the number of particles,
- $d_{time}$ is the time during which the observables are averaged,
- $\epsilon$ is the strength parameter mentioned in Eq. 1,
- $\alpha$ is the smoothing parameter of the observables, used in the equation :

$$\tilde{\Delta}_j(t) = \alpha \int_0^{\infty} \Delta_j(t-\tau) e^{-\alpha\tau} d\tau \tag{2}$$

  so that $\tilde{\Delta}_j(t)$ replaces $\Delta_j(t)$ in Eq. 1.

The choice of these parameters is crucial, each galaxy requiring an appropriate set.

## 3 Results

A preliminary model has been obtained on the nearby disky galaxy NGC 3377. In Fig. 2 we illustrate the fit of the photometry and LOSVDs obtained with the SAURON instrument on this galaxy, for a given mass (MGE) distribution. The maps are globally well reproduced ($I$, $\sigma$, $h_3$). Some remaining discrepancy in the velocity field is probably due to an observed asymmetry in the kinematics of this galaxy (tilt between the photometric and kinematic minor-axis). The relatively high $h_4$ observed level is not reached by the model : this may be partly related to a template mismatch effect which mostly affects the even moments of the LOSVD.

## 4 Conclusion and perspectives

Although the results shown here are very preliminary, they illustrate the possibility of fitting the kinematics, which is an important step in the development of such a code. It has also been designed to adjust complex data such as the one provided by 3D spectrographs. A few improvements are now being implemented :

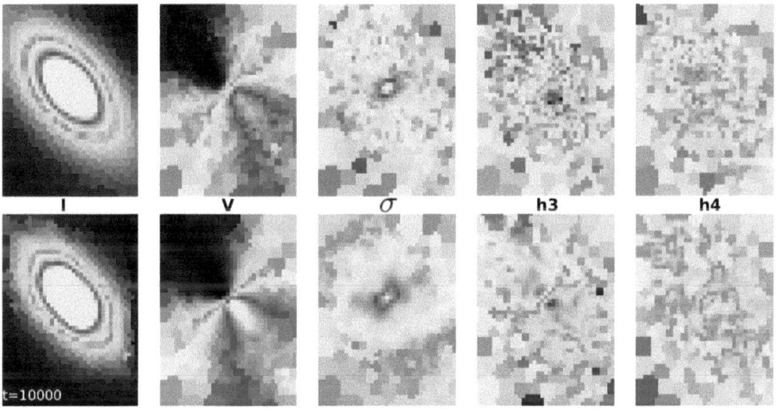


**Fig. 2.** $I, V, \sigma, h_3, h_4$ from the observations (top) and the model (bottom) of NGC 3377

- Self-consistency : it is technically easy to replace the current integration module with a self-consistent scheme. The implications of a continuous weight changing remains however unclear and have to be examined.
- Addition of a central black hole : this would allow the use of higher spatial resolution data of nearby galaxies.
- Stellar populations : this new class of observables would allow us to link the dynamics and chemistry of the stellar component.

## References


[1] Binney, J., MNRAS, **363**, 937 (2005)
[2] Bissantz, N., Debattista, V. P., Gerhard, O., ApJ, **601**, L155 (2004)
[3] Cappellari, M., Copin, Y., MNRAS, **342**, 345 (2003)
[4] de Zeeuw, P. T., Bureau, M., Emsellem, E., Bacon, R. et al: MNRAS, **329**, 513 (2002)
[5] Emsellem, E., Monnet, G., Bacon, R., A&A, **285**, 723
[6] Emsellem, E., Cappellari, M., Peletier, R. F. et al: MNRAS, **352**, 721 (2004)
[7] Monnet, G., Bacon, R., Emsellem, E., A&A, **253**, 366 (1992)
[8] Naab, T., Burkert, A., Hernquist, L., ApJ, **523**, L133 (1999)
[9] Press, W. H., Schechter, P., ApJ, **187**, 425 (1974)
[10] Schwarzschild, M., ApJ, **232**, 236 (1979)
[11] Springel, V., Yoshida, N., White, S. D. M., New Astronomy, **6**, 79 (2001)
[12] Syer, D., Tremaine, S., MNRAS, **282**, 223 (1996)

# Internal Kinematics of Galaxies: 3D Spectroscopy on Russian 6m Telescope

A. V. Moiseev

Special Astrophysical Observatory, Russian Academy of Sciences, 369167 Russia
moisav@sao.ru

**Summary.** We consider some results on the gas and stellar kinematics of nearby galaxies recently obtained on the 6m telescope of SAO RAS using panoramic spectroscopy methods. The circumnuclear regions of the galaxies were observed with integral-field spectrograph MPFS. The large-scale ionized gas kinematics was studied with the scanning Fabry-Perot interferometer (FPI) in the multi-mode focal reducer SCORPIO. The main attention is devoted to kinematically decoupled regions in the galaxies: bars, spirals, polar disks and rings.

## 1 Introduction

The circumnuclear regions within the inner kiloparsec of disk galaxies turn out to be decoupled on the basis of the dynamic characteristics. Using the technique of panoramic spectroscopy makes it possible to study in detail the differences in kinematics of the stellar and gaseous components. I briefly review the following types of kinematically decoupled regions in galactic disks:

- non-circular motions caused by dynamical effects (spirals, bars, colliding rings);
- circular rotation in different planes or directions (polar rings, counter-rotating disks);
- non-circular motions caused by violent star formation (bubbles, high-velocity clouds).

## 2 Instrumentation

The study of two-dimensional kinematics of galaxies at the 6m telescope of SAO RAS with the scanning FPI was started by our colleagues from Marseille Observatoire (J. Boulesteix et al.) in cooperation with team in SAO in the first half of 1980s. The observations were made with the system CIGALE. Then the IPCS was replaced by a CCD and in 2000 the first observations with a new multimode focal reducer SCORPIO (3) were carried out. Today the FPI mode of SCORPIO provides a $6' \times 6'$ field of view with a spectral resolution 1-2.5Å in the H$\alpha$, [N II], [S II] and [O III] emission lines. So, using

FPI data we create monochromatic images as well as fields of line-of-sight velocities in these lines (Fig. 1).

The integral field spectrograph MPFS is based on the idea by Victor Afanasiev et al., i.e. the combination of a lens array with a bundle of fibers (1). A first version of the spectrograph was developed in 1990. The current variant of the MPFS became operational at the 6 m telescope in 1998 (2). Fibers transmit light from $16 \times 16$ square elements of the galaxy image to the slit of the spectrograph, together with 16 additional fibers that transmit the sky background taken away from the galaxy. The angular scale is $1''$ per spaxel and the spectral resolution is $4 - 12$Å. Using MPFS for kinematics of galaxies allows us to build velocity fields of ionized gas as well as of their stellar component.

Descriptions of our spectrographs are available at SAO RAS web page: http://www.sao.ru/hq/lsfvo/"

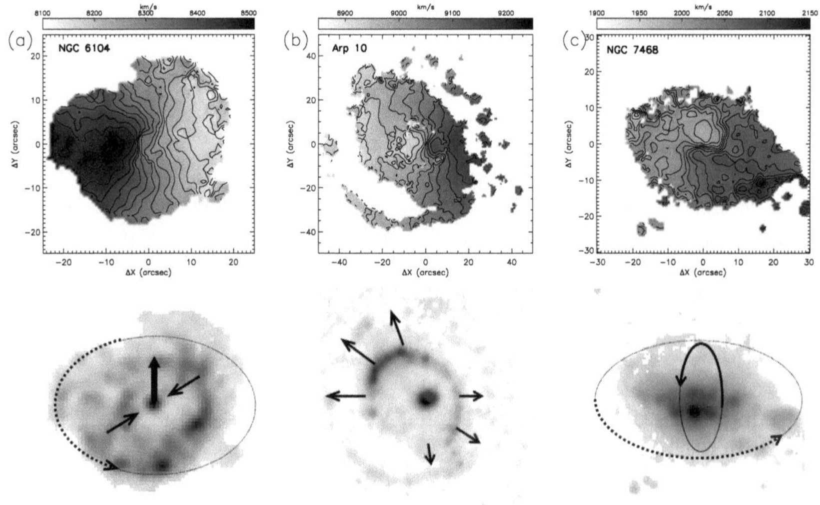


**Fig. 1.** FPI velocity fields of ionizing gas and Hα images with the sketch of gas motions: radial inflow along a bar and nuclear jet outflow in Sy NGC 6104 (a, (11)); expansion of the emission ring in Arp 10 (b, (8)); polar disk in NGC 7468 (c, (9))

# 3 Examples of the Objects

## 3.1 Barred and Double-barred Galaxies

The values of non-circular motions in barred galaxies can be estimated by the detailed analysis of their velocity fields. For instance, the Sy galaxy

NGC 6104, where the ionized gas radial inflow motions along the bar co-exist with nuclear outflow caused by jet-clouds interactions, is shown in Fig. 1a.

Recently we have observed a sample of candidate double-barred galaxies and suggest that these objects are, in fact, galaxies with very different circumnuclear structure (7). The majority of the observed morphological and kinematic features in the sample may be explained without the secondary bar hypothesis. Three cases of inner polar disks, one counter-rotating gaseous disk and seven nuclear disks nested in large-scale bars were found in this work.

### 3.2 Counter-rotation

The form of the motion of gas clouds may noticeably differ from the rotation of the stellar component even in 'quiet' galaxies without an AGN. The characteristic case is the lenticular galaxy NGC 3945 (Fig. 2a). The velocity field of the stars in the circumnuclear disk shows the normal circular rotation. For the gas velocity field the situation is more complicated. In $r < 6''$ (0.5 kpc), the line-of-sight velocities of ionized gas are at the maximum amplitude to those of the stars but opposite in sign. At larger distances the direction of rotation of gas changes abruptly and coincides with the stellar component of rotation. The large-scale FPI velocity field confirms the fact of normal gas rotation at large radii, up to 11 kpc. Therefore, we have detected a circumnuclear disk of ionized gas rotating in the opposite direction with respect to the stellar component, (7). This is probably attributable to a merger of an accreted gaseous cloud with the corresponding direction of angular momentum (4).

### 3.3 Collisional Rings

Collisional ring galaxies represent a class of objects in which nearly circular density waves are driven into a disk as a result of an almost face-on collision with another galaxy. We have observed with FPI the peculiar galaxy Arp 10, which has two rings (an inner and an outer one) and an extended outer arc. The H$\alpha$ velocity field shows evidence for significant radial motions in both outer and inner galactic rings. We fit a model velocity field taking into account the regular rotation and projection effects. The expansion velocity of the NW part of the outer ring exceeds 100 km/s, whereas it attains only 30 km/s in the SE part (8). Therefore, the asymmetric shape of the outer ring (Fig. 1b) may be caused by a systematic difference in the ring expansion velocity and collisional origin of the rings.

### 3.4 Polar Rings and Nuclear Polar Disks

Recently the team in St.Petersburg University (V. Hagen-Thorn, L. Shaliapina, V. Yakovleva), in a collaboration with us, have attempted

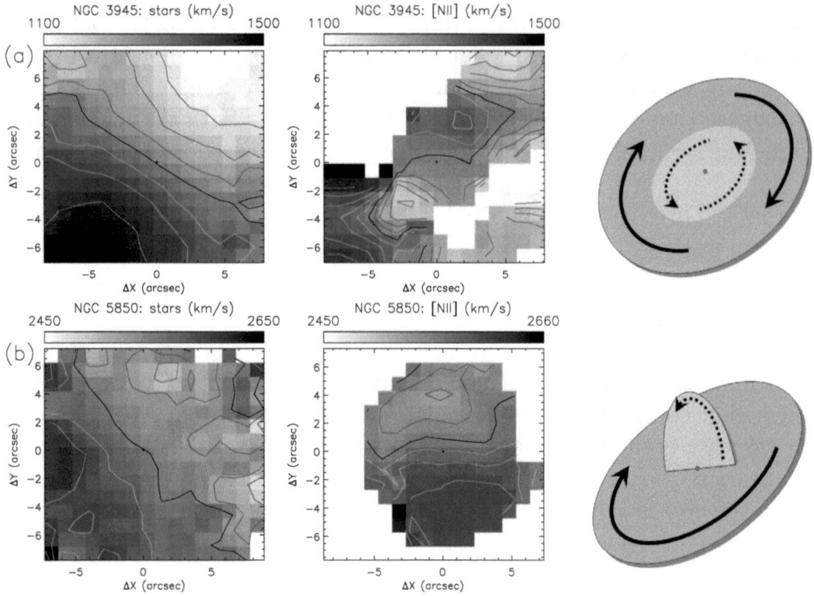


**Fig. 2.** MPFS data (from (7)): the velocity fields of stars and ionized gas for the circumnuclear region of NGC 3945 (a, counter-rotated disk) and NGC 5850 (b, inner polar disk). The right panels show the sketch of motions of gas and stars.

to observe with FPI the gas kinematics in candidate polar-ring galaxies. Interesting results were obtained. For example, they detected an inner gaseous disk whose rotation plane is almost perpendicular to the plane of the 'main' galactic disk in NGC 7648 (9). Fig. 1c shows the sharp turn of isovelocities in the galactic circumnuclear region. The central collision of NGC 7468 with a gas-rich dwarf galaxy and their subsequent merging seem to be responsible for the formation of the disk.

In the barred galaxy NGC 5850, the direction of rotation measured from the stellar component coincides with the line of nodes of the disk, whereas in the ionized gas it differs by more than 50° and coincides with the position angle of the central isophotes. Such a behavior is typical for a disk inclined to the galactic plane. A more reasonable assumption is that the gas, at $r < 1$ kpc, rotates in a polar plane with respect to the global galactic disk (7). In this case, the polar gaseous disk lies orthogonal to the major axis of the bar. It is remarkable that similar polar mini-disks inside the large-scale bars or the three-axial bulge have already been detected in about twenty galaxies (see (10) and contribution by O. Sil'chenko in this volume).

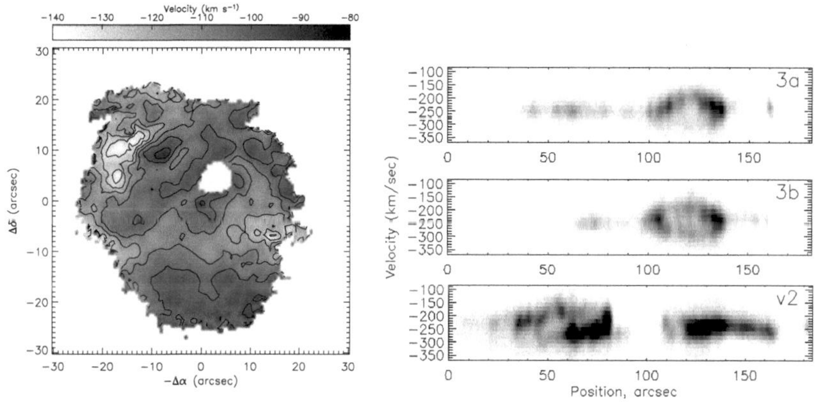


**Fig. 3.** Gas kinematics in dwarf galaxies: (left) – velocity field of VIIZw403 (6), (right) – the position velocity diagrams through star forming regions in IC 1613. The remarkable velocity ellipsoids correspond to expansion of H$\alpha$ envelopes.

### 3.5 Star Formation in Dwarf Galaxies

A burst of star formation also produces non-circular gas motions triggered by the combined effect of stellar winds and supernova explosions in rich stellar groupings. In dwarf galaxies the formation of giant multi-shell complexes around stellar groupings can proceed unhindered. See, for example, observations with SCORPIO-FPI of the nearby irregular galaxy IC1613 (5): the authors refined the expansion velocities of individual shells of the ionized and neutral gas (Fig. 3). In such galaxies the main part of line-of-sight velocities connects with an expansion of HII regions, frequently without any rotation (like VIIZw403, see Fig. 3)

This work was partially supported by the RFBR grant No. 05-02-16454. I also thank the Russian Science Support Foundation and the Organizing Committee of the workshop.

## References

[1] V.L. Afanasiev, V.V. Vlasyuk, S.N. Dodonov, O.K. Sil'chenko: preprint of SAO RAS, **54**, 1 (1990)
[2] V.L. Afanasiev, S.N. Dodonov, A.V. Moiseev: In: *Stellar dynamics: from classic to modern*, ed by L.P. Osipkov, I.I. Nikiforov (Saint Petersburg 2001) pp 103-109
[3] V.L. Afanasiev, A.V Moiseev: Astr. Lett., **31**, 193 (2005) (astro-ph/0502095)
[4] F. Bertola, L.M. Buson, W.W. Zeilinger: ApJ, **401**, L79, (1992)

[5] T.A. Lozinskaya, A.V. Moiseev, N.Yu. Podorvanyuk: Astr. Lett., **29**, 77 (2003)
[6] T.A. Lozinskaya, A.V. Moiseev, V.Yu. Avdeev, O.V. Egorov: Astr. Lett., accepted (2006)
[7] A.V. Moiseev, J.-R. Valdés, V.H. Chavushyan: A&A, **421**, 433 (2004)
[8] A.V. Moiseev, D. Bizyaev, E.I. Vorobyov: AAS Meeting 205, #26.05 (2005) (astro-ph/0501601)
[9] L.V. Shaliapina, A.V. Moiseev, V.A. Yakovleva, V.A. Hagen-Thorn, O.Yu. Barsunova: Astr. Lett., **30**, 583 (2004) (astro-ph/0411457)
[10] O.K. Sil'chenko, V.L. Afanasiev: AJ, **127**, 2641 (2004)
[11] A.A. Smirnova, A.V. Moiseev, V.L. Afanasiev: Astr. Lett., submitted (2006)

# SAURON Observations of Sa Bulges: The Formation of a Kinematically Decoupled Core in NGC 5953

J. Falcón-Barroso[1], R. Bacon[2], M. Bureau[3], M. Cappellari[1], R. L. Davies[3], P. T. de Zeeuw[1], E. Emsellem[2], K. Fathi[4], D. Krajnović[3], H. Kuntschner[5], R. M. McDermid[1], R. F. Peletier[6] and M. Sarzi[3,7]

[1] Sterrewacht Leiden, Leiden, The Netherlands.
jfalcon@strw.leidenuniv.nl
[2] CRAL - Observatoire de Lyon, Lyon, France.
[3] University of Oxford, Oxford, United Kingdom.
[4] Rochester Institute of Technology, New York, USA.
[5] European Southern Observatory, Garching, Germany.
[6] Kapteyn Astronomical Institute, Groningen, The Netherlands.
[7] University of Hertfordshire. Hatfield. United Kingdom.

**Summary.** We present results from our ongoing effort to understand the nature and evolution of nearby galaxies using the SAURON integral-field spectrograph. In this proceeding we focus on the study of the particular case formed by the interacting galaxies NGC 5953 and NGC 5954. We present stellar and gas kinematics of the central regions of NGC 5953. We use a simple procedure to determine the age of the stellar populations in the central regions and argue that we may be witnessing the formation of a kinematically decoupled component (hereafter KDC) from cold gas being acquired during the ongoing interaction with NGC 5954.

## 1 Interacting galaxies and the formation of KDCs

The interaction between galaxies lies at the heart of the current lore of hierarchical formation models. While most of the recent numerical simulations are able to make predictions about how such interactions affect the outer regions of galaxies (e.g. Combes et al. 1995; Moore et al. 1999), it is only in the last years that they have started to achieve sufficient resolution to provide clues of their effects on the inner regions of galaxies (e.g. Naab & Burkert 2001). Signatures of counter-rotation in either the stellar or gas components of galaxies have often been used to argue for the external origin of material and therefore interaction between galaxies (e.g. Bertola et al. 1988). In merger remnants, direct evidence for past interactions is the presence of a KDC in the inner regions of the galaxy.

One of the major goals of the SAURON survey is to study the structural and kinematical properties of such KDCs, as well as their frequency. Using that information, it is possible to step back in time and infer the conditions that led to the final configuration of the merger remnant. Previous papers in

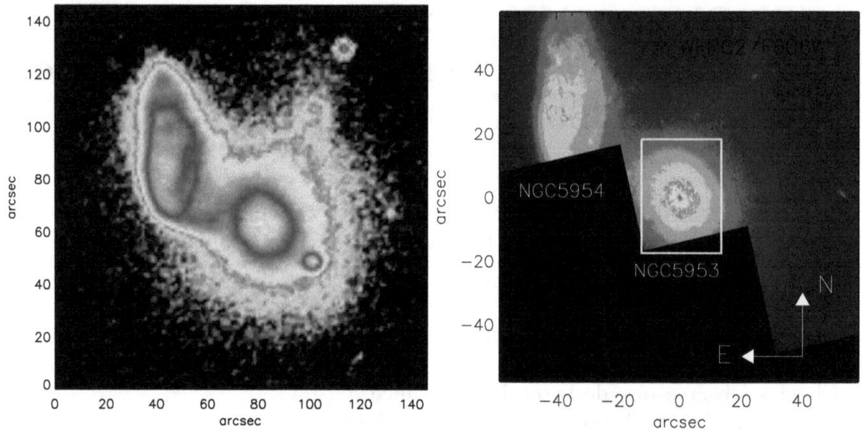


**Fig. 1.** The interacting pair formed by NGC 5953 and NGC 5954. Left: *B*-band DSS image (logarithmically scaled) showing the bridge of material between the two galaxies. Right: High spatial resolution HST/WFPC2-F606W image. Overlaid on NGC 5953 is the SAURON field-of-view covering the inner 4×3 kpc of the galaxy.

the SAURON series, concentrated on E and S0 galaxies, have shown that KDCs are common in early-type galaxies (e.g. de Zeeuw et al. 2002; Emsellem et al. 2004). In general these are old KDCs that are thought to have formed early in the evolution of the host galaxies (Kuntschner et al. 2006). By zooming in onto the nuclear regions, however, one discovers that there is also a large fraction of young decoupled components in the centres (see McDermid et al., these proceedings).

Among the 24 Sa spiral bulges in the SAURON survey, we were able to detect kinematically decoupled components in up to 50% of the galaxies, most of them aligned with the galaxies photometric major axis, and likely the end result of secular evolutionary processes (e.g. Wozniak et al. 2003). Only two cases display strongly misaligned or counter-rotating KDCs (i.e. NGC 4698 and NGC 5953, see Falcón-Barroso et al. 2006). NGC 4698 is a well studied galaxy, where the KDC may be the result of an intermediate-size merger (Bertola et al. 1999). The case of NGC 5953 is particularly interesting in the survey because it is the only example where we are witnessing an interaction that is currently taking place (see Figure 1), rather than observe the end product of a past merger.

## 2 The interacting galaxy NGC 5953

The system formed by the interacting pair NGC 5953/NGC 5954 has been extensively studied in the past. Most of the work has focused on the active nature of NGC 5953's nucleus (Jenkins 1984; Reshetnikov 1993; Gonzalez

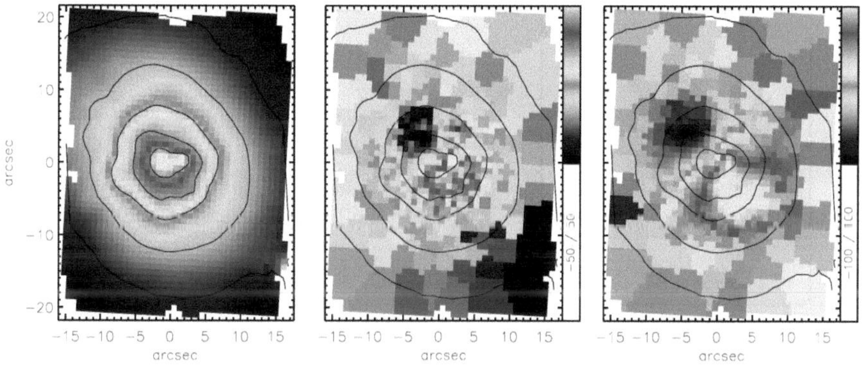


**Fig. 2.** Kinematics of NGC 5953. From left to right: i) SAURON reconstructed intensity, ii) stellar velocity, iii) H$\beta$ ionised-gas velocity. Levels for the kinematics are indicated in a box on the right-hand side of each map, and are expressed in km s$^{-1}$.

Delgado & Perez 1996). Rampazzo et al. (1995) combined broad-band imaging and long-slit spectroscopy to produce a detailed photometric and kinematic analysis of the interaction between the two galaxies. More recent results make use of a Fabry-Pérot spectrograph to map the two-dimensional gas kinematics using the [N II] emission line (Hernández et al. 2003; see also Fuentes-Carrera et al. in these proceedings).

In Figure 2, we present the reconstructed intensity map and the stellar and gas kinematics of the central regions of NGC 5953. The SAURON observations reveal the presence of a kinematically decoupled component in the central regions, where the stars are counter-rotating with respect to the bulk of the galaxy. The ionised-gas velocity map of the central component is consistent with that of the stars in the inner parts, and it is in good agreement with that of Hernández et al. (2003). While the kinematic major axis of the stellar component (both the inner and outer parts) appears to be aligned with respect to the photometric major axis of the galaxy, the kinematics of the ionised-gas in the outer parts shows signs of non-axisymmetry.

In addition to the stellar and the gas kinematics, we have also determined the [O III]/H$\beta$ ratio over the SAURON field-of-view. This ratio is a good indicator of young metal-rich stellar populations where star formation is very intense (Kauffmann et al. 2003). In Figure 3 (left panel) the ratio map exhibits very low values at the location of the kinematically decoupled component and high values in the very centre of the galaxy. These results are similar to those of Yoshida et al. (1993) in the inner regions (i.e. [O III]/H$\beta$=0.14-0.41) and are consistent with a star-forming region surrounding an active galactic nucleus.

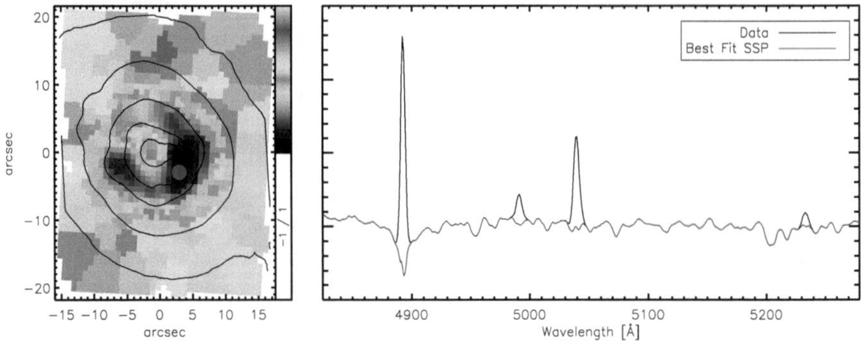


**Fig. 3.** A young KDC in NGC 5953. From left to right: i) SAURON [O III]/H$\beta$ ratio map (in logarithmic scale), ii) SAURON spectrum (black solid line) at the location marked in the left panel (red dot), with the best single stellar population model (t=3 Gyr) overlaid (red solid line).

## 3 A recently formed KDC in NGC 5953

Despite the fact that the gas and stars in the inner parts of the galaxy appear to be kinematically decoupled from the main body and despite the extremely low [O III]/H$\beta$ ratio in the same location, there is no direct proof that both the star-forming gas and stellar decoupled component are related, although this is likely. To establish this link, we need to estimate the age of the stars in that region and determine whether it is indeed consistent with a young stellar population.

Here we follow the simple procedure outlined in Vazdekis (1999) to measure the age of the underlying stellar population. Briefly, the idea is to find the single stellar population (SSP) model from a library of model templates that best matches the overall spectrum in the kinematically decoupled region. Figure 3 (right panel) shows the result of this experiment. The best-fit SSP model has an age of ~3 Gyr. By making the simple assumption that the galaxy is made by two populations (i.e. young and old), the luminosity-weighted age of 3 Gyr measured here sets an upper limit to the age of the young population. The true age of this population is however difficult to assess given that different fractions of young stars on top of an old population can lead to the same luminosity-weighted age. Nevertheless, given the quality of the fit, it seems certain that young stars are present at the location of the KDC.

The combination of these different observations suggests that at least part of the gas in NGC 5953 was acquired from the nearby interacting galaxy NGC 5954, settled to the disk plane with a direction of rotation opposite to that of the stars in the main disk, and itself started to form stars. The kinematic and population information set strong constraints on the formation timescale of the KDC we observe in this galaxy, and suggest that we are indeed witnessing the early stage of its formation.

## References

[1] Bertola F., Buson L. M., Zeilinger W. W., 1988, Nature, 335, 705
[2] Bertola, F., Corsini, E. M., Vega Beltran, J. C., et al., 1999, ApJ, 519, L127
[3] Combes F., Rampazzo R., Bonfanti P. P., Pringniel P., Sulentic J. W., 1995, A&A, 297, 37
[4] de Zeeuw, P.T., et al., 2002, MNRAS, 329, 513
[5] Emsellem, E., et al., 2004, MNRAS, 352, 721
[6] Falcón-Barroso, J., et al., 2006, MNRAS, submitted
[7] Gonzalez Delgado R. M., Perez E., 1996, MNRAS, 281, 781
[8] Hernández-Toledo H. M., Fuentes-Carrera I., Rosado M., Cruz-González I., Franco-Balderas A., Dultzin-Hacyan D., 2003, A&A, 412, 669
[9] Jenkins C. R., 1984, ApJ, 277, 501
[10] Kauffmann, G., et al., 2003, MNRAS, 346, 1055
[11] Kuntschner, H., et al., 2006, MNRAS, 369, 497
[12] Moore B., Ghigna S., Governato F., Lake G., Quinn T., Stadel J., Tozzi P., 1999, ApJ, 524, L19
[13] Naab T., Burkert A., 2001, ApJ, 555, L91
[14] Rampazzo R., Reduzzi L., Sulentic J. W., Madejsky R., 1995, A&A, 110, 131
[15] Reshetnikov V. P., 1993, A&A, 280, 400
[16] Yoshida M., Yamada T., Kosugi G., Taniguchi Y., Mouri H., 1993, PASJ, 45, 761
[17] Vazdekis A., 1999, ApJ, 513, 224
[18] Wozniak H., Combes F., Emsellem E., Friedli D., 2003, A&A, 409, 469

# Characterizing Stellar Populations in Spiral Disks

M. Mollá[1], S. Cantin[2], C. Robert[2], A. Pellerin[3] and E. Hardy[4]

[1] CIEMAT, Avda. Complutense 22, Madrid (Spain) mercedes.molla@ciemat.es
[2] Université Laval and Observatoire du mont Mégantic, Québec (Canada)
[3] Space Telescope Science Institute, 3700 San Martin Drive, Baltimore (USA)
[4] National Radio Astronomy Observatory, Casilla El Golf 16-10, Santiago (Chile)

**Summary.** It is now possible to measure detailed spectral indices for stellar populations in spiral disks. We propose to interpret these data using evolutionary synthesis models computed from the Star Formation Histories obtained from chemical evolutionary models. We find that this technique is a powerful tool to discriminate between old and young stellar populations. We show an example of the power of Integral Field spectroscopy in unveiling the spatial distribution of populations in a barred galaxy.

## 1 Introduction

The information obtained for a spiral galaxy may come from two sources: the gas or the stars. It is usually obtained from emission or absorption lines respectively (see (6) for a review). In the case of spiral galaxies, the spectra of the HII regions, which will display emission lines of H$\alpha$, [O III] etc., provide information on the chemical abundances in the gas and/or the *present* Star Formation Rate (SFR). The atomic and/or molecular gas densities may also be estimated from observations. These data represent the present time state of a galaxy.

It is currently possible to perform spectrophotometric observations in different wavelength bands in order to obtain the radial distribution of surface brightness, colors, or even of spectral indices that quantify the absorption lines of the stellar disk. These observations measure the added light of all the stellar generations still present in the disk of the galaxy, thus providing information on the *average* properties of the disk over the galaxy life time.

The question then is how to use these two types of data, emission and absorption, to determine the history of a spiral galaxy. What are the possible evolutionary paths followed by a galaxy in order to reach the present observed state of its youngest generation while displaying the observed average properties?

## 2 Observing Spiral Disks

The measurement of the stellar absorption line indices of a galaxy is important since it provides information on the averaged properties of the various stellar populations and thus represents a powerful tool in breaking the age-abundance degeneracy (see next section). Historically observations of stellar absorption lines in spiral disks have been difficult to obtain because the flux level from the sky is usually very close to that emitted by the stellar populations under study, especially in the outer disk regions. The solution has been to obtain narrow-band images with filters specially designed for each galaxy (1) or to perform Fabry-Perot interferometry with a Taurus-Tunable filter (10). Both methods are conceptually similar, although the instrumentation is different. The idea resides in measuring fluxes within wavelength bands corresponding to specific spectral indices, as close in time as possible with the observations of the corresponding continuum bands. The galaxy disk is divided in concentric rings and the flux is then integrated azimuthally so that the effective signal-to-noise for the specific absorption line increases. A more modern approach is the use of a 3D spectroscopic instrument such as OASIS (2) or SAURON (4) which offer high spatial and spectral resolution. These instruments allow separation of the gas and young star emission line regions from the older phases of star formation.

## 3 The Degeneracy Problem : Theoretical Models

The information from the gas phase, such as density, abundance, actual star formation rate, is usually analyzed using chemical evolutionary models. They describe how the proportion of heavy elements present in the interstellar medium (ISM) increases, starting from primordial abundances, when stars evolve and die. Modern codes solve numerically the equation system used to describe a scenario based on initial conditions for the total mass of the region, the existence of infall or outflow of gas, and the initial mass function (IMF). Stellar mean-lifetimes and yields, known from stellar evolutionary tracks, are also included. Finally, a star formation rate is assumed. These many inputs, which greatly influence the final results of a model, imply that a large number of parameters must be chosen. The *best* model for a galaxy will of course be the one which reproduces the observational data as closely as possible. The resulting star formation history (SFH) and the age-metallicity relation (AMR) might be taken as the basic relations to describe the evolution of the galaxy.

Chemical evolutionary models, however, suffer from the well known *uniqueness* problem: it is usually impossible to describe the evolution of a galaxy only by knowing the final state or present day abundances, since more than one evolutionary scenario can be used to reach the present situation. A solution is feasible however if the number of observations is larger than the number of parameters used to constrain the model.

On the other hand, data related to the stellar phase, such as the radial distribution of the surface brightness or colors, are usually analyzed through (evolutionary) synthesis models – see (3) for a recent and updated evolutionary synthesis model – based on a single stellar population (SSP) created by an instantaneous burst of star formation. These codes compute the spectral energy distribution (SED), $S_\lambda$, colors and spectral absorption indices emitted by a SSP of metallicity $Z$ and age $\tau$, from the sum of spectra of all stars created and distributed along an HR diagram, convolved with an initial mass function. This SED is a characteristic of each SSP of a given age and metallicity.

These models suffer, in turn, from a *degeneracy* problem: the old and metal-poor SSP's show a similar SED to the young and metal-rich ones. A possible solution is to use absorption spectral indices, since some of them depend on $Z$ while some others, particularly the Balmer and other hydrogen lines, depend only on $\tau$. Thus, with two orthogonal indices, it is possible to find the parameters ($\tau$, $Z$) for a SSP. This technique was first developed and used to study globular clusters and elliptical galaxies, for which it was assumed that the star formation occurred in a short and early burst. It seems possible from this scheme to find when the burst took place, based on the age of the stellar population, and also what is the metallicity of the stars.

## 4 Applying Synthesis Models to Spiral Disks

When the star formation does not occur in a single burst, as it seems to be the case in spiral galaxies where the star formation is continuous, the characteristics found using an SSP model represent only averaged values. These are weighted by the luminosity of each stellar generation, since the final SED corresponds to the one emitted by the sum of different SSP's when a convolution with the SFR, $\Psi(t)$, is done :

$$F_\lambda(t) = \int_0^t S_\lambda(\tau, Z)\Psi(t')dt', \qquad (1)$$

where $\tau = t - t'$.

We must find, e.g. using a least square method, the best superposition of these SSP's which will fit the data. We must estimate the proportion of each SSP defined by an age and a metallicity which best reproduces the observations. This would give the SFH, $\Psi(t)$, and AMR (i.e. $Z(t)$). However, this method, in addition to being costly in computing-time, may produce some unphysical solutions. Due to this the SFH is usually assumed, not computed, e.g. an exponentially decreasing function of time. This also requires some hypotheses about the shape and the intensity of the SFR or, to avoid a bias, the adoption of many functions. This last option leads to an excessively large number of models. A second point, usually forgotten, is that $S_\lambda(\tau, Z) = S_\lambda(\tau, Z(t'))$, that is, the metallicity changes with time since stars continue to form. Therefore, it is not clear which $Z$ must be selected at each time step without knowing how it is evolving in the galaxy. Usually, only one $Z$ is used for the whole integration.

We suggest here a method to explore spiral galaxies, which are very well studied objects from the chemical evolutionary point of view. Since a large number of emission data is available, it is possible to compute and constrain chemical evolutionary models. The SFR and the AMR obtained as a result of these models may be used as an input for Eq. 1. If the SED, colors or absorption indices are well fitted simultaneously with the present–time data, we then have a good characterization of the evolutionary history of the galaxy.

We applied our technique to three Virgo galaxies for which the radial distribution of the spectral indices Fe5270 and Mg2 had been observed (1; 9). We first computed the multiphase chemical evolutionary models for these galaxies, using a code that was developed for the Solar Neighbourhood (5) and later applied to the Milky Way Galaxy (MWG) and to other spiral disks (see (7) and references therein). As we explained in (9), the radial distribution of the spectral indices were predicted using Eq.1 where the SFH and AMR result from the best chemical evolution model.

In order to study the problems of uniqueness and degeneracy of these models, we performed 500 different models for one of the three Virgo galaxies: NGC 4303 (8). From a statistical analysis using a $\chi^2$ method, we determine that only $\sim$20 ( a 4%) of these models can reproduce the present day data. The evolutionary synthesis code was then applied to these 20 models to calculate the radial distribution of the Fe5270 and Mg2 spectral indices. Only $\sim$6 (a 1%) of them were able to fit simultaneously both sets of observations. Furthermore, these latter models defined a very small region in the parameter space. This implies that the SFR and AMR obtained for each radial region of the galaxy can be known with a high level of confidence. Our technique is therefore quite powerful in estimating the evolutionary history of a spiral galaxy when both emission and absorption line data are available.

## 5 The Barred Galaxies NGC 4900 and NGC 5430

We have observed two barred galaxies NGC 4900 and NGC 5430 with OASIS, an optical integral-field spectrograph at the Canada-France-Hawaii Telescope (2). About 1000 spectra in the wavelength ranges from 4700 to 5500Å and from 6270 to 7000Å have been collected with a spatial resolution of 0.80″ within the central 12″ × 17″ of each galaxy. As shown in Figure 1, these data allow us to create detailed maps for the different spectral indices showing various morphologies. When compared with model predictions, the values for the H$\beta$, $Fe5270$ and $Fe5335$ spectral indices allow us to locate the gas and

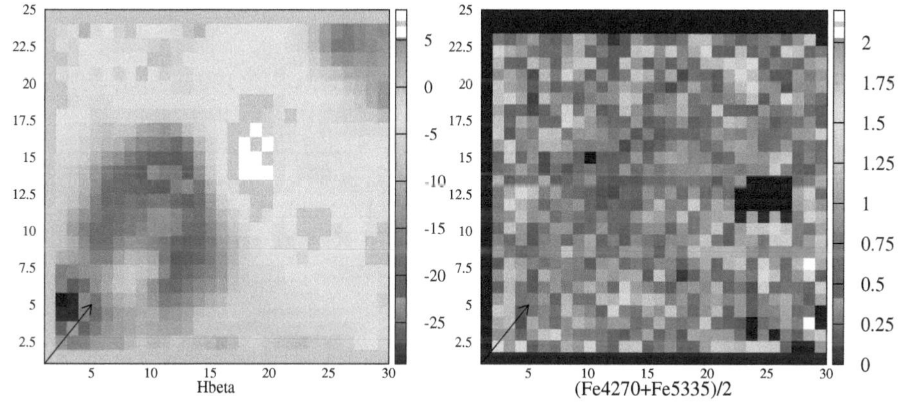


**Fig. 1.** Maps of spectral indices H$\beta$ (left) and $< Fe >= (Fe5270 + Fe5335)/2$ (fight) in units of Å. One element in $x$ and $y$ covers 100 pc.

the stellar populations. Thus, in NGC 4900, the youngest stellar population (as described by H$\beta$, left panel of Fig.1) is located parallel to the galaxy large scale bar (marked by the arrow), while the older stars, as described by Fe5270 and Fe5335 spectral indices, are distributed around the bar.

## 6 Conclusions

1. Measurements of spectral indices in spiral disks are now possible (1; 10), either through narrow-band imaging, mostly with a Fabry-Perot interferometer and a Taurus Tunable filter, or using integral field spectroscopy.
2. The use of chemical evolutionary tracks and evolutionary synthesis models to interpret these data results in a powerful tool to determine with precision the evolutionary history of spiral galaxies.
3. The properties of spatially-resolved spiral regions, such as the central region of barred galaxies, observed using 3D spectroscopy, may be related to other processes, e.g. to a bar which may produce new star forming bursts. These effects may be quantified using our combined evolutionary models.

## References

[1] Beauchamp, D., & Hardy, E. 1997, AJ, 113, 1666
[2] Cantin, S., Robert. C., Pellerin, A. & Mollá, M. 2005, New Astronomy Review, in press

[3] Delgado, R. M. G., Cerviño, M., Martins, L. P., Leitherer, C., & Hauschildt, P. H. 2005, MNRAS, 357, 945
[4] Falcón-Barroso, J., et al. 2004, MNRAS, 350, 35
[5] Ferrini, F., Matteucci, F., Pardi, C., & Penco, U. 1992, ApJ, 387, 138
[6] Henry, R. B. C., & Worthey, G. 1999, PASP, 111, 919
[7] Mollá, M., & Díaz, A. I. 2005, MNRAS, 358, 521
[8] Mollá, M., & Hardy, E. 2002, AJ, 123, 3055
[9] Mollá, M., Hardy, E., & Beauchamp, D. 1999, ApJ, 513, 695
[10] Ryder, S. D., Fenner, Y., & Gibson, B. K. 2005, MNRAS, 358, 1337

# The Stellar Populations of E and S0 Galaxies as Seen with SAURON

H. Kuntschner[1], E. Emsellem[2], R. Bacon[2], M. Bureau[3], M. Cappellari[4], R. L. Davies[3], T. de Zeeuw[4], J. Falcón-Barroso[4], D. Krajnović[3], R. M. McDermid[4], R. F. Peletier[5] and M. Sarzi[3]

[1] Space Telescope European Coordinating Facility, Garching, Germany
hkuntsch@eso.org
[2] Centre de Recherche Astronomique de Lyon, Lyon, France
[3] University of Oxford, Oxford, UK
[4] Leiden Observatory, Leiden, The Netherlands
[5] Kapteyn Astronomical Institute, Groningen, The Netherlands

**Summary.** We present selected results from integral-field spectroscopy of 48 early-type galaxies observed as part of the SAURON survey. Maps of the H$\beta$, Fe5015, Mg $b$ and Fe5270 indices in the Lick/IDS system were derived for each of the survey galaxies. The metal line strength maps show generally negative gradients with increasing radius roughly consistent with the morphology of the light profiles. Remarkable deviations from this general trend exist, particularly the Mg $b$ isoindex contours appear to be flatter than the isophotes of the surface brightness for about 40% of our galaxies without significant dust features. Generally these galaxies exhibit significant rotation. We infer from this that the fast-rotating component features a higher metallicity and/or an increased Mg/Fe ratio, as compared to the galaxy as a whole.

We also use the line strength maps to compute average values integrated over circular apertures of one effective radius, and derive luminosity weighted ages and metallicities. The lenticular galaxies show a wide range in age and metallicity estimates, while elliptical galaxies tend to occupy regions of older stellar populations.

## 1 The SAURON Survey

We are carrying out a survey of the dynamics and stellar populations of 72 representative nearby early-type galaxies and spiral bulges based on measurements of the two-dimensional kinematics and line strengths of stars and gas with SAURON, a custom-built panoramic integral-field spectrograph for the William Herschel Telescope, La Palma (Bacon et al. (1)). The goals and objectives of the SAURON survey are described in de Zeeuw et al. (2), which also presents the definition of the sample. The full maps of the stellar kinematics for the 48 elliptical (E) and lenticular (S0) galaxies are given in Emsellem et al. (3). The morphology and kinematics of the ionised gas emission are presented in Sarzi et al. (4). Here we summarize selected results of the absorption line strength measurements, which are described more fully in Kuntschner et al. (5).

## 2 Mg *b* Isoindex Contours Versus Isophotes

One of the most interesting aspects of integral-field spectroscopy is the capability to identify two-dimensional structures. For the first time we can use this in connection with line strength indices and compare isoindex contours with the isophotal shape. One might expect the index to follow the light in slowly-rotating giant elliptical galaxies, since the stars are dynamically well mixed. However, one could imagine that dynamical substructures with significantly different stellar populations could leave a signature in line strength maps which are sensitive to e.g., metallicity.

The Mg *b* index is the most stringent feature in our survey and potential differences between isoindex contours and isophotes should be most apparent. Indeed we find a number of galaxies in our survey where the isoindex contours clearly do not follow the isophotes, e.g., NGC 4570 and NGC 4660 (see Fig. 1). Both galaxies show fast-rotating components along the direction of enhanced Mg *b* strength. In order to further study such a possible connection, we determine the best fitting simple, elliptical model for each of the reconstructed images and Mg *b* maps in our survey (excluding all galaxies with significant dust absorption). For an example of this procedure see Fig. 2.

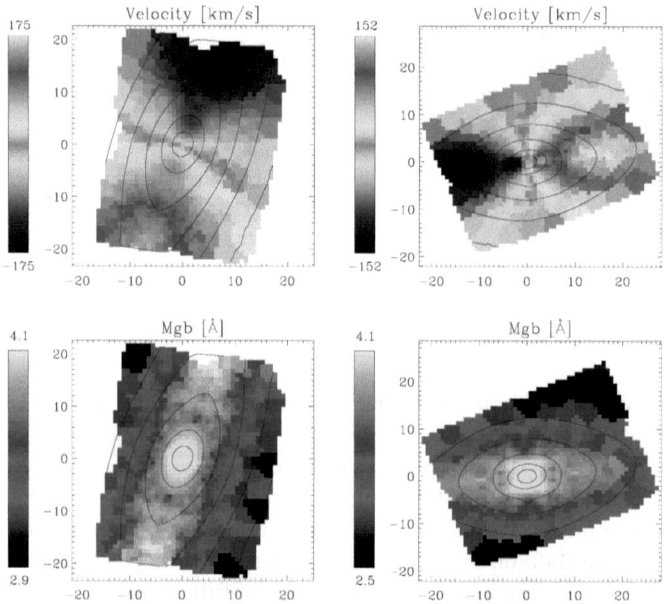


**Fig. 1.** The velocity maps and Mg *b* maps are shown for NGC 4570 (left) and NGC 4660 (right). Both galaxies show fast-rotating components along the direction of enhanced Mg *b* strength. For presentation purposes the Mg *b* maps are symmetrized. The x- and y-axis are given in arcsec; North is up and East to the left.

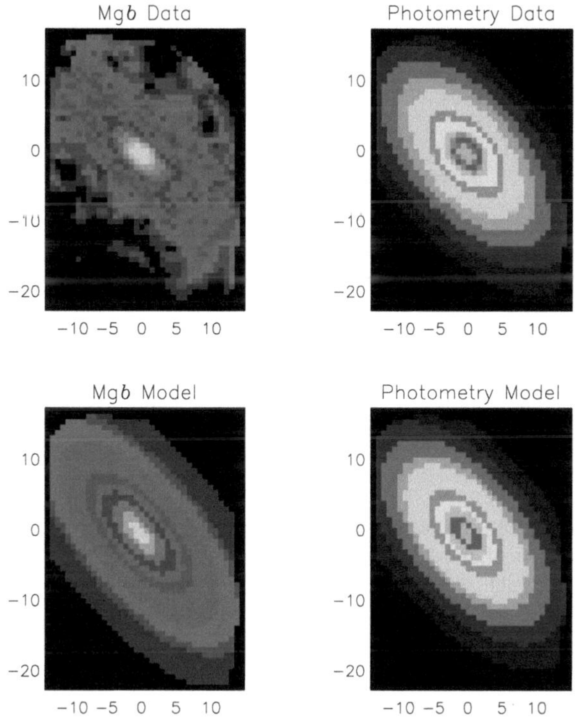


**Fig. 2.** *Top row:* The interpolated Mg *b* map and the reconstructed image of NGC 3377. *Bottom row:* The elliptical models with constant position angle and ellipticity were fitted to the Mg *b* map and the reconstructed image. The best fitting ellipticity for the isophotes and the Mg *b* map is $0.473\pm0.003$ and $0.573\pm0.022$, respectively. The x- and y-axis are given in arcsec; North is up and East to the left.

The final results of the ellipse fitting are shown in Fig. 3. Applying a $2\sigma$ error cut, 16 out of 41 galaxies appear to have more flattened Mg *b* contours than the isophotes. Most of these galaxies show a high degree of rotational support in the direction of enhanced Mg *b* strength. Thus, the flattened Mg *b* distribution suggests that the fast-rotating components in these galaxies exhibit a stellar population different from the main body. The enhanced Mg *b* strength can be interpreted to first order as higher metallicity and/or increased [Mg/Fe] ratio.

## 3 Stellar Populations

In order to provide a global measurement of the stellar populations for each galaxy we derive a central averaged spectrum from all data available within

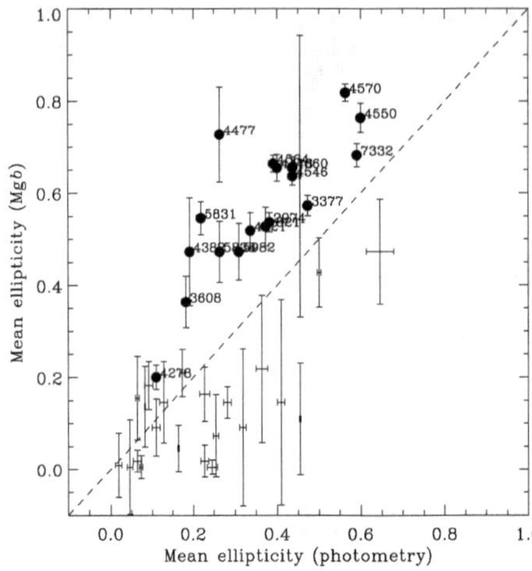


**Fig. 3.** Comparison of the average ellipticity of constant Mg *b* strength with the best fitting elliptical model of the isophotes. Errors are evaluated by a Monte-Carlo simulation. All galaxies which are more than $2\sigma$ above the one-to-one line are indicated by filled circles and their NGC numbers.

one effective radius $R_e$. Since our line strength maps do not always cover the full area of one effective radius we apply aperture corrections. However, even the galaxies with the smallest coverage (NGC 4486 and NGC5846) feature line strength data out to about 30% of $R_e$ (corresponding to a field coverage of $\approx 60''$) and corrections remain small ($< 8\%$). The median coverage of the line strength maps in our survey is $0.8 R_e$

Estimates of the luminosity weighted age and metallicity of early-type galaxies can be inferred from index-index diagrams. Generally a metallicity sensitive index is plotted against an age sensitive index. In order to minimize the influence of abundance ratio variations on our metallicity estimates we use a composite index [MgFe50]$'$. This index is constructed such that a mean metallicity is measured rather than a metallicity biased by non-solar abundance ratios.

Fig. 4 shows an H$\beta$ *vs* [MgFe50]$'$ diagram. Overplotted are the model predictions from Thomas, Maraston & Bender (6). The lenticular galaxies in our sample show a wide range in age and metallicity estimates. The elliptical galaxies however, tend to occupy regions of older stellar populations. There is weak evidence that ellipticals in the field may on average have experienced some "frosting" of young stars yielding lower luminosity weighted age estimates. It is worth noticing that luminosity weighted metallicities rarely exceed solar metallicity when taking an average over one effective radius.

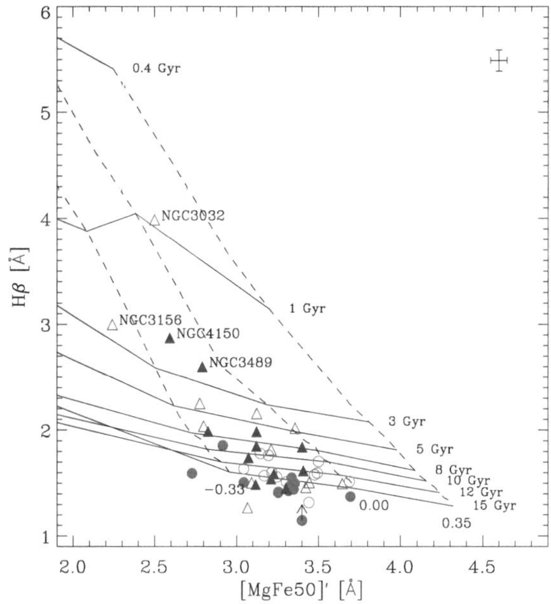


**Fig. 4.** Central, one $R_e$ aperture H$\beta$ *versus* [MgFe50]′ line strength measurements. The filled and open circles are cluster and field ellipticals, respectively; filled and open triangles represent cluster and field S0s, respectively. Overplotted are models by Thomas, Maraston & Bender (6). The error bar in the right upper corner gives the mean offset error to the Lick/IDS system.

## 4 Conclusions

The results presented here are the outcome of the first comprehensive survey of the line strength distributions of nearby early-type galaxies with an integral-field spectrograph. This data set demonstrates that many nearby early-type galaxies display a significant and varied structure in their line strength properties. This structure can be of a latex subtle nature as seen in the deviations of Mg *b* isoindex contours compared to the isophotes of galaxies.

The 2D coverage of the line strength allows us to connect the stellar populations with the kinematical structure of the galaxies and thus further our knowledge of the star-formation and assembly history of early-type galaxies.

## References


[1] Bacon R., Copin Y., Monnet G., Miller B. W., Allington-Smith J. R., Bureau M., Carollo M. C., Davies R. L., Emsellem E., Kuntschner H., Peletier R. F., Verolme E. K., de Zeeuw P. T., MNRAS, **326**, 23 (2001)

[2] de Zeeuw P. T., Bureau M., Emsellem E., Bacon R., Carollo C. M., Copin Y., Davies R. L., Kuntschner H., Miller B. W., Monnet G., Peletier R. F., Verolme E. K., MNRAS, **329**, 513 (2002)
[3] Emsellem E., Cappellari M., Peletier R. F., McDermid R. M., Bacon R., Bureau M., Copin Y., Davies R. L., Krajnović D., Kuntschner H., Miller B. W., de Zeeuw P. T., MNRAS, **352**, 721 (2004)
[4] Sarzi M., Falcón-Barroso J., Davies R. L., Bacon R., Bureau M., Cappellari M., de Zeeuw P. T., Emsellem E., Fathi K., Krajnović D., Kuntschner H., McDermid R. M., Peletier R. F., 2006, MNRAS, accepted, astro-ph/0511307
[5] Kuntschner H., Emsellem E., Bacon R., Bureau M., Cappellari M., Davies R. L., de Zeeuw P. T., Falcón-Barroso J., Fathi K., Krajnović D., McDermid R. M., Peletier R. F., Sarzi M., 2006, MNRAS, submitted
[6] Thomas D., Maraston C., Bender R., MNRAS, **339**, 897 (2003)

# VIMOS-IFU Spectroscopy of Extragalactic Star Forming Complexes

N. Bastian[1], E. Emsellem[2], M. Kissler-Patig[3] and C. Maraston[4]

[1] University College London, Department of Physics and Astronomy, Gower Street, WC1E 6BT, London, UK bastian@star.ucl.ac.uk
[2] CRAL–Observatoire de Lyon, , 9 avenue Charles André, 69561 Saint-Genis-Laval Cedex, France
[3] European Southern Observatory, Karl-Schwarzchild-Strasse 2 D-85748 Garching b. München, Germany
[4] University of Oxford, Denys Wilkinson Building, Keble Road, Oxford, OX13RH, UK

**Summary.** We present *VLT-VIMOS-IFU* spectroscopy of the overlap region in the merging Antennae galaxies (NGC 4038/39). This regions contains six complexes of young massive clusters, three of which display strong Wolf-Rayet features, indicating extremely young ages ($\sim 3-7$ Myr). The high area normalised star formation rates within these young complexes places them in the regime of starburst galaxies, thereby justifying the designation of *localised starbursts*. The ionised gas surrounding the complexes is expanding at speeds of $20-40$ km/s. This slow expansion can be understood as a bubble, caused by the stellar winds and supernovae within the complexes, expanding into the remnant of the progenitor giant molecular cloud. We also find that the complexes are likely to lose a large fraction of their extended material to the surrounding regions, however the central regions may merge into a single massive star cluster.

## 1 Introduction

Recent studies of massive young cluster (YMC) populations have shown that young clusters are often not isolated, but tend to be part of extended groupings or complexes (13). This implies that clusters do not form in isolation, but as part of a larger hierarchy of structure formation, extending from stellar to kiloparsec scales (3).

In order to investigate the role that these complexes of young clusters have in the hierarchy of star formation in galaxies, as well as their impact on the surrounding ISM and GMCs, we have obtained *VLT-VIMOS-IFU* spectroscopy of two cluster-rich regions in the merging galaxies NGC 4038/39 (the Antennae). The positions of our two pointings are shown in Fig. 1 superimposed on a *HST-WFPC2* V-band image. The Antennae make an almost ideal site for this type of study as it hosts a large young cluster population which has been studied in great detail (e.g. (12)), and shows groupings on multiple spatial scales.

In addition to the important role that complexes play in the hierarchy of star formation in galaxies, they may also be sites of heavy merging of star clusters allowing the formation of an extremely massive central star cluster (5; 7).

The spectra obtained (covering 4100 – 6100 Å) allow us to determine the age, extinction, star formation rates, metallicity and expansion velocities of the gas within each of the complexes. Additionally, we supplement the IFU spectra with *HST-WFPC2* imaging in order to measure the structural parameters of the complexes. The full analysis of this dataset, along with details on data reduction are presented in Bastian et al. 2005.

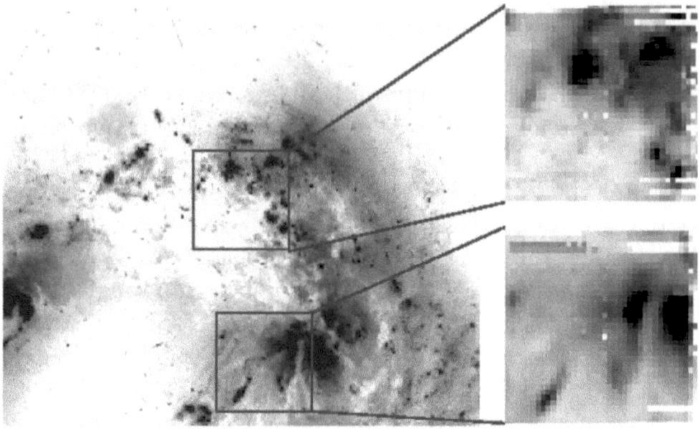

**Fig. 1.** Location of the two VIMOS pointing marked on a *HST-WFPC2* V-band mosaic image. Fields 1 and 2 are the top and bottom right panel, respectively.

## 2 Wolf-Rayet Features and Age Dating of the Complexes

In three of the complexes in Field 1 (i.e. in the overlap region) we detected a strong Wolf-Rayet (WR) features, namely the "blue bump", which is a combination of broad N III, C IV, and He II lines. This feature shows the existence of very young hot stars, and implies an age (at least of the most recent star formation episode within the complex) of 3 – 7 Myr.

The spatial resolution of the observations allow us to find where strong Wolf-Rayet features are present within the field of view. To do this, we fit a polynomial to the spectral region around the 'blue bump', and then subtract the continuum. We then sum the remaining flux in that region, and reconstruct the spatial image. The results are shown in the right panel of Fig. 2, where bright colours represent strong Wolf-Rayet features and dark colours represent regions where little or no features are detected. Here we see that Complexes 4, 5, and 6 are the only regions of strong WR features in this field. From the fact that these three complexes show WR features, we can conclude that they (or at least the massive stars within the complex) have the same age within $\sim 4$ Myr (9), and all have ages less than 10 Myr. This strongly suggests that the formation of these complexes, or at least the most recent round of star formation within them, was triggered at about the same time. We will return to this point in § 4.

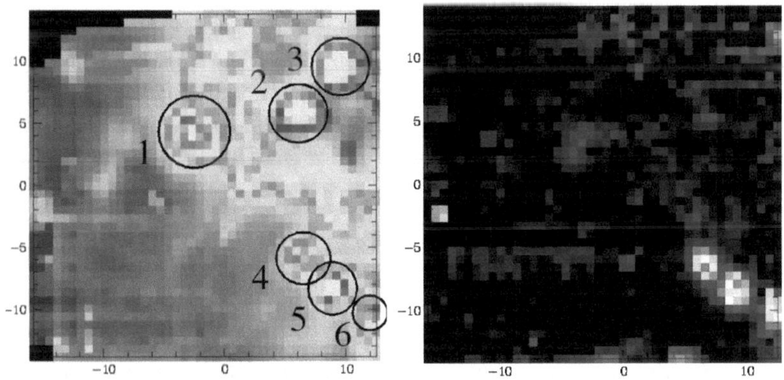


**Fig. 2. Left:** The reconstructed image of Field 1 (the top right panel of Fig. 1) of the VIMOS-IFU datacube. The six complexes are labeled. **Right:** The spatial distribution of Wolf-Rayet features in Field 1. Brighter/lighter colours indicates strong Wolf-Rayet features. Note that Complexes 4, 5, 6 all show strong WR features. One spaxel (spatial pixel) in all VIMOS reconstructed images corresponds to 63 pc.

# 3 Expansion Velocities and Star Formation Rates

In order to estimate the velocity dispersion of the gas (a measure of the random motions of the gas), we have fit the width of the H$\beta$ and [O III]$\lambda\lambda$4959, 5007 emission lines after subtraction of the best-fitting stellar background template. The results are shown in the right panel of Fig. 3, where the dark (blue) and bright (white) regions represent velocity dispersions of 20 and 100 km/s, respectively. Note that all of the complexes, but in particular the youngest complexes (i.e. Complexes 4, 5 and 6) all show low velocity dispersions.

These low values of the gas velocity dispersion can be understood as bubbles, originating from a cluster wind, which is made up from the collection of winds from hot stars and supernovae. If one assumes that the winds/SNe act collectively to form a cluster wind, then one can estimate the speed of expansion of the resulting bubble using equations 12.12 and 12.19 in (8). For this we assume that the ambient medium that the bubble is expanding into has a typical density of a GMC in the Antennae (0.3 $M_\odot/pc^3$ (11)) and a wind kinetic luminosity of between $10^{39}$ and $10^{42}$ ergs/s (10). We further assume that the wind has been blowing constantly for the past 3-7 Myr. With these assumptions (and assuming that the driven bubble is in the energy or momentum 'snowplow' phase) the bubble outflow velocity is expected to be between 10 and 35 km/s. These values are in excellent agreement with the observed velocity dispersions.

Additionally, we can estimate the star formation rates inside the complexes using the de-reddened H$\beta$ flux (the extinction in this line was determined using the H$\beta$/H$\gamma$ flux ratio and assuming Case B recombination). We find area normalised star formation rates of a 1–7 $M_{\odot}/yr/kpc^2$, which is comparable to starburst galaxies, thereby justifying the label *localised starbursts.* As star formation is highly non-uniform (i.e. taking place within complexes) we would also expect the metal distribution in the ISM to be patchy, as recently observed in the Antennae (4).

## 4 Groupings of Complexes and Their Fate

We noted in § 2 that Complexes 4, 5, and 6 have the same age (within $\sim$ 4 Myr) and are all located in a fairly compact region. The left panel in Fig. 3 shows a continuum subtracted *HST-WFPC2* H$\alpha$ image of Field 1. Note that Complexes 4, 5 and 6 share a common ionised envelope, again arguing for a similar age of each of the complexes. Also, these three complexes have very similar metallicities (1). The size of the ionised region surrounding these complexes is $\sim$ 735 pc (assuming a distance of 19.2 Mpc), which corresponds to the size scale of large GMCs or GMC association in the Antennae. The age difference between the complexes (derived from the star clusters within them, see (1)) is incompatible with the distance between the complexes divided by the local sound speed. Thus, it is likely that an external perturbation (e.g. a large scale shock or gravitational perturber) triggered the collapse of a whole large GMC/GMC association, forming the three individual complexes.

Finally, by comparing the measured size of each complex to their estimated tidal radii we can investigate their future evolution. This is of particular interest as a number of recent simulations have suggested that cluster complexes may be sites of heavy merging of star clusters, allowing for the formation of extremely massive clusters (5; 7). Current observations also suggest such a formation mechanism for the most massive clusters (6). As an initial assumption we suppose that the complexes are in a rotating disk. Taking the galactic rotation velocity at the position of the complexes (13) and their mass from photometric estimates, we find that most of the complexes are much larger than their tidal radii, suggesting that they will loose a substantial amount of matter to their surroundings. However, due to the steep density profile of matter within the complexes (2) we do expect the central parts of the complexes to be sites of heavy merging. Thus, while the complexes as a whole appear unbound, the observed high central densities support the Kroupa (7) and Fellhauer & Kroupa (5) scenario.

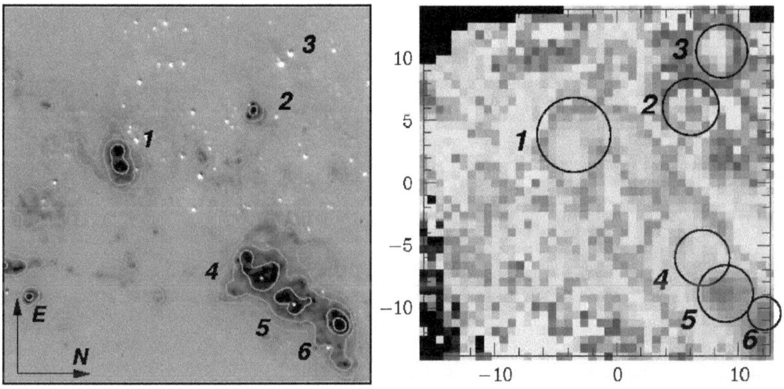


**Fig. 3. Left:** Continuum-subtracted Hα HST-WFPC2 image of Field 1, with the complexes labeled. Note the shared H II envelope around Complexes 4, 5, and 6. **Right:** The velocity dispersion of the gas as measured from the combination of Hβ and the [O III]λλ4959, 5007 lines. The colour coding extends from 0 to 100 km/s, with dark (blue) representing the lowest velocity dispersions and white representing the highest. The measurements have been corrected by subtracting in quadrature the measured width of the skyline at λ5577. Note that the gas at the position of the complexes has the lowest velocity dispersion in the region (20-40 km/s).

## References


[1] Bastian, N., Emsellem, E., Kissler-Patig, M., & Maraston, C. 2005, A&A, in press (astro-ph/0509249)
[2] Bastian, N., Gieles, M., Efremov, Yu.N., & Lamers, H.J.G.L.M. 2005, A&A, 443, 79
[3] Elmegreen, B. G., & Efremov, Y. N. 1996, ApJ, 466, 802
[4] Fabbiano, G. Baldi, A., King, A.R. et al. 2004, ApJL, 605, L21
[5] Fellhauer, M. & Kroupa, P. 2005, MNRAS, 359, 223
[6] Kissler-Patig, M., Jordán, A., & Bastian, N. 2005, A&A, in press (astro-ph/0512360)
[7] Kroupa, P. 1998, MNRAS, 300, 200
[8] Lamers, H.J.G.L.M. & Cassinelli, J.P. 1999, In: Introduction to Stellar Winds, Cambridge University Press
[9] Norris, R.P.F.: The Role of Massive Stars in Young Starburst Galaxies. PhD Thesis, University College London (2003)
[10] Tenorio-Tagle, G., Silich, S., Rogríguez González, A., & Muñoz-Tuñón, C. 2005, ApJL, 628, 13
[11] Wilson, C.D., Scoville, N., Madden, S.C., & Charmandaris, V. 2003, ApJ, 599, 1049
[12] Whitmore, B.C., Zhang, Q., Leitherer, C., et al. 1999, AJ, 118, 1551
[13] Zhang, Q., Fall, M., Whitmore, B.C. 2001, ApJ, 561, 727

# Analyzing Non-circular Motions in Spiral Galaxies Through 3D Spectroscopy

I. Fuentes-Carrera[1], M. Rosado[2] and P. Amram[3]

[1] Instituto de Astronomi a, Geofísica e Ciências Atmosféricas, Universidade de São Paulo, rua do Matão 1226 , Cidade Universitaria, 05508-900, São Paulo, SP, Brazil isaura@astro.iag.usp.br

[2] Instituto de Astronomía, UNAM, A.P.70-264, Ciudad Universitaria, México D.F., 04510, México margarit@astroscu.unam.mx

[3] Laboratoire d'Astrophysique de Marseille, 2 Place Le Verrier, 13248 Marseille Cedex 04 France amram@oamp.fr

**Summary.** 3D spectroscopic techniques allow the assessment of different types of motions in extended objects. In the case of spiral galaxies, thes type of techniques allow us to trace not only the (almost) circular motion of the ionized gas, but also the motions arising from the presence of structure such as bars, spiral arms and tidal features. We present an analysis of non-circular motions in spiral galaxies in interacting pairs using scanning Fabry-Perot interferometry of emission lines. We show how this analysis can be helpful to differentiate circular from non-circular motions in the kinematical analysis of this type of galaxies.

## 1 Kinematical Information from 3D Spectroscopy: Radial Velocity Fields

Scanning Fabry-Perot (FP) observations allow us to build a line profile for each pixel on extended objects such as galaxies. The intensity profile found along the scanning process contains information about the monochromatic emission and the continuum emission of the object. Radial velocity maps are computed considering the Doppler shift of the observed emission-line.

## 2 Decomposition of Observed Radial Velocities

In the general case, the observed radial velocity $V_{obs}$ at a certain point $(R, \theta)$ on the plane of the galaxy has the following decomposition:

$$V_{obs} = V_{syst} + \Big[ (V_{circ} + V_{tan}) \times \cos\theta + V_{rad} \times \sin\theta \Big] \times \sin i + \quad V_{\perp} \times \cos i \quad (1)$$

where $i$ is the inclination angle between the plane of the galaxy and the plane of the sky, $V_{syst}$ is the systemic velocity of the galaxy, $V_{tan}$ is the tangential velocity component in the plane of the galaxy additional to the circular velocity component $V_{cir}$, $V_{rad}$ is the radial velocity component in the plane of the galaxy, and $V_{\perp}$ is the velocity component perpendicular to the plane of the galaxy. See Figure-1 for more details.

In practice, the assumption is made that $V_{cir}$ is much greater than the other velocity components so that the observed radial velocity at each point of the galaxy can be transformed into a the circular velocity via the following equation:

$$V_{circ}(R) \sim \left( \frac{V_{obs}(\rho, \phi) - V_{syst}}{\cos\theta \sin i} \right) \tag{2}$$

This equation is then used to build the rotation curve of a galaxy following the procedure described in (1).

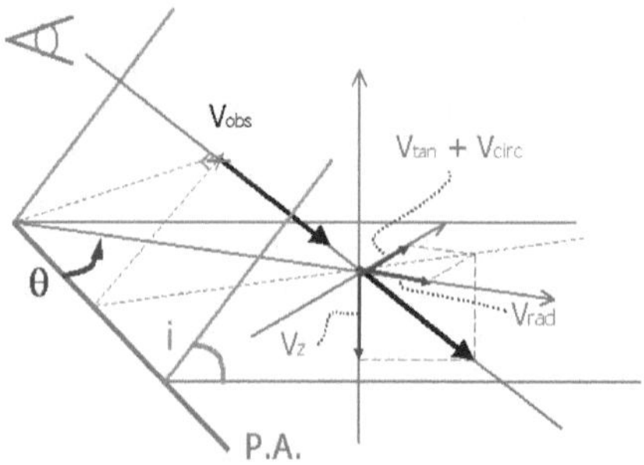


**Fig. 1.** Decomposition of the observed radial velocity $V_{obs}$ with respect to the plane of the galaxy $P$. $V_{rot}$ is the rotation velocity around the center of the galaxy, $V_{rad}$ is the radial velocity from the center of the galaxy and $V_{\perp}$ is the velocity perpendicular to the plane of the galaxy

# 3 Evaluating Non-Circular Motions

## 3.1 Rotation Curves vs. Velocity Fields

Two dimensional kinematical fields of disk galaxies allow us to study the motion of the gas all over the galaxy. These motions can be matched with structures found through photometric methods in order to determine to which extent the gas is following pure circular motion around the center of the galaxy and to which extent there are important contributions from non-circular velocities. In this way we can associate particular features on the rotation curves of galaxies with the presence of certain morphological features (Figure-2).

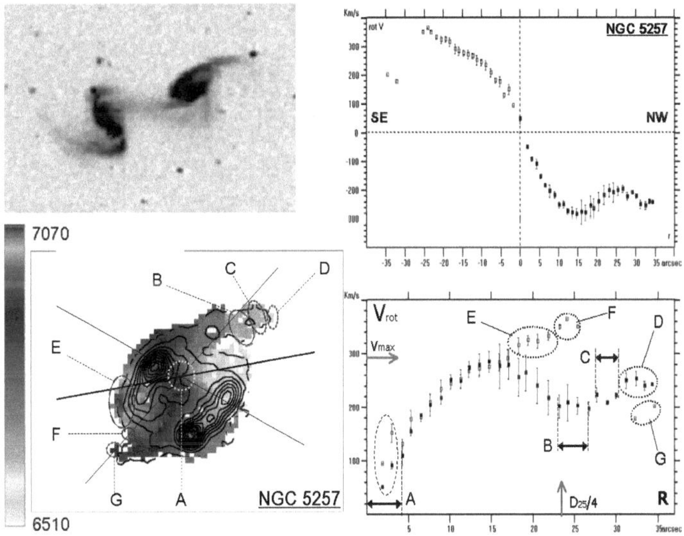


**Fig. 2.** The top left panel shows the interacting pair Arp 240 (NGC 5256/57). The bottom left panel shows the velocity field of NGC 5257 with the monochromatic (H$\alpha$) isophotes superposed, The right panels show the rotation curve of the galaxy, approaching and receding sides separately (*top*) and their superposition (*bottom*). Letters shown in the curve are associated to the regions shown in the middle panel and indicate features such as a central bar (region A), the spiral arms (region B, C, E and F) and the beginning of a tidal bridge (region G) and its probable counterpart (region D) -Figures taken from (3)

### 3.2 FWHM

The velocity dispersion for each pixel can also be computed from the $FWHM$ of the profile at each pixel deconvolved by the instrumental profile. Depending on the inclination of the galaxy, the velocity dispersion at each pixel may be associated to motions of the gas perpendicular to the plane of the galaxy (in the case where the galaxy is almost face-on, $i \sim 0°$) or to non-circular motions in the plane of the galaxy (if the galaxy is almost edge-on, $i \sim 90°$) in the approximation that the ISM is optically thin (Figure-3 ). For intermediate values of $i$, velocity dispersion is associated to both types of motion.

### 3.3 Residual Velocities

Residual velocity fields are obtained by subtracting the observed radial velocity field from a *ideal* radial velocity field. This field is the radial velocity field of a galaxy in with uniform circular motions only so that

$$V_{ideal}(R) = V_{syst} + \left(V_{circ}(R) \times \cos\theta \times \sin i\right) \tag{3}$$

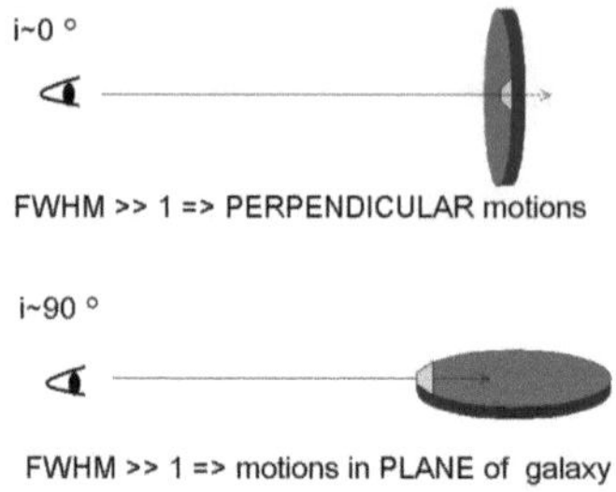


**Fig. 3.** Analysis of non-circular motions using the FWHM values for particular inclinations of a disk galaxy

| θ (deg) | 0° | | 90° | | 180° | | 270° | |
|---|---|---|---|---|---|---|---|---|
| $V_{res}$ | + | − | + | − | + | − | − | + |
| $V_{tan}$ | − | + | Negligible | | + | − | Negligible | |
| $V_{rad}$ | Negligible | | − | + | Negligible | | + | − |

**Fig. 4.** Analysis of non-circular motions from the sign of residual velocities when velocities perpendicular to the plane of the galaxy $V_\perp$ are negligible

The *ideal* velocity field is constructed from the observed rotation curve by assigning to all the points of the galaxy at a certain radius, the rotational velocity from the rotation curve at that particular radius. These fields prove to be very useful to evaluate the validity of the kinematical parameters chosen to compute the rotation curve of a disk galaxy (4). They also give the opportunity to detect and analyze non-circular motions of the gas in each galaxy since points where the residual velocity $V_{res} \sim 0$ correspond to regions in the galaxy where circular motions predominate. For points where $V_{res} > 0$ or $V_{res} < 0$, the interpretation becomes more complicated. By definition the residual velocity $V_{res} = V_{ideal} - V_{obs}$ at each point is given by

$$V_{res} = -\Big[ V_{tan} \times \cos\theta + V_{rad} \times \sin\theta \Big] \times \sin i - V_\perp \times \cos i \qquad (4)$$

If $V_\perp \simeq 0$, the sign of $V_{res}$ is a combination of the values of $V_{tan}$, $V_{rad}$ and $\theta$. For instance along the minor axis, $\theta = 90°,\ 270°$ implies $V_{tan}$ contribution is insignificant. For $\theta = 90°$, $V_{rad}$ has the opposite sign of $V_{res}$, while for $\theta = 270°$, $V_{rad}$ has the same sign as $V_{res}$. Along the major axis $V_{rad}$ contribution is insignificant so that for $\theta = 180°$, $V_{tan}$ has the same sign as $V_{res}$, while for $\theta = 0°$, $V_{tan}$ has the opposite sign of $V_{res}$. This analysis is shown in Figure-4.

The analysis of non-circular motions from the residual velocity field is complemented with information from the velocity dispersion field. For instance for low inclinations, $V_{\perp}$ can be neglected in equation 4 for regions displaying small velocity dispersion values. An example of this analysis is presented in Figure-5 and thoroughly explained in (3).

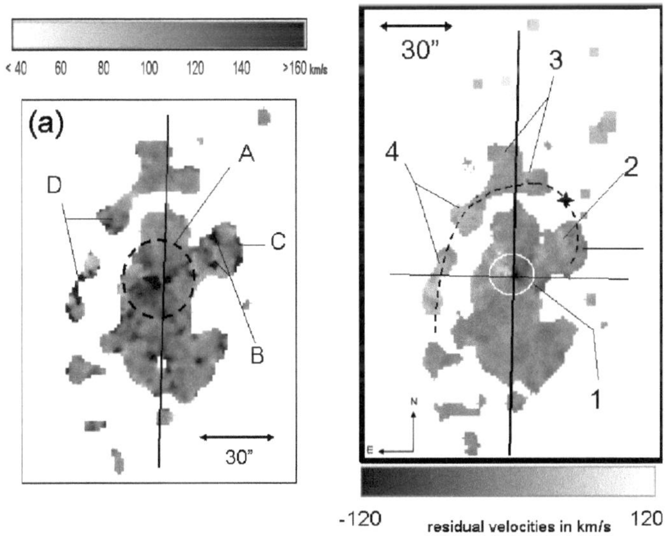


**Fig. 5.** *Left panel* FWHM map of the spiral galaxy NGC 5426 in the interacting pair Arp 271. Letters show regions with large values. Since $i = 60$ deg for this galaxy. these values can be associated to motions perpendicular to the plane of the galaxy. *Right panel* Residual velocity map for the same galaxy. Numbers show regions with important contributions of non-circular velocities that can be associated to the passage of gas through the spiral density wave. Black star indicates the possible location of the co-rotation radius. This analysis is explained in detail in (3).

## References

[1] P. Amram, C. Balkowski, J. Boulesteix et al, A&A, **310**, 737 (1996)
[2] I. Fuentes-Carrera: Cinematica y dinamica de pares de galaxias en interaccion. PhD Thesis, UNAM, Mexico (2003)
[3] I. Fuentes-Carrera, M. Rosado, P. Amram et al, A&A, **415**, 451 (2004)
[4] P.J. Warner, M.C.H. Wright & J.E. Baldwin, MNRAS, **163**, 163 (1973)

# Harmonic Analysis of the H$\alpha$ Velocity Field of NGC 4254

L. Chemin[1,2], O. Hernandez[1], C. Balkowski[2], C. Carignan[1] and P. Amram[3]

[1] Université de Montréal, Dépt. de Physique, C.P. 6128 succ. centre-ville, Montréal (QC), CANADA H3C 3J7 chemin@astro.umontreal.ca
[2] Observatoire de Paris, section Meudon, GEPI, CNRS-UMR 8111 & Université Paris 7, 5 Pl. Janssen, 92195, Meudon, France
[3] Observatoire Astronomique Marseille-Provence, 2 Pl. Le Verrier, 13248, Marseille Cedex 4, France

The ionized gas kinematics of the Virgo Cluster galaxy NGC 4254 (Messier 99) is analyzed by an harmonic decomposition of the velocity field into Fourier coefficients. The aims of this study are to measure the kinematical asymmetries of Virgo cluster galaxies and to connect them to the environment. The analysis reveals significant $m = 1, 2, 4$ terms which origins are discussed.

## 1 Introduction

Galaxies in clusters are sensitive to environmental effects like the cluster tidal field, gravitational encounters with other galaxies, galaxy mergers, ram pressure stripping and accretion of gas (see e.g. (6), (10)). Such external events dramatically affect their structure, triggering internal perturbations like bars or oval distortions (e.g. (1)), spirals, warps (5) or lopsidedness (2). Their kinematics is also disturbed, as revealed by long-slit spectroscopy 1-D rotation curves (Rubin et al. 1999).

High-resolution H$\alpha$ velocity fields were obtained for 30 Virgo cluster galaxies (3) in order to study the degree of perturbation of their 2-D kinematics and the influence of the environment on the kinematics. The harmonic analysis is a powerful tool to detect kinematical anomalies, as already shown on HI velocity fields (9). This technique is applied to the H$\alpha$ velocity field of NGC 4254 (Figure 1).

## 2 The Virgo Atlas : Observations

The data acquisition and reduction are described in details in (3). Data cubes of 30 galaxies were obtained in the H$\alpha$ emission-line at the Observatoire du mont Mégantic (Canada), the Observatoire de Haute-Provence, the ESO 3.6-m telescope and the Canada-France-Hawaii-Telescope between 2000 and

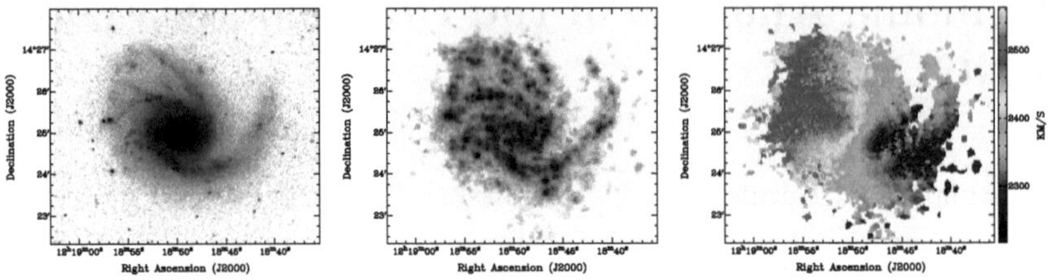

**Fig. 1.** $H-$band image, H$\alpha$ emission and velocity maps of NGC 4254. The NIR image is from (4).

2005. The instrument is composed of a focal reducer and a Fabry-Perot interferometer coupled with a photon-counting system. The spectral sampling of the observations varies between 7 and 16 km s$^{-1}$, the spatial sampling between 0.42" and 1.61", and the field-of-view between 3.6' and 13.7'. The typical total exposure time per spectral channel is ∼4 minutes.

## 3 Results of the Harmonic Decomposition and Analysis

A tilted-ring model is first fitted to the velocity field to derive the inclination, position angle, systemic velocity and kinematical centre. The velocity field is then expanded into Fourier coefficients by fitting $v_{\rm obs} = c_0 + \sum_m c_m \cos(m\Psi) + s_m \sin(m\Psi)$ (9). $\Psi$ corresponds to the angle in the plane of a ring, the coefficient $c_0$ to the systemic velocity of a ring, the first order term $c_1$ to the rotation curve and all other terms to non-circular motions. Since no warping of the optical disk is clearly detected in this galaxy, the inclination ($i$) and kinematic position angle ($P.A.$) can be kept constant as a function of radius during the fitting. The Fourier coefficients are computed up to the 4-th order. Figure 2 shows the results of the decomposition inside a radius of $R = 150$". The results are :

- a large variation of the $c_0$ term at small radius which is accompanied by significant non-zero $c_2$ and $s_2$ terms. This likely indicates the effect of a lopsided potential ($m = 1$ perturbation).
- a nearly constant value of $7 \pm 2$ km s$^{-1}$ inside 100" for the $s_1$ term. This feature does not disappear when $i$ and/or $P.A.$ are allowed to vary. It could be due to elliptical streaming in a $m = 2$ perturbing potential and/or to a radial inflow (considering trailing spiral arms).
- large variations of the $c_4$ and $s_4$ terms at large radius ($R > 100$"). At these radii, the emission is dominated by many HII regions in the northern and western arms and no evident $m = 3$ or $m = 5$ modes are detected. The origin of these asymmetries still remains to be explained.

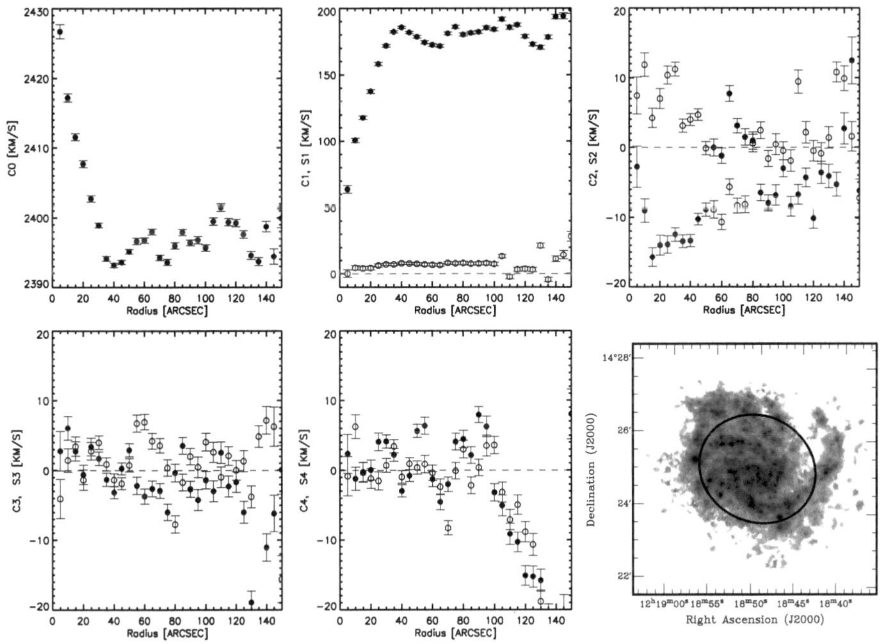


**Fig. 2.** Harmonic coefficients $c_m$ and $s_m$ (full and open symbols resp.) up to $m = 4$. The ellipse in the Hα image displays the projection of the $R = 100"$ annulus.

This work is in progress and we plan to investigate which event has created the perturbations of the potential of NGC 4254. It could be due to the accretion of gas into the disk plane (7).

## References

[1] Bournaud F., & Combes F. 2002, A&A, 392, 83
[2] Bournaud F., et al., 2005, A&A, 438, 507
[3] Chemin L., et al., 2005, MNRAS, in press, astro-ph/0511417
[4] Gavazzi G., et al., 2003, A&A, 400, 451
[5] Huang S., & Carlberg R. G., 1997, ApJ, 480, 503
[6] Moore B., Lake G., Katz N., 1998, ApJ, 499, L5
[7] Phookun B., Vogel S. N., & Mundy L. G., 1993, ApJ, 418, 113
[8] Rubin V. C., Waterman A. H., & Kenney J. D. P., 1999, AJ, 118, 236
[9] Schoenmakers R. H. M., Franx M., & de Zeeuw P. T., 1997, MNRAS, 292, 349
[10] Vollmer B., et al., 2001, ApJ, 1561, 708

# The NIR Tully-Fisher Relation in Different Environments: the Importance of 3D Observations

I. Fuentes-Carrera[1], C. Mendes de Oliveira[1], M. Rosado[2], P. Amram[3], H. Plana[4] and O. Garrido[3,5]

[1] IAG-USP, rua do Matão 1226 , Cidade Universitaria, 05508-900, São Paulo, SP, Brazil isaura@astro.iag.usp.br,oliveira@astro.iag.usp.br
[2] IA-UNAM, A.P.70-264, Ciudad Universitaria, México D.F., 04510, México margarit@astroscu.unam.mx
[3] LAM, 2 Place Le Verrier, 13248 Marseille Cedex 04 France amram@oamp.fr, olivia.garrido@oamp.fr
[4] LATO/DCET/UESC, Rodovia Ilhéus-Itabuna, km 16, CEP 45662-000 Ilhéus, BA, Brazil plana@uesc.br
[5] Observatoire de Paris-Meudon, 5 Place Jules Janssen, 92195 Meudon, France

## 1 Introduction

Recent results show that the Tully-Fisher (TF) relation for interacting galaxies has a larger scatter than for isolated ones (8),(10). These results were derived from observing techniques that do not take fully into account the (most probable) perturbed nature of these galaxies. We present preliminary results of a study of the TF relation in the near-infrared (NIR) for galaxies in binary systems and in compact groups using 3D observing techniques.

## 2 Why 3D Spec?

Interacting galaxies might be perturbed by their companions. Tidal features or distorted spiral arms can be responsible for perturbations to the circular velocity of a galaxy inducing an error in the estimation of the maximum rotation velocity ($V_{max}$) used to derive the TF relation. 3D techniques such as Fabry-Perot (FP) interferometry allow us to trace the internal motions of a galaxy to distinguish non-circular from circular motions (4),(5).

## 3 The Samples

Our sample of Hickson compact groups (HCGs) consists of 9 groups with a total of 25 galaxies. The TF in the B-band for this set of galaxies was previously studied by (12), finding that, with a few exceptions, galaxies in compact groups follow the *standard* TF relation. The sample of interacting galaxy pairs consists of 11 pairs in an early stage of interaction: both galaxies are distinguishable but already display interacting features such as tidal tails and bridges and perturbed spiral arms. For the control sample of isolated galaxies we will use isolated galaxies from the GHASP survey (6).

# 4 Observations and Reductions

The galaxies in all samples were observed using a scanning FP interferometer in order to derive the H$\alpha$ emission-line velocity maps. Data reduction was done using the ADHOCw software[6] developed by J. Boulesteix. Details on the observation and data reduction are given in (1), (11) and (13), for the HCGs and in (3), (4) and (7), for the interacting pairs.

# 5 The Data

## 5.1 Absolute Magnitudes

The NIR is fairly independent of current star formation providing a measure of current stellar content that is, presumably, more closely tied to the total mass of a galaxy. Uncertainties due to interstellar absorption are also greatly reduced. We use the isophotal K20 magnitude given by 2MASS. The K magnitude was chosen following the result of (2) showing this magnitude produces the NIR-TF with the least scatter for galaxies in clusters. Magnitudes have not been corrected by absorption, nor internal or Galactic extinction.

## 5.2 Rotation Velocities

The rotation curve of each galaxy was derived from the 2D velocity field as described in (5). As a first approach, the maximal rotation velocity of each galaxy $V_{max}$ was taken to be the velocity of the furthest emitting point from both sides of the curve, that did not display non-circular or peculiar velocities.

# 6 Preliminary Results

Figure-1 presents the preliminary TF for galaxies for which $V_{max}$ has already been determined taking into account the effects of non-circular motions. Globally galaxies in pairs and in HCGs seem to follow the same trend along the diagram except for two galaxies whose K magnitudes were unusually low: H100D and UGC 11283. These values are probably due to an error in the 2MASS data, since these galaxies show no peculiar behavior in the B band.

[6]www.oamp.fr/adhoc/adhocw.htm

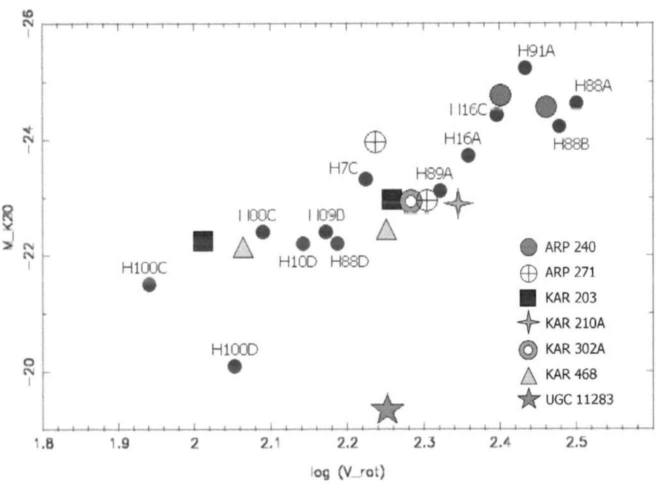


**Fig. 1.** Preliminary NIR-TF plot for galaxies in pairs and compact groups. Black dots show galaxies in HCGs. The name of the galaxy is also shown. Symbols show galaxies in pairs. Symbols of same type indicate galaxies belonging to the same pair. Magnitude values have not been corrected by absorption.

Once we determine the Vmax of all the galaxies in the sample, the resulting NIR-TF relations will be compared to the NIR-TF derived for isolated galaxies from the GHASP survey

## References

[1] P. Amram, H. Plana, C. Mendes de Oliveira et al, A&A, **402**, 865 (2003)
[2] N. Bouche & S. Schneider, PASP, Conf. Ser. **218**, 111 (2000)
[3] I. Fuentes-Carrera: Cinemática y dinámica de pares de galaxias en interacción. PhD Thesis, UNAM, México (2003)
[4] I. Fuentes-Carrera, M. Rosado, P. Amram et al, A&A, **415**, 451 (2004)
[5] I. Fuentes-Carrera, M. Rosado & P. Amram *-these proceedings*
[6] O. Garrido, M. Marcelin, P. Amram et al, MNRAS, **362**, 127 (2005)
[7] H. Hernandez-Toledo, I. Fuentes-Carrera, M. Rosado et al, A&A, **412**, 669 (2003)
[8] S.J. Kannappan, D.G. Fabricant & M. Franx, AJ, **123**, 2358 (2002)
[9] I.D. Karachentsev, S.N. Mitronova, V.E. Karanchentseva et al, A&A, **396**, 431 (2002)
[10] I. Marquez, J. Masegosa, M. Moles et al A&A, **393**, 389 (2002)
[11] C. Mendes de Oliveira, H. Plana, P. Amram et al, AJ, **507**, 691 (1998)
[12] C. Mendes de Oliveira, P. Amram, H. Plana et al, AJ, **126**, 2635 (2003)
[13] H. Plana, P. Amram, C. Mendes de Oliveira et al, AJ, **125**, 1736 (2003)

# 3D Kinematics of Local Galaxies and Fabry-Perot Database

O. Garrido[1], P. Amram[2], C. Balkowski[1], M. Marcelin[2] and I. Jegouzo[1]

[1] Observatoire de Paris, section de Meudon, GEPI, 5 place Jules Janssen, 92195 Meudon, France
[2] Laboratoire d'Astrophysique de Marseille, Observatoire de Marseille, 2 place Le Verrier, 13248 Marseille cedex 04, France

## 1 The GHASP program

### 1.1 Definition

GHASP (Gassendi HAlpha survey of SPirals) is a survey led at the 1.93m telescope of the Observatoire of Haute-Provence (France) using a scanning Fabry-Perot and a photon-counting system (1). GHASP mapped the distribution and the kinematics of ionized gas, observing the H$\alpha$ line, for 220 nearby (v$\leq$10 000 km/s) Spirals and Irregulars located in low-density environments. High spatial (2 arcsec) and spectral (0.15A) resolution 2D and 3D data have been derived : continuum map, H$\alpha$ map, velocity field, position-velocity diagram, rotation curves and data cubes. By now, data for 96 galaxies have been published (2), (3) ,(4), (5) enabling to study the distribution of dark halos and the kinematical properties of galaxies. More information about GHASP and the total list of observed galaxies is available on : **http://www-obs.cnrs-mrs.fr/interferometrie/GHASP/ghasp.html**. GHASP provides a reference sample of 2D velocity fields for nearby galaxies covering the plane : absolute magnitude- morphological type.

### 1.2 Results

In (5), GHASP galaxies have been differentiated into isolated and softly interacting galaxies thanks to the determination of a parameter which measures the degree of interaction suffered by a galaxy. For this purpose, the parameter has been evaluated for all the companions located within a box of twenty times the diameter of the considered galaxy, and in the range $\pm$700 km/s in radial velocity.

No strong difference has been remarked between the two groups : the extension of the ionized gas, proportional to R25, is similar; the inner slope of the rotation curves is correlated with the central concentration of light; the kinematical asymmetry increases when considering later type, less massive and bluer galaxies. We note that the slope of the Tully-Fisher relation is smaller for isolated galaxies. The only strong difference concerns the outer slope of the rotation curves : no decreasing outer slope was found in the softly-interacting group and the outer slope for these galaxies is not correlated with the type or the kinematical asymmetry, contrarily to isolated galaxies. This shows that faint interactions can only affect the outer optical part of the galaxies (in terms of modification of fundamental or kinematical properties) whereas the radio parts of softly-interacting galaxies can already show strong disturbances.

## 2 Fabry-Perot database

### 2.1 The content

Fabry-Perot data are most often obtained with instrument visitors and then are not available to the community. FP data are useful to compare with simulations, data obtained at other wavelengths and 3D data obtained for distant galaxies (6) which suffer from low spatial resolution. A FP database is actually under construction and will be completely operational at the end of the year 2005. It is already visible at http://fpdatabase.obspm.fr.

Two kinds of data can be downloaded :
-rough data in .fits format for user who wants to do all the reduction process : calibration and observation cubes.
-reduced data ready to be used in .fits formats : continuum map, monochromatic map, velocity field, position-velocity diagram, rotation curve and calibrated data cube (see Fig.1).
All the data are corrected from the geometric distortion and have WCS astrometry. In a first time, about a hundred of galaxies (observed by GHASP) will be available in the archive; after, around 400 galaxies should be available including galaxies in high density environments (clusters, pairs, groups). We also plan to present the data obtained for the H$\alpha$ survey of the Magellanic Clouds (7) and of the Milky Way (8) under the form of mosaics covering large fields of view.

### 2.2 The tools

Researches by catalog name, fundamental parameters, position or SQL researches will be possible. To begin, the downloads and the visualization of the associated images will be possible for 2D data, whereas only downloads will be available for 3D data. Visualization's tools for 2D and 3D data will be developed in the archive. One of the goals is to adopt for Fabry-Perot data the formats defined by the Virtual Observatory. To begin, we project to provide a SIA (Simple Image Access) service. As soon as the 3D OV format will be chosen, the database will also provide transactions with SSA (Simple Spectrum Access). This will enable to exploit our data with OV tools.

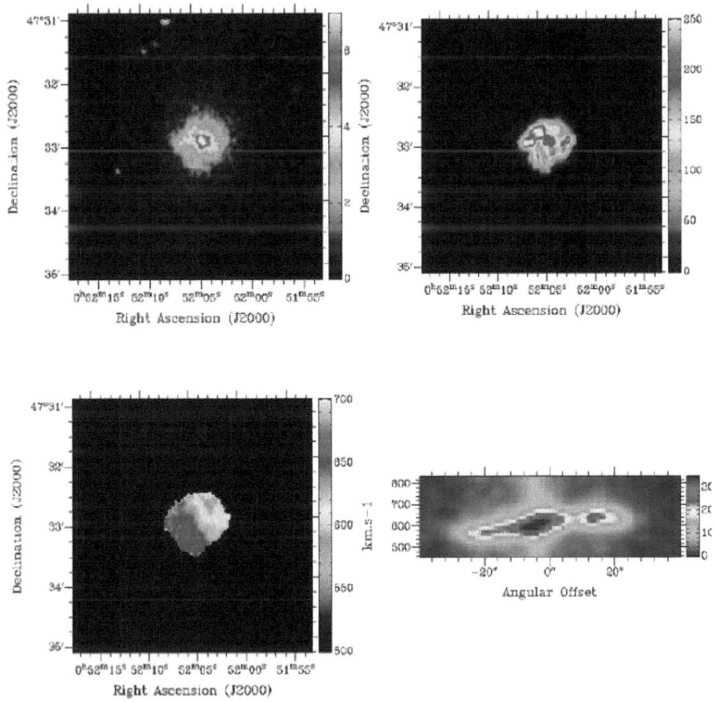


**Fig. 1.** Examples of maps provided by the Fabry-Perot archive for UGC 528. Up (left to right) : continuum map and H$\alpha$ map. Down : velocity field and position-velocity diagram. The field of view is 4x4 with a spatial resolution around 2.5 arcsec and a spectral resolution around 1.5nm.

## References


[1] J-L. Gach, O. Hernandez, J. Boulesteix et al: PASP **114**, 1043 (2002).
[2] O. Garrido, M. Marcelin, P. Amram, J. Boulesteix: A &A, **387**, 821, (2002).
[3] O. Garrido, M. Marcelin, P. Amram, O. Boissin: A &A, **399**, 51 (2003).
[4] O. Garrido, M. Marcelin, P. Amram: MNRAS, **349**, 225 (2004).
[5] O. Garrido, M. Marcelin, P. Amram et al: MNRAS, **362**, 127 (2005).
[6] H. Flores, M. Puech, F. Hammer et al: A&A, **420**, L31 (2004).
[7] P. Ambrocio-Cruz, A. Laval, M. Rosado et al.: A&A, **127**, 2145 (2004).
[8] D. Russeil, C. Adami, P. Amram et al.: A&A, **429**, 497 (2005).

# Validation of Stellar Population and Kinematical Analysis of Galaxies

M. Koleva[1,2], N. Bavouzet[3], I. Chilingarian[2,4] and P. Prugniel[2,3]

[1] Department of Astronomy, Sofia University, Bulgaria
[2] CRAL, Observatoire de Lyon, France
[3] GEPI, Observatoire de Paris, France
[4] Sternberg Astronomical Institute, Moscow, Russia

**Summary.** 3D spectroscopy produces hundreds of spectra from which maps of the characteristics of stellar populations (age-metallicity) and internal kinematics of galaxies can be derived. We carried out simulations to assess the reliability of inversion methods and to define the requirements for future observations. We quantify the biases and show that to minimize the errors on the kinematics, age and metallicity (in a given observing time), the size of the spatial elements and the spectral dispersion should be chosen to obtain an instrumental velocity dispersion comparable to the physical dispersion.

**Key words:** kinematics, stellar populations of galaxies, error analysis

## 1 Analyzing the Data

Recently it became possible to derive simultaneously internal kinematics and characteristics of the stellar population of galaxies ((1)).

We use a simple parametric procedure ((2)) to fit the moments of the line-of-sight velocity distribution (LOSVD), v, $\sigma$, etc and the parametrized star formation history (SFH), containing either single stellar population (SSP) or several star bursts, using models computed with PEGASE.HR ((3)). Before running the inversion procedure one needs to: (a) determine variations of the spectrograph's line-spread function with wavelength and over the field of the IFU; (b) inject this information into template spectra. We stress that the population parameters are constrained by the absorption lines and not by the shape of the continuum which may be affected by internal extinction or calibration uncertainties.

## 2 Validation of the Population Pixel Fitting

The questions we address are:

1. Does the method return unbiased estimates of kinematics, age ($t$) and metallicity ($Z$) at any signal-to-noise ratio (SNR)?
2. What are the degeneracies between different parameters?

3. What SNR and spectral resolution are required to obtain a given precision?
4. Are various models of stellar populations consistent?

*Biases and degeneracy.* We have performed extensive Monte-Carlo simulations to fit SSP or SFH containing two bursts with SNR ranging from 1 to 500 $pix^{-1}$ (at R=10000; pix=0.2Å). As long as the grid of models is fine enough for performing good interpolation, our method is not biased down to SNR=5 $pix^{-1}$. The main degeneracy is naturally between age and metallicity, but Fig 1 presents also the degeneracy between $\sigma$ and $Z$: a discrepancy toward higher $Z$ (sharp lines) is compensated by lower $\sigma$. This latter degeneracy is the strongest coupling between the kinematical and population parameters. An error of 1 dex on the metallicity results in an error of about 25% on the velocity dispersion. This can introduce a significant systematics in the determination of the mass-to-light ratio of galaxies.

*Relation between precision and SNR.* The precision essentially depends on the total SNR integrated over the whole wavelength range: a lower SNR per pixel can be balanced by a larger number of pixels. When the wavelength range is shortened to exclude the blue region containing $H_\beta$ line and bluer (but keeping the same total SNR), the precision on kinematics and metallicity is not seriously affected, but the precision on age becomes twice worse.

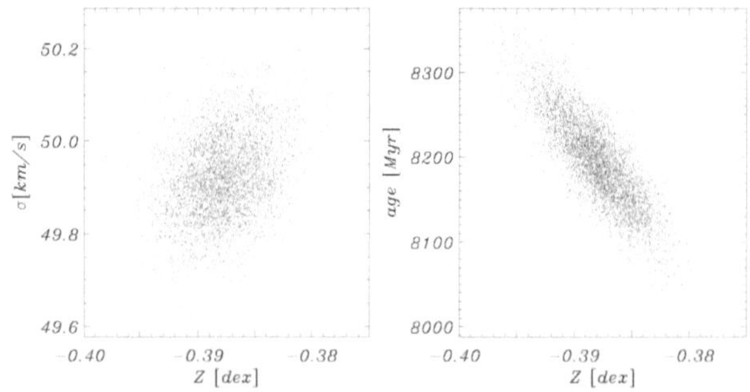


**Fig. 1.** Metallicity-velocity dispersion and Age-metallicity degeneracies. Monte-Carlo simulations with PEGASE.HR spectra, SNR $= 100\,\mathrm{pix}^{-1}$, $\sigma = 50\,\mathrm{km\,s}^{-1}$

*Relation between resolution and precision.* The precision on kinematical and population parameters depends on $(\sigma^2 + \sigma_{\rm ins}^2)$, where $\sigma_{\rm ins}$ is the instrumental velocity dispersion. The resolution is an important parameter for the precise determination of the kinematics. But it has weaker influence over the errors on age and metallicity (see the relations below). When $\sigma < 0.4\,\sigma_{\rm ins}$, the different minimization strategies with which we experimented generally become unstable to measure the internal kinematics.

## 3 Conclusion

The method based on the pixel fitting with PEGASE.HR templates can efficiently constrain kinematics and stellar population. We summarize below the relations between the errors on the population characteristics (radial velocity, velocity dispersion, population age and metallicity), the total signal-to-noise ratio, and the resolution (for the spectral range 4000 – 6800Å):

$$\begin{aligned} err_{\mathrm{v}} &\approx 5 \times SNR^{-1} \times (\sigma^2 + \sigma_{\mathrm{ins}}^2)^{2/3} \\ \sigma \times err_{\sigma} &\approx 5 \times SNR^{-1} \times (\sigma^2 + \sigma_{\mathrm{ins}}^2)^{6/5} \\ err_t/t &\approx 54 \times SNR^{-1} \times (\sigma^2 + \sigma_{\mathrm{ins}}^2)^{1/14} \\ err_Z &\approx 12 \times SNR^{-1} \times (\sigma^2 + \sigma_{\mathrm{ins}}^2)^{1/7} \end{aligned}$$

These relations are useful to select the observational setup: what is the optimal compromise between the spectral dispersion and the size of the spatial elements that minimizes the errors in a given observing time? Considering only the sources of uncertainties modeled in our simulations, it appears that the best precision on the internal kinematics will be obtained when $\sigma_{\mathrm{ins}} \approx 0.8\,\sigma$. For the best precision on the parameters of the stellar population, it is preferable to maximize the total SNR with a lower dispersion. A single setup with a resolution matching the velocity dispersion is in general a good choice.

The template mismatch due to abundance effects and the uncertainties in the modeling of a stellar populations are probably the source of significant biases on the parameters of the stellar population. In particular, we can guess that if Balmer lines are not in the wavelength range, the determination of age will be extremely sensitive to abundance mismatch. We have inverted observations of globular clusters ((4)) and simulated spectra from Bruzual & Charlot ((5)). For old populations the estimates appear too young and metal-rich. We are investigating the origin of this problem that may be connected with the bias found by Prieto et al. ((6)) in the Elodie library.

## References

[1] Ocvirk, P., Pichon, C., Lancon, A., Thiebaut, E.: MNRAS **365**, 374 (2006)
[2] Chilingarian, I., Prugniel, P., Sil'Chenko, O., Afanasiev, V.: IAU Colloquium 198, 105 (2005)
[3] Le Borgne, D., Rocca-Volmerange, B., Prugniel, P., et al.: A&A, 425, 881 (2004)
[4] Rose, J. A., et al.: ApJS **169**, 138 (2005)
[5] Bruzual, G., Charlot, S.: MNRAS **344**, 1000 (2003)
[6] Allende Prieto, C., Beers,T. C., Wilhelm, R., et al.: Ap.J. **636**, 804 (2006)

# Non-Gravitational Motions in Galaxies

D. Tamburro, H.-W. Rix and F. Walter

Max-Planck-Institut für Astronomie, Königstuhl 17, D-69117 Heidelberg, Germany tamburro@mpia.de

## 1 Summary

We present a preliminary study of non–gravitational dynamics in galaxies using radio observations of the 21 cm line of atomic hydrogen (HI). Radio spectral line HI observations are complementary to optical 3D observations as they trace the atomic gas phase which is the dominant phase of the interstellar medium in most galaxies. Radio spectral line observations also have several advantages compared to optical/NIR 3D spectrographs: large fields of views (FOV) can be imaged instantaneously at high spatial resolution (e.g., high–resolution VLA observations can cover a FOV of 30′ at a spatial resolution of 7″). Another advantage is the velocity resolution: resolutions of $<5\,\mathrm{km\,s^{-1}}$ can be achieved routinely in HI spectral line observations, allowing for a detailed analysis of the dynamics of the interstellar medium.

We have obtained VLT imaging using the VIMOS integral field unit (IFU) for a number of southern galaxies that are part of the SINGS sample (SINGS: The Spitzer Infrared Nearby Galaxy Survey, Kennicutt et al. 2003). The VLT data will be used to combine the kinematics of the ionized gas in the center with the larger scale kinematics of the HI, which is partly missing in the very center since it is ionized by radiation from stars (or has turned into molecular gas). We have started analyzing HI data for the same galaxies that have been obtained as part of THINGS (THINGS: The HI Nearby Galaxy Survey, Walter et al. 2004). The main goal is combine the VLT and HI observations to derive the non-gravitational motions in these systems. These motions are of particular interest, as they can be used to constrain the energetic feedback into the ISM of the winds of young massive stars or SNe explosions that are able to provide the energetic "feed back". In order to understand galaxy evolution it is crucial to quantify this energy injection and to understand to what extent it regulates star formation.

We have developed routines in IDL to extract the non-gravitational motions of galaxies from a datacube. For every spatial element (x,y) of the data cube the line profiles have been fitted with a single Gaussian, $G(v) = a\,\exp[-(v-v_m)^2/2\,\sigma^2]$, from which we obtain the peak value $a(x,y)$, the main velocity $v_m(x,y)$ and the velocity dispersion $\sigma(x,y)$. Some of the line profiles within the spiral arms are strongly non-Gaussian, asymmetric and often with multiple peaks; reminiscent of the effects expected by winds of massive stars and multiple SNe explosions. The velocity field has been fitted by $\chi^2$ minimization with a rotating disk model, assuming circular orbits and retaining only those pixels with S/N > 3. This simple model uses the equation:

$$v(x, y) = v_{\rm sys} + v_c \sin i \cos\phi \qquad (1)$$

where $v_{\rm sys}$ is the systemic velocity; $i$ the disk inclination; $\phi$ is a function of the position angle, of $i$, and of the center coordinates $(x_0, y_0)$; and $v_c$ is the rotation curve, for which we adopted the equation: $v_c(r) = GM\, r^a/(r + r_s)^{a+b}$, where the fitted parameters $M$, $r_s$, $a$ and $b$ define the maximum rotation velocity, the scale radius, the central slope and asymptotic slope respectively. Error bars are obtained with the Monte Carlo method. Most of the deviations of the rotation curve from the model derived by the $\chi^2$ fit are produced by the wiggles due to the spiral arms.

A better description of the kinematics would be achieved by a harmonic expansion of the velocity map on individual rings as Schoenmakers et al. (1997):

$$v(r, \phi) = c_0(r) + \sum_{n=1}^{N} [c_n(r)\, \cos(n\phi) + s_n(r)\, \sin(n\phi)] \qquad (2)$$

Star winds and SN explosions reflect on higher order harmonic modes. We plan to quantify the small-scale deviations which are not caused by gravity by using such harmonic expansions in the future.

## References

[1] Kennicutt, R. C., et al. 2003, PASP, 115, 928
[2] Schoenmakers, R. H. M., Franx, M., & de Zeeuw, P. T. 1997, MNRAS, 292, 349
[3] Begeman, K. G. 1989, A& A, 223, 47
[4] Walter, F., Brinks, E., de Blok, W. J. G., et al. 2005, ASP Conf. Ser. 331: Extra-Planar Gas, 331, 269

# Studying the Roots of the Supergalactic Wind in NGC 1569 with Gemini GMOS/IFU

M.S. Westmoquette[1], K. M. Exter[2], L. J. Smith[1] and J. S. Gallagher III[3]

[1] University College London, Gower Street, London, WC1E 6BT, U.K. msw@star.ucl.ac.uk
[2] Instituto de Astrofísca de Canarias, C/Via Lactea s/n, E38200, La Laguna (Tenerife), España katrina@iac.es
[3] Department of Astronomy, University of Wisconsin-Madison, 5534 Sterling, 475 North Charter St., Madison WI 53706, U.S.A. jsg@astro.wisc.edu

**Summary.** We present Gemini North GMOS/IFU observations of the roots of the NGC 1569 super-galactic wind. This forms part of a larger and on-going study of the formation and collimation mechanisms of galactic winds. Four IFU pointings cover the base of the outflows originating from the central super-star clusters; we present a preliminary analysis for two positions. The choice of single-slit mode and the R831 grating (dispersion = 0.34 Å $pix^{-1}$) allowed us to cover the spectral range 4740-6860Å through 500 object fibres. We observe the nebular diagnostic lines of H$\alpha$, H$\beta$, [O III]$\lambda\lambda$4959, 5007, [N II]$\lambda\lambda$6548, 6583, [S II]$\lambda\lambda$6716, 6731. The data were reduced using the standard Gemini pipeline included with IRAF, and corrected for differential atmospheric refraction. We find emission lines clearly composed of multiple components—in most cases a bright narrow feature (intrinsic FWHM = 50 km $s^{-1}$) superimposed on a fainter broad component (FWHM $\leq$ 300 km $s^{-1}$). We discuss the origin of these components, in the context of the wind/interstellar medium interaction.

## 1 Introduction

NGC 1569 is classified as a post-starburst dwarf-irregular galaxy at a distance of $2.2 \pm 0.6$ Mpc (4), and is therefore the closest example of a starburst, and an excellent analogue to high-redshift starburst galaxies. It has undergone a recent episode of star-formation that is thought to have ended 5-10 Myr ago and lasted ~100 Myr (1). Previous studies of deep H$\alpha$ images find many disrupted filaments and shells (3). A possible H I bridge between NGC 1569 and a nearby H I cloud has also been observed (8) and has led some authors to speculate that an interaction with this cloud may have been the trigger for the current starburst event.

NGC 1569 hosts several Super Star Clusters (SSCs); the two brightest are designated NGC 1569A (found to have two components; ~5-8 Myr (5)) and NGC 1569B (~10 Myr); and together with cluster 30, they provide a significant fraction (20-25%) of the total optical and near-infrared light emitted from the central region of the galaxy (7). It has been suggested that the combined supernova-driven wind from these central clusters have blown several kiloparsec-scale "superbubbles" with radial velocities of as much as $\pm 200$ km s$^{-1}$ (2), at least one of them having "blown out" (6). Very broad wings have also been found on the H$\alpha$ emission-line profile at the location of super star-cluster A (2). This is suggestive of recent supernova activity, and provides evidence that the outflow is already very fast close to the central sources.

## 2 Observations & Data Reduction

We obtained Gemini North GMOS observations of NGC 1569 with four IFU pointings covering the base of the galactic wind outflow. We opted for the single-slit mode and R831 grating, giving a dispersion of 0.34 Å pix$^{-1}$ over the spectral range 4740-6860Å. In this configuration there are 500 object fibres arranged in a rectangle (each having a diameter of 0.2", covering 3".5 × 5") separated by 1 arcmin from a block of 250 sky fibres. We observe the nebular lines of H$\alpha$, H$\beta$, [O III]$\lambda\lambda$4959, 5007, [N II]$\lambda\lambda$6548, 6583, [S II]$\lambda\lambda$6716, 6731 with a spectral resolution of 60 km s$^{-1}$ at H$\alpha$. The data were reduced using the standard Gemini pipeline included with IRAF which involves bias correction and flat-fielding, extracting and collapsing each aperture as projected on the CCD, throughput correcting, and photometric calibration. The data were then converted into 3D cube format using the E3D Vistool by interpolating the hexagonal arrangement of spaxels into congruent squares. Correction for differential atmospheric refraction was performed using a custom script making use of the IRAF STSDAS Drizzle task.

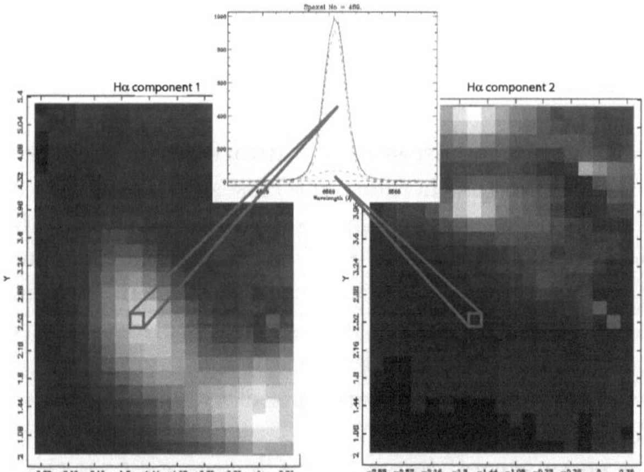


**Fig. 1.** Position 2 H$\alpha$ line flux maps. The panels show the spatial distribution of flux in each emission component. Also illustrated is how these maps are produced from fitting Gaussians to the combined profile for each spaxel.

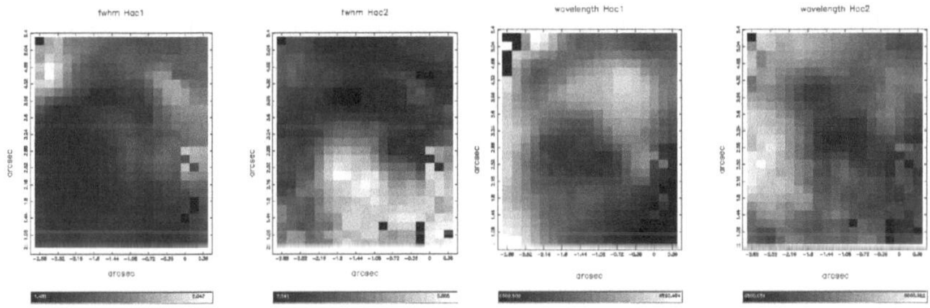


**Fig. 2.** Position 2 H$\alpha$ emission line maps. The two panels on the left represent the intrinsic FWHM for component 1 (black-white=30–70 km s$^{-1}$) and component 2 (black-white=100–300 km s$^{-1}$); the panels on the right show the heliocentric radial velocities for comp. 1 (black-white=-110 – -90 km s$^{-1}$) and comp. 2 (black-white = -110 – -75 km s$^{-1}$). Note the region of white in the comp 2 FWHM map corresponds with the bright blob of gas emitting the narrow component 1 (see Fig 1).

## 3 Analysis & Discussion

We used a graphical curve-fitting program written in IDL to automate the fitting of multiple Gaussian components to each of the emission lines observed. The decision to fit extra components was determined by setting a threshold (determined by visual inspection of the fits) on the reduced chi-squared value output for each fit. We were then able to analyse the properties of each component separately, allowing us to disentangle flux, velocity and line width information for each component. Fig 2 shows some of the results for the fits to IFU position 2.

This spatially resolved spectroscopy of the outflow roots shows for the first time the difference between the slow moving, cooler gas emitting strongly in the bright ionized regions, and the faint, fast moving hotter gas that fills the space in-between. We see broad-line emission flowing away from the gas blobs which may represent the roots of the superwind, before it has had a chance to be collimated into a bi-polar structure.

## References

[1] Greggio L., Tosi M., Clampin M., de Marchi G., Leitherer C., Nota A., Sirianni M., 1998, ApJ, 504, 725
[2] Heckman T.M., Dahlem M., Lehnert M.D., Fabbiano G., Gilmore D., Waller W.H., 1995, ApJ, 448, 98
[3] Hunter, D.A., Hawley W.N., Gallagher J.S., 1993, AJ, 106, 5
[4] Israel, F.P. 1988, A&A, 194, 24

[5] de Marchi, G., Clampin, M., Greggio, L., Leitherer, C., Nota, A., Tosi, M. 1997, ApJ, 479, L27
[6] Martin C.L., ApJ, 1998, 506, 222
[7] Origlia L., Leitherer C., Aloisi A., Greggio L., Tosi M., 2001, 122, 815
[8] Stil J.M., Israel F.P., 1998, A&A, 337, 64

# Part IV

# Lensing Studies

# Gravitational Lensing Studied with Integral Field Spectroscopy

L. Wisotzki

Astrophysikalisches Institut Potsdam, An der Sternwarte 16, D-14482 Potsdam, Germany, lwisotzki@aip.de

**Summary.** I briefly recall some basic features of gravitational lensing, chiefly in the context of multiple image formation which is the most widely observed phenomenon of 'galaxy lensing'. In such cases, the angular scales involved are suitably matched to the sizes of existing integral field units, and therefore the impact of 3D spectroscopy has been largest. I review some recent advances of studying galaxy-scale gravitational lensing with integral field units, highlighting the investigation of small-scale structure in both lenses and sources.

## 1 Introduction

Gravitational lensing comes in many flavours: weak and strong lensing; lensing by stars, galaxies or galaxy clusters; lensing detected through high-resolution imaging or through photometric time series. Integral field spectroscopy as a relatively new observational technique, has not yet produced a major impact in the field of lensing studies. For many lensing applications, this is likely to remain so for a while. This brief review focuses on a few cases where 3D spectroscopy could already make a contribution. It also recalls a few basics of gravitational lensing, setting the scene for other presentations contained in these proceedings. The discussion is limited exclusively to the phenomena associated with 'strong lensing', and mainly to the particular case of individual galaxies acting as lenses. For a comprehensive introduction to the field, the reader is referred to the recent compendium by Kochanek, Schneider, & Wambsganss (7)[1] which provides also several references to other texts and tutorials.

## 2 Some Basics

Light travels on geodesic curves, which can be bent when passing near mass concentrations. There are several ways to describe this phenomenon, for instance in terms of a variable effective diffractive index as in geometrical optics, or as wavefront distortion due to a propagation speed that depends on the local potential. The crossing time for different geometrical paths differ, due

[1]available online at http://www.astro.uni-bonn.de/~peter/SaasFee.html

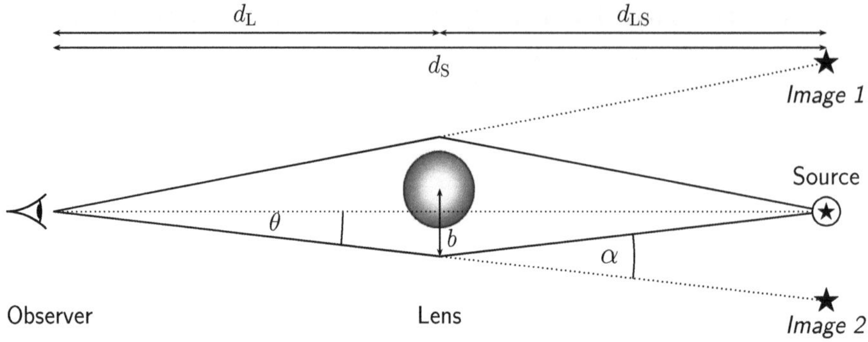


**Fig. 1.** Schematic layout and notation convention, for the case of image splitting of a background source into apparently two *mirage* images.

to two effects: a geometric time delay because of varying path lengths; and a purely relativistic (or 'Shapiro') time delay. By Fermat's principle, those paths are favoured where the arrival time is stationary. If the lens is sufficiently massive, there can be multiple solutions to this condition, in which case a compact background source is split into multiple images. Furthermore, background sources can be geometrically distorted and magnified (causing an amplification of total flux), and this magnification may depend on the source size. In case of significant transverse motions, the magnification can also change with time, leading to variability.

For a point mass, the deflection angle $\alpha$ is given by the simple relation

$$\alpha = \frac{4GM_L}{c^2}\frac{1}{b}$$

where $M_L$ ist the lens mass and $b$ is the impact parameter of the considered light ray (see Fig. 1). The lensing geometry relates the deflection angle $\alpha$ in the lens system to the angle $\theta$ as seen from the observer. In the specific case of a point mass lens perfectly aligned with a background source, the arrival time has a minimum at

$$\theta = \sqrt{\frac{4GM_L}{c^2}\frac{d_{LS}}{d_L\, d_S}} \equiv \theta_E$$

where the linear scale is set by the angular diameter distances $d_L$, $d_S$, and $d_{LS}$. In this case, the source is deformed into an 'Einstein ring' with radius $\theta_E$ (the 'Einstein radius'). More generally speaking, $\theta_E$ gives a characteristic angular scale also for non-point mass lenses: $\pi\theta_E^2$ is roughly the cross-section for strong gravitational lensing effects to become detectable, and multiple images have typical separations of the order $2\theta_E$. In a scenario with a distant quasar at $z_S \simeq 2.0$ being lensed by a galaxy located about halfway, $z_L \simeq 0.4$, this number becomes $\theta_E \sim 0.''73 \times (M_L/10^{11}\, M_\odot)^{1/2}$, thus is of order $1''$ for galaxy-scale lenses, and of order 1′ for cluster-scale lenses. It is also instructive to estimate the corresponding linear scales of one Einstein radius projected into the source plane, $\theta_\mathrm{E} d_\mathrm{S} \sim 6\,\mathrm{kpc} \times (M_\mathrm{L}/10^{11}\,M_\odot)^{1/2}$. A source is effectively point-like for lensing purposes if its physical size is $R_\mathrm{S} \ll \theta_\mathrm{E} d_\mathrm{S}$; conversely, if $R_\mathrm{S} \gg \theta_\mathrm{E} d_\mathrm{S}$, the source will be largely unaffected by lensing effects. This becomes important in the discussion of the possible effects of microlensing by stars in the lensing galaxy: compact regions in a background quasar can be susceptible to microlensing; while more extended parts of the quasar host galaxy are only affected by 'macrolensing' due to the overall lens potential.

Modelling of gravitational lensing usually starts with two assumptions: (i) deflection angles are small; (ii) the size of the lens along the optical axis is small and can be neglected – the 'Thin Lens Approximation'. A lens model is then fully specified by the geometric layout and the projected surface mass density distribution in the lens plane, $\Sigma(\boldsymbol{\theta})$, where $\boldsymbol{\theta} = (\theta_x, \theta_y)$ specifies the location in the lens plane. The surface mass distribution is related to the 2-dimensional potential $\phi(\boldsymbol{\theta})$ by

$$\Sigma(\boldsymbol{\theta}) \propto \nabla^2 \phi(\boldsymbol{\theta}) \,.$$

Deflected light rays from a source originally at $\boldsymbol{\theta}_\mathrm{S}$ and passing the lens plane at a position $\boldsymbol{\theta}$ suffer a time delay $\tau$ to reach the observer:

$$\tau(\boldsymbol{\theta}) = \frac{1+z_\mathrm{lens}}{c} \frac{d_\mathrm{L}\, d_\mathrm{S}}{d_\mathrm{LS}} \left( \frac{1}{2} |\boldsymbol{\theta} - \boldsymbol{\theta}_\mathrm{S}|^2 - \phi(\boldsymbol{\theta}) \right) .$$

By Fermat's principle, images form at stationary points of the time delay surface:

$$\nabla_\theta(\tau) = \nabla_\theta \left( \frac{1}{2} |\boldsymbol{\theta} - \boldsymbol{\theta}_\mathrm{S}|^2 - \phi \right) = 0 \,.$$

The magnification (distortion) of each image is

$$\mu = \frac{\text{image area}}{\text{source area}} = \left( \frac{\partial \boldsymbol{\theta}_\mathrm{S}}{\partial \boldsymbol{\theta}} \right)^{-1} .$$

Two simple rules can be distilled: (i) image *positions* are determined by the *gradient* of the time delay surface; (ii) image *magnifications* are determined by the *curvature* of the time delay surface. Consider for example the important case of a galaxy as lens. On the one hand there is the overall gravity action of stars, gas and dark matter combined; the gradient of this potential, and therefore of the time delay surface, determines the image splitting on arcsec scales. This is often called 'macrolensing'. On the other hand, individual stars in the lensing galaxy can locally change the curvature of $\tau(\boldsymbol{\theta})$ by a substantial amount, leading to possibly drastic changes of magnification. The same stars do not affect the gradient on scales much beyond their Einstein radii which is of order of $\mu$arcsec, and the image *positions* are therefore very nearly those obtained from macrolensing alone.

The number of images arising in a given lensing configuration depends mainly on the degree of symmetry in the lens and on the alignment of lens and source. A perfect alignment of source and a symmetric lens leads to a complete Einstein ring. A slight breaking of the symmetry produces five images, one of which (the central image) is highly demagnified and usually obscured by the lens itself, so this yields effectively a quadruple system. A somewhat higher degree of asymmetry produces three images. Again the central image is highly demagnified, effectively leaving a double system. If the source lies outside the Einstein radius of the lens, there is only one image which is then only weakly distorted and magnified. More insights into qualitative features of strong gravitational lensing can be found in the highly instructive paper by Saha & Williams (18).

The detailed modelling of lens mass distributions has reached a high level of sophistication. However, in the case of quasar lensing where the images of the source are usually point-like, the number of independent observable constraints is small. In double systems these are ideally 2 coordinate pairs, 1 flux ratio $\mu_1/\mu_2$ (the absolute magnifications are not observable), and 1 time delay difference $\Delta\tau = \tau_1 - \tau_2$ (again, the time delays themselves are not observable). If the lensing galaxy is detected, 1 coordinate pair and the lens redshift may be added. Furthermore, measuring the time delay requires a big observational effort and may take years, whereas the flux ratios can be affected by microlensing. Thus, all except the simplest mass models are severely underconstrained in double systems. This is somewhat better in quads where up to 5 coordinate pairs, 3 independent flux ratios and 3 differential time delays can be observed. Here one can at least allow for a quadrupole term in the potential, including a possible tidal contribution from the lens environment (companion, group or cluster), and still have a few degrees of freedom for parameter estimation.

Several software packages for the modelling of gravitational lenses are freely available, of which I recommend two very different ones. The state-of-the-art in detailed modelling is represented by C. Keeton's `gravlens` and `lensmodel` programs[2] (6). On the other hand, P. Saha has designed an intuitive interactive modelling tool called `SimpLens`[3] (19) which is particularly useful for pedagogic purposes.

## 3 Integral Field Spectroscopy of Lensed Systems

As a rule, cluster lens systems make poor targets for integral field spectroscopy as they have overall sizes of $\sim 1'$, much too large for most existing IFUs (but see the contribution by Sanchez et al., these proceedings). However, it was recently demonstrated that individual lensed and distorted galaxies make excellent targets for 3D spectroscopy (22). I will not discuss cluster

[2] available at `http://www.physics.rutgers.edu/~keeton/gravlens/`

[3] available at `http://ankh-morpork.maths.qmul.ac.uk/~saha/astron/lens/`

lenses any further as this topic is well represented in these proceedings (see contributions by Soucail et al., Covone et al., and Swinbank et al.).

Systems where the lensing potential is dominated by a single galaxy show typical image separations of up to a few arcsec. This is well matched to the field of view of most existing integral field spectrographs, and consequently the existing literature on IFS studies of gravitational lenses deals chiefly with galaxy lens systems, predominantly multiply imaged quasars. Galaxy-galaxy lenses are still much rarer in numbers, although they are currently increasing at a great rate, mainly due to the application of ingeneous new search techniques (e.g., 3; 26).

What are the main astrophysical issues that can be tackled with integral field spectroscopy (IFS) of such systems? One can group these into four broad categories. Firstly, there can be little doubt that IFS should be the method of choice for straightforward follow-up spectroscopic confirmation of a system identified by its lens-like morphological appearance. In a single shot one can obtain redshifts of all components of the background source, and if it is bright enough, also of the lens (provided that the IFU data can be adequately deblended; see below). This has been exploited most notably in the above-mentioned survey for galaxy-galaxy lenses (Bolton et al. 3; also Bolton et al. 2). For galaxy-quasar lenses this is more an expectation than a description of the present situation, as most lensed quasars known to date were verified by slit spectroscopy (exceptions are APM 08279+5255: Ledoux et al. 9 and HE 0435−1223: Wisotzki et al. 27).

The second category where IFS studies can contribute is the dynamics of the lensing galaxies. At least in those cases where the lensing galaxy is at reasonably low redshift – a prominent example is the 'Einstein Cross' Q 2237+0307 (5) with $z_L = 0.04$, and a more recent case is ESO 325-G004 (21) – one can directly compare and possibly combine the constraints on the mass distribution from lensing with morphological constraints and those from velocity or velocity dispersion fields. An example of what can be done already without the spectroscopic constraints is the mass model of Q 2237+0307 by Trott & Webster (23). But as much as IFS is concerned, this is more a preview for the near future than an assessment of what has already been done.

The third topical category is the detailed comparison of spectra between the components of a lens systems. Some of the first IFS observations targeted the Einstein Cross for precisely this purpose (1; 4), trying to measure the flux ratios not only in broad-band fluxes, but also separately for various emission lines and the continuum. The Einstein Cross has ever since remained popular as a target for newly commissioned IFUs, as witnessed by the studies of Mediavilla et al. (10), Metcalf et al. (12), and Wayth et al. (25). Other lensed QSOs studied with the IFS technique were SBS 0909+532 (15), HE 0435−1223 (27), HE 0047−1756 (30), and UM 673 (28). A comparison of lensed QSO spectra can provide evidence for extinction in the lens and constrain the shape of the extinction curve at substantial redshifts (15; 31; 11).

Another very important application is the search for small-scale structure in both sources and lenses, which will be discussed in Section 4.

Finally, IFS observations should also allow major progress in the study of lensed quasar host galaxies. HST revealed that many lensed QSOs show prominent host galaxies stretched out into partial or almost complete Einstein rings, especially in the near infrared (see Peng et al. 16 for several examples; Kochanek et al. 8 for a particularly beautiful one). Getting spectra of these should be feasible soon; the main obstacle to overcome is the high redshifts of the QSOs, demanding infrared rather than optical instruments. An exception is the reported detection of extended C III] line emission from a lensed QSO host galaxy by Mediavilla et al. (10) and Motta et al. (14), using INTEGRAL

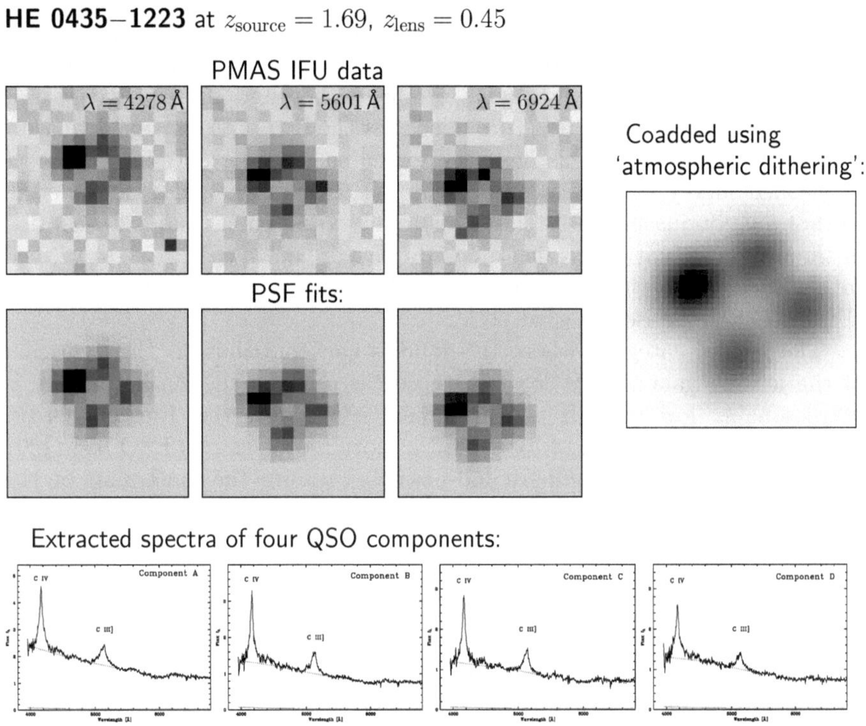


**Fig. 2.** Integral field spectroscopic data of the quadruple QSO HE 0435−1223, obtained with PMAS on the Calar Alto 3.5 m telescope (27). The upper right panels show monochromatic slices at different wavelengths; notice the centroid shift due to differential atmospheric refraction. Below we show the corresponding synthesised images after simultaneous PSF fitting to the entire datacube. The resultant extracted spectra of the four components are shown in the bottom. The right-hand panel shows a supersampled coadded image using the technique of 'atmospheric dithering' as explained in the text.

on the William Herschel Telescope. However, recent GMOS IFU data (25) could not confirm the feature.

*Note on Deblending.* In order to be able to compare spectral properties of multiply imaged QSO components, the images need to be accurately deblended. This is by no means trivial. In many of the observations reported above, the components showed considerable mutual overlap. Furthermore, the effects of differential atmospheric refraction are not to be underestimated. Figure 2 shows monochromatic slices of a data cube of the quadruple QSO HE 0435−1223, extracted at different wavelengths. Atmospheric dispersion caused centroid shifts of more than 1 arcsec (two full spaxels in this case) between different wavebands. Simply extracting spectra by adding up columns from the data cube would yield completely wrong results. On the other hand, the effect can even be employed to one's advantage as the gradual centroid shift provides a natural subpixel dithering. The top-right panel in Fig. 2 demonstrated how 'atmospheric dithering' can assist in supersampling the coarse spaxels of the original PMAS monochromatic image slices into a much higher quality broad-band image.

In the most commonly employed extraction strategy, narrow-band images are constructed that are centered on emission lines or continuum regions; these images are afterwards treated separately (e.g., 12; 25). A more comprehensive deblending approach was developed by us, implementing an iterative multicomponent fitting loop over all spatial layers of a data cube. A simple early version of this scheme was based on the assumption that all components could be approximated by analytical functions (27; 28). Figure 2 shows the outcome of that process, applied to HE 0435−1223: high S/N spectra of all 4 QSO components are obtained, yielding accurately deblended emission line data and – because of the fully preserved spectrophotometric properties – also continuum slopes. A similar but much more powerful algorithm has now been presented by Sánchez et al. (20, see also Sanchez et al., these proceedings).

## 4 Macro-, Micro-, and Mesolensing

The principal empirical constraints on mass models of lenses are the positions and fluxes of individual components. While lens models are highly successful in following the astrometric constraints down to the measurement limit of a few milli-arcseconds, they often enough fail dismally in reproducing the relative brightnesses, i.e. the magnification ratios (e.g., 13). In several cases, this failure is generic to all smooth lens models. For example, the two bright components A1 and A2 of the bright quadruple system PG 1115+080 constitute a close pair of images that should have almost equal magnifications in about any smooth lens model, yet they differ by $\sim 0.5$ mag (for this and several other examples, see the CASTLeS website[4]). Differential extinction may explain a few of these cases, but in most lensing galaxies there is no evidence for significant dust extinction. The conclusion is that the lensing potential cannot be smooth, but must have 'substructure' of some sort. We have seen above that if the subunits have masses $M \ll M_{\rm L}$ of the lens as a whole, the deflection angles can be expected to follow the overall 'macrolensing' potential, while the magnifications can experience major modifications due to local perturbations of the potential, in agreement with the observed behaviour.

Two types of substructure are being discussed. Microlensing due to individual stars of typical masses $M_\star \sim 0.1 \ldots 1\,M_\odot$ is characterised by Einstein radii of the order of 1 $\mu$arcsec, which projected into the source plane corresponds to only $\theta_{\rm E} d_{\rm S} \sim 10^{-2}$ pc. This is similar to theoretical expectations for the size of the accretion disk in quasars, and in fact the timescales of observed microlensing events in the Einstein Cross have already provided empirical upper limits to the source size (24; 32). On the other hand, the broad- and particularly the narrow-line regions in QSOs are supposedly much larger and should not be much affected by microlensing. Thus, the flux ratio discrepancies should show up mainly in the optical/UV continuum, and much less in the emission line flux ratios if to be explained by microlensing.

Another possibility is lens substructure in the form of dark matter clumps, with masses in the range $\sim 10^5 \ldots 10^8\,M_\odot$. The presence of such clumps is predicted by CDM structure formation simulations, but if they exist, they must be largely invisible, as dwarf satellite galaxies fail to add up in sufficient numbers. Such 'dark matter only' entities could nevertheless be detected through their lensing action. The signature of a single clump in the above mass range can be easily estimated: its Einstein radius will be around 10–100 pc, thus the magnification ratios of continuum source and broad line region should be identical, while an extended narrow-line region with diameter $> 1$ kpc will be unaffected. Being intermediate between 'macro-' and 'microlensing', this phenomenon has been dubbed 'milli-' or 'mesolensing'. (I wonder whether cluster lensing should now be rebaptised as 'kilolensing'.)

There is now ample evidence that microlensing by stars affects the spectra of lensed quasars significantly. A major difference between continuum and emission line flux ratios was detected first in the double QSO HE 1104−1805 (29). Several other cases followed, and in a recent tally (28) we estimated that more than 50 %, possibly even most of all macrolensed QSOs also suffer from significant microlensing. Many of these systems, however, were doubles and thus too poorly constrained to be useful for real testing of the mass models. But for the quadruple system HE 0435−1223 we could show with IFS observations that indeed the broad emission line flux ratios are much closer to the model predictions (27). A similar result was obtained for the Einstein Cross by Metcalf et al. (12) and Wayth et al. (25). For both objects,

[4] `http://www.cfa.harvard.edu/castles`

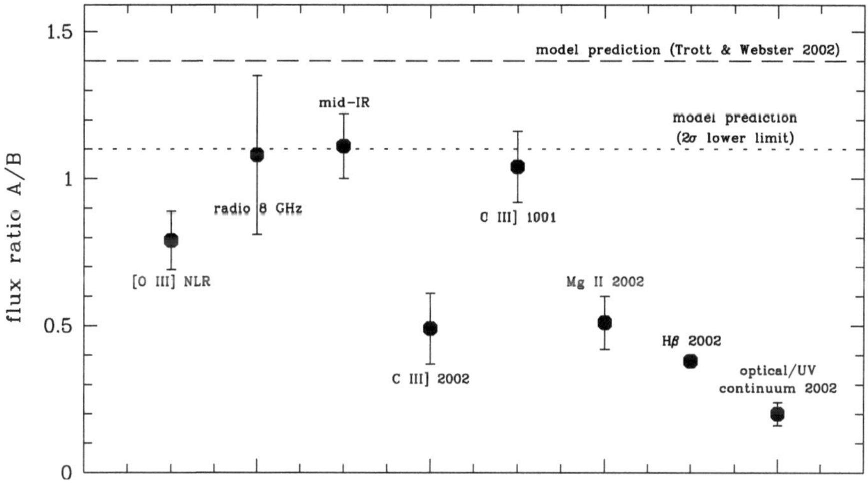


**Fig. 3.** Flux ratios between components A and B in the Einstein Cross Q 2237+0307, measured at various wavelengths and for different emission lines (collected from Racine (17), Metcalf et al. (12), Wayth et al. (25), and references therein). The dotted line denotes the $2\sigma$ lower limit for the prediction from the best-fitting mass model of the lensing galaxy (23).

however, the agreement between models and BLR flux ratios is still far from satisfactory, introducing the question whether both micro- *and* mesolensing may have to be considered. Further support for the mesolensing interpretation comes from anomalous flux ratios found also in radio observations of some systems; as the radio emitting region is resolved to be $\sim 40$ mas, these cannot be affected by microlensing.

The Einstein Cross is particularly interesting as it is so extensively studied. Furthermore, the expected time delay between images is of order of hours, much shorter than timescales of intrinsic variability, so that single-epoch observations are sufficient. Figure 3 shows a synopsis of measured flux ratios $A/B$ taken from the literature. A remarkable trend is the increase from right to left, which could physically be interpreted as a trend of increasing source size and therefore decreasing impact of microlensing. The difference in the C III] points between 1991 and 2002 were interpreted by Wayth et al. (25) as signature of a microlensing event acting on the BLR in 2002. Another remarkable feature is the lower value obtained for the narrow [O III] $\lambda$5007 line by Metcalf et al. (12), which they took as evidence for some moderate mesolensing.

I conclude this very superficial review of lensing constraints on lens substructure by expressing that the present situation is certainly very uncertain. Most likely both effects, micro- and mesolensing, exist and can be observed. Disentangling them is difficult, but possible. Integral Field Spectroscopy in the optical and near-infrared will play a key role, as it is undoubtedly the most efficient observational technique for this purpose.

## References


[1] Adam, G., Bacon, R., Courtes, G., et al. 1989, A&A, 208, L15
[2] Bolton, A. S., et al. 2005, ApJL, 624, L21
[3] Bolton, A. S., et al. 2006, ApJ, 638, 703
[4] Fitte, C., Adam, G. 1994, A&A, 282, 11
[5] Huchra, J., Gorenstein, M., Kent, S., et al. 1985, AJ, 90, 691
[6] Keeton, C. 2001, ApJ, submitted, astro-ph/0102340
[7] Kochanek, C., Schneider, P., Wambsganss, J. 2004, Gravitational Lensing: Strong, Weak & Micro, Proceedings of the 33rd Saas-Fee Advanced Course
[8] Kochanek, C. S., et al. 2006, ApJ submitted, astro-ph/0508070
[9] Ledoux, C., Theodore, B., Petitjean, P., et al. 1998, A&A, 339, L77
[10] Mediavilla, E., Arribas, S., del Burgo, C., et al. 1998, ApJL, 503, L27+
[11] Mediavilla, E., Muñoz, J. A., Kochanek, C. S., et al. 2005, ApJ, 619, 749
[12] Metcalf, R. B., et al. 2004, ApJ, 607, 43
[13] Metcalf, R. B., Zhao, H. 2002, ApJL, 567, L5
[14] Motta, V., Mediavilla, E., Muñoz, J. A., Falco, E. 2004, ApJ, 613, 86
[15] Motta, V., Mediavilla, E., Muñoz, J. A., et al. 2002, ApJ, 574, 719
[16] Peng, C. Y., et al. 2006, ApJ submitted, astro-ph/0509155
[17] Racine, R. 1992, ApJL, 395, L65
[18] Saha, P., Williams, L. L. R. 2003, AJ, 125, 2769
[19] Saha, P., Williams, L. L. R. 2004, AJ, 127, 2604
[20] Sánchez, S. F., et al. 2006, New Astronomy Review, 49, 501
[21] Smith, R. J., Blakeslee, J. P., Lucey, J. R., Tonry, J. 2005, ApJL, 625, L103
[22] Swinbank, A. M., Smith, J., Bower, R. G., et al. 2003, ApJ, 598, 162
[23] Trott, C. M., Webster, R. L. 2002, MNRAS, 334, 621
[24] Wambsganss, J., Schneider, P., Paczynski, B. 1990, ApJL, 358, L33
[25] Wayth, R. B., O'Dowd, M., Webster, R. L. 2005, MNRAS, 359, 561
[26] Willis, J. P., Hewett, P. C., Warren, S. J. 2005, MNRAS, 363, 1369
[27] Wisotzki, L., et al. 2003, A&A, 408, 455
[28] Wisotzki, L., et al. 2004, AN, 325, 135
[29] Wisotzki, L., Köhler, T., Kayser, R., Reimers, D. 1993, A&A, 278, L15
[30] Wisotzki, L., Schechter, P. L., Chen, H.-W., et al. 2004, A&A, 419, L31
[31] Wucknitz, O., Wisotzki, L., Lopez, S., Gregg, M. D. 2003, A&A, 405, 445
[32] Wyithe, J. S. B., Webster, R. L., Turner, E. L. 2000, MNRAS, 318, 762

# 3D Spectroscopy as a Tool for Investigation of the BLR of Lensed QSOs

L. Č. Popović

Astronomical Observatory, Volgina 7, 11160 Belgrade, Serbia
lpopovic@aob.bg.ac.yu**

**Summary.** Selective amplification of the line and continuum source by microlensing in a lensed quasar can lead to changes of continuum spectral slopes and line shapes in the spectra of the quasar components. Comparing the spectra of different components of the lensed quasar and the spectra of an image observed in different epochs one can infer the presence of millilensing, microlensing and intrinsic variability. Especially, microlensing can be used for investigation of the unresolved broad line (BLR) and continuum emitting region structure in active galactic nuclei (AGN). Therefore the spectroscopic monitoring of selected lensed quasars with 3D spectroscopy open new possibility for investigation of the BLR structure in AGN. Here we discuss observational effects that may be present during the BLR microlensing in the spectra of lensed QSOs

## 1 Introduction

Gravitational lensing is in general achromatic (the deflection angle of a light ray does not depend on its wavelength); however, the wavelength-dependent geometry of the different emission regions may result in chromatic effects (see (9), and references therein). Studies aimed at determining the influence of microlensing on the spectra of lensed quasars (hereafter QSOs) need to account for the complex structure of the QSO central emitting region. Since the sizes of the emitting regions are wavelength-dependent, microlensing by stars in a lens galaxy will lead to a wavelength-dependent magnification. The geometries of the line and the continuum emission regions are in general different and there may be a variety of geometries depending on the type of AGN (i.e. spherical, disc-like, cylindrical, etc.). Observations and modeling of microlensing of the broad-line region (BLR) of lensed QSOs are promising, because the study of the variations of the broad emission-line shapes in a microlensed QSO image could constrain the size of the BLR and the continuum region.

Our knowledge of the inner structure of quasars is very limited and largely built on model calculations. Continuum-line reverberation experiments with

**The work is supported by the Ministry of Science and Environment Protection of Serbia through the project "Astrophysical Spectroscopy of Extragalactic Objects" and by Alexander von Humboldt Foundation through "Rückkehrstipendium".

low-redshift QSOs tell us that the broad-emission line region (BLR) is significantly smaller than earlier assumed, and it is typically several light days up to a light year across (e.g., (4)). It means that the BLR radiation could be significantly amplified due to microlensing by (star-size) objects in an intervening galaxies (2). Hence, gravitational lensing can provide an additional method for studying the inner structure high-redshift quasars for several reasons:

(i) the extra flux magnification, from a few to 100 times, provided by the lensing effect enables us to obtain high signal-to-noise ratio (S/N) spectra of distant quasars with less observing time;

(ii) the magnification of the spectra of the different images may be chromatic (as was noted in (11),(12),(13),(9)) because of the line and continuum emitting region are different in sizes and geometrically complex and/or complex gravitational potential of lensing galaxy; (iia) consequently, microlensing events lead to wavelength-dependent magnifications of the continuum that can be used as indicators of their presence (11),(9);

(iii) gravitational microlensing can also change the shape of the broad lines (see (8)(2)(9)), the deviation of the line profile depends on the geometry of the BLR.

Finally, the monitoring of lensed QSOs in order to investigate the effect of lensing on the spectra can be useful not only for constraining the unresolved structure of the central regions of QSOs, but also for providing insight to the complex structure of the lens galaxy.

## 2 Structure of the central part of AGNs and probability of the BLR microlensing

According to the standard model of AGNs, a QSO consists of a black hole surrounded by a (X-ray and optical) continuum emitting region probably with an accretion disk geometry, a broad line region (BLR) and a larger region, narrow line region (NLR) that can be resolved in several nearby AGNs (e.g., (5)). The physics and structure of the NLR have been investigated using the observations of the region in nearby AGNs, while the physics and structure of the BLR cannot be investigated by direct observations. Our knowledge of the inner structure of quasars is very limited and largely built on model calculations. Continuum-line reverberation experiments (in the UV/optical spectral band) with low-redshift QSOs tell us that the broad-emission line region (BLR) is significantly smaller than earlier assumed, and it is typically several light days up to a light year across (e.g., (4)).

Consequently, one can expect that the magnification of the BLR (or a part of the BLR) and continuum emission due to microlensing. In Fig. 1 (left), the cumulative probabilities for microlensing of the BLR and continuum as a function of the ERR (in units of the BLR dimensions) are given. As one can see in Fig. 1 (left), there is a global correlation between the BLR and

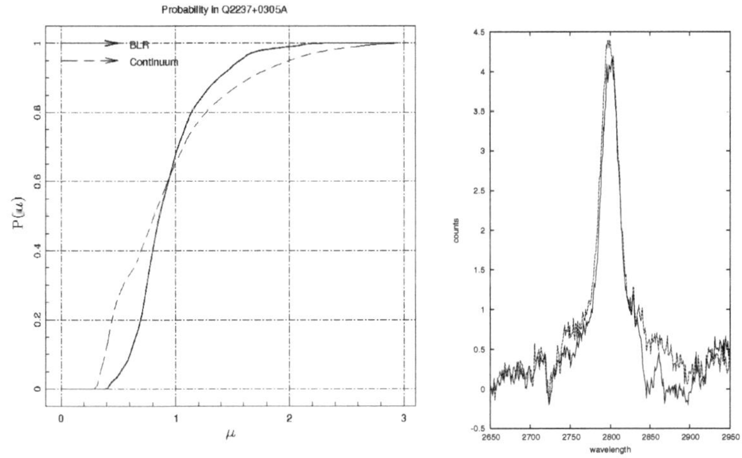


**Fig. 1.** *Left*: Accumulate probability of the magnification of the BLR, in solid line, and the continuum, in dashed line (1). *Right*: Comparison of the line shape of Mg II$\lambda$=2798 Å of SDSS J0924+0219 observed at two epochs; 15/01/2005 – solid line, and 01/02/2005 – dashed line (3)

continuum microlensing probabilities, i.e one can expect that the variation in the line profile of a QSO should be also seen as amplification in the continuum. We should note here that the emission of the BLR can be partially amplified due to microlensing, i.e. one part of the BLR can be affected by microlens that can result in amplification of the only one part of broad lines (see Fig 1, right)

## 3 3D spectroscopy as tool for the BLR investigation

A systematic search for microlensing signatures in the spectra of lensed quasars should be performed with 3D spectroscopy. One of examples is the Cosmological Monitoring of Gravitational Lenses (COSMOGRAIL, see (3)). To detect microlensing, simultaneous observations of spectra of images of a lensed QSO at several epochs, preferably separated by the time-delay, are needed (see (9)).

Taking into account the redshift of lensed QSOs it is needed to obtain the spectra from 3500 to 9000 Å (covering also, the broad C IV, CIII] and Mg II lines which are emitted from the BLR region) of a sample multi-imaged QSOs. Concerning the estimation of the BLR dimensions (see (4)), one should select a sample of lensed QSOs where the BLR microlensing might be expected. To find the possible microlensing one can apply the method given by (9) comparing the spectra (in the continuum and in the broad lines) of different components in order to detect the difference caused by microlensing or/and millilensing. Using previous theoretical estimates of line shape variations due

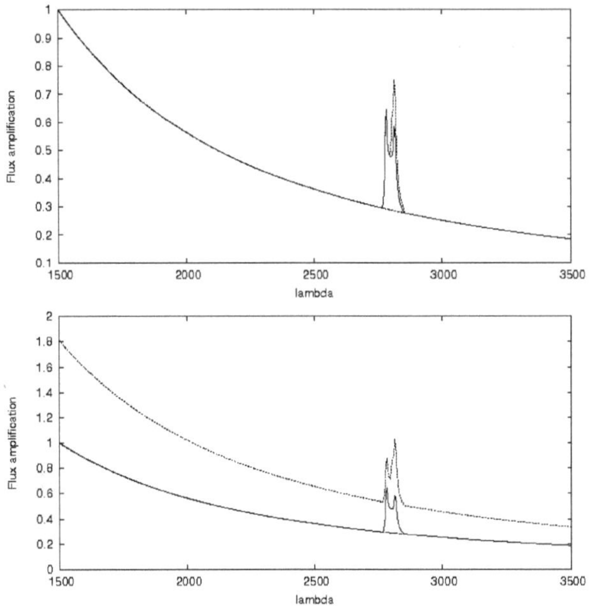


**Fig. 2.** The simulation of the variation in Mg II$\lambda$ = 2798Å and the continuum between 1500 Å and 3500 Å due to microlensing by straight fold caustic. The microlensing may affect only line profile (top) as well as the line profile and the continuum (down).

to microlensing ((8),(2),(1),(9)) the observed spectra can be fitted with the theoretical line profile assuming different geometries. From this one will be able to estimate the geometry and dimension of the BLR. Also, comparing difference in amplification of the continuum, C IV and Mg II lines one will be able to conclude about differences between high and low ionized line emitting regions and compare them with the size and geometry of the continuum emission region.

## 4 Microlensing of the BLR/continuum emission

Let us discuss the expected variability in the line and continuum shapes due to microlensing. Recently, the broad line variability that may be caused by microlensing were observed by (10),(3). Concerning the theoretical predictions (see (8),(2)) one can expect that different parts of a line can be amplified and that the line should change in intensity as well as in the shape during the microlensing event. Also, the continuum should vary (6) and there is a correlation between the continuum and line amplification (see Fig. 1, left). Also, it is expected that the continuum emission is wavelength dependent (9). But, all of these effects might not be seen at the same time of a microlensing event. As an example in Fig. 2 (top) we present simulation of microlensing of an emission from the accretion disk, taking that the continuum is coming from the inner part of the disk (from 50 Rg to 200 Rg) and the Mg II line from the outer part (from 200 Rg to 1200 Rg). The used model of the disk is the same as it is given in (7), but with $i = 14°$ and for the UV radiation. In some cases the continuum can remain constant, while the line can be amplified. It corresponds to the case where only the outer part of the disk is microlensed (the part that emits the line). Such amplification of a line without amplification of the continuum can be seen only for a limited period of time. But during a complete microlensing event the continuum should be also amplified (see Fig 2, down). Consequently, to register the microlensing presence, one should monitor lensed QSOs using 3D spectroscopy in order to have observations of all images at the same time from different epochs.

## References


[1] C. Abajas, E. Mediavilla, J.A. Muñoz, L. Č. Popović: Mem. S.A.It.S (accepted) (2005)
[2] C. Abajas, E. Mediavilla, J.A. Muñoz, L. Č. Popović, A. Oscoz: ApJ **576**, 640 (2002)
[3] A. Eigenbrod, F. Courbin, S. Dye, et al.: A&A, sent (astro-ph/0510641)
[4] S. Kaspi, P. S. Smith, H. Netzer, et al.: ApJ **533**, 631 (2000)
[5] J. H. Krolik: 'Active galactic nuclei' Prinston, N. J. (1999)
[6] G. F. Lewis, R. A. Ibata: MNRAS **348**, 24L (2004).
[7] L. Č. Popović, P. Jovanović, E. G. Mediavilla, et al.: ApJ (accepted, astro-ph/0510271)
[8] L. Č. Popović, E. G. Mediavilla, J. A. Muñoz: A&A **378**, 295 (2001).
[9] L. Č. Popović, G. Chartas: MNRAS **357**, 135 (2005).
[10] G. T. Richards, R.C. Keeton, P. Bartosz, P. et al.: ApJ **610**, 679 (2004).
[11] J. Wambsganss, B. Paczynski: AJ **102**, 86 (1991).
[12] L. Wisotzki, T. Becker, L. Christensen: A&A **408**, 455 (2003).
[13] O. Wucknitz, L. Wisotzki, S. Lopez, M. D. Gregg: A&A **405**, 445 (2003).

# A VIMOS-IFU Survey of $z \sim 0.2$ Massive Lensing Galaxy Clusters: Constraining Cosmography

G. Soucail[1], G. Covone[2,3] and J.-P. Kneib[3]

[1] OMP, Laboratoire d'Astrophysique de Toulouse-Tarbes, 14 Avenue Belin, 31400 Toulouse, France, soucail@ast.obs-mip.fr
[2] INAF – Osservatorio Astronomico di Capodimonte, Naples, Italy
[3] OAMP, Laboratoire d'Astrophysique de Marseille, France

**Summary.** We present an integral field spectroscopy survey of rich clusters of galaxies aimed at studying their lensing properties. Thanks to knowledge of the spectroscopic characteristics of more than three families of multiple images in a single lens, one is able in principle to derive constraints on the geometric cosmological parameters. We show that this ambitious program is feasible and present some new results, in particular the redshift measurement of the giant arc in A2667 and the redshift confirmation of the counter-image of the radial arc in MS2137–23. Prospects for the future of such a program are presented.

## 1 Cosmological Applications of Gravitational Lensing

Gravitational lensing has been well studied for over 15 years and many extensive reviews are available in the literature (1). So we simply state here the basic lensing equation, which corresponds to a mapping of the image plane on the source plane

$$\boldsymbol{\beta} = \boldsymbol{\theta} - \boldsymbol{\nabla}\psi(\boldsymbol{\theta}) \qquad\qquad \psi(\boldsymbol{\theta}) = \frac{2}{c^2}\frac{D_{LS}}{D_{OS}D_{OL}}\varphi(\boldsymbol{\theta})$$

where $\varphi(\boldsymbol{\theta})$ is the Newtonian gravitational potential of the deflecting lens. The distances are angular-distances between the lens, the source and the observer. The Einstein radius is defined as

$$R_E = \sqrt{\frac{4GM}{c^2}\frac{D_{OL}D_{LS}}{D_{OS}}} = 2.6'' \left(\frac{\sigma^2}{c^2}\right)\frac{D_{LS}}{D_{OS}}$$

and represents the radius of the circular image in the ideal case of a source perfectly aligned behind a point source of mass $M$ or behind a singular isothermal sphere with velocity dispersion $\sigma$. For any lensing mass, it represents the typical angular scale of the lensed images. Numerically, it is about $1 - 3''$ for a galaxy lens and $10 - 30''$ for a cluster of galaxies. In practice, gravitational lensing has many astrophysical applications, related to either the determination of the lensing potential and the mass distribution in galaxies and clusters of galaxies, or the study of the distant galaxies thanks to the magnification of their images (2). But it also depends on the geometrical cosmological parameters and represents an alternative possibility to constrain these parameters, independently from other methods like supernovae or CMB fluctuations. This method was explored recently in detail (3). It requires several sets of multiple images from different sources at different redshifts through the same lens. For each source the lens equation can be written as

$$\boldsymbol{\theta}_S = \boldsymbol{\theta}_I - \frac{2\sigma_0^2}{c^2} \boldsymbol{f}(\boldsymbol{\theta}_I, \theta_C, \alpha, \ldots) \times E(\Omega_M, \Omega_\Lambda, z_L, z_S)$$

where the function $f$ includes all the characteristics of the lensing potential and $E = D_{LS}/D_{OS}$ includes the cosmological dependence [$(\Omega_M, \Omega_\Lambda)$ or $(\Omega_M, w)$ for a Euclidean Universe] as well as the lens and source redshifts. Provided $z_L$ and $z_S$ are well known for two sets of multiple images, one can in principle solve the lens equation for both sets and constrain the E-term, with a well-defined degeneracy between the two geometrical cosmological parameters used (3). However the cosmological dependence in the E-term is weak and better constraints require more than two sets of multiple images, spread in redshift as well as a very accurate mass determination in the lens (Fig. 1).

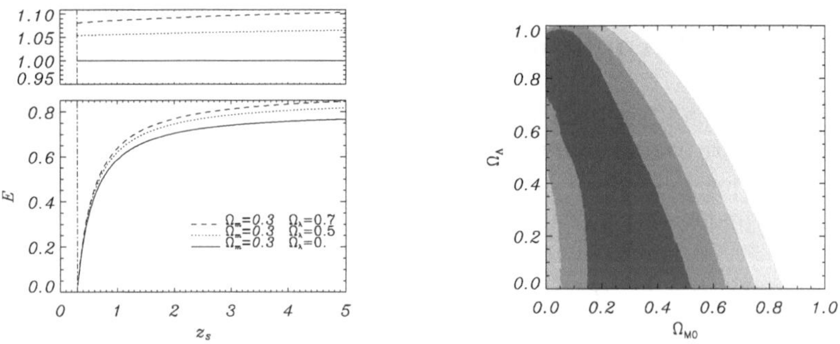


**Fig. 1.** Left: Evolution of the E-term with redshift for different cosmological models, for a lens at $z = 0.3$. Right: Confidence levels obtained from the modeling of the cluster-lens Abell 2218 (5)

This method was first attempted in the cluster Abell 2218 ($z_L = 0.17$,), a strong gravitational lens close to the ideal case: it displays at least 4 sets of multiple images with redshifts ranging from $z_{S1} = 0.702$ to $z_{S4} = 5.576$. An accurate mass model was developed (4), including 2 major mass components located on the two brightest galaxies, as well as the contribution of about 17 galaxy masses scaled by their luminosity. This model prescription was constrained more accurately by including the 4 families of multiple images, and the likelihood distribution in the $(\Omega_M, \Omega_\Lambda)$ plane finally displayed the expected degeneracy (Fig. 1), reinforcing the validity of the mass model (5).

## 2 Observations of Massive Lensing Clusters with VIMOS/IFU

In order to tighten the likelihood distribution in the $(\Omega_M, \Omega_\Lambda)$ plane we need to add more cluster-lenses with similar numbers of multiple images but different lens redshifts or mass distributions. This is the main motivation of a survey of massive cluster lenses started two years ago at ESO and using the IFU mode of VIMOS on the VLT. This instrument configuration has many advantages, thanks to its wide field of view of $1' \times 1'$ in order to get as many redshifts as possible, both on cluster galaxies and on lensed arcs with curved shapes. The high density of objects in the central cores of clusters is therefore mapped in a single shot, allowing both the study of the dynamics of the cluster galaxies and the redshift determination of the brightest multiple arcs.

**Table 1.** Summary of the VIMOS/IFU observations of the survey

| Cluster | redshift | # pointings | Run | $T_{exp}$ (ksec) |
|---|---|---|---|---|
| Abell 1689 | 0.184 | 4 | May-June 2003 | 59.4 |
| Abell 2390 | 0.233 | 3 | June 2003 | 32.4 |
| AC 114 | 0.312 | 2 | June 2004 | 21.6 |
| Abell 2667 | 0.233 | 1 | June 2003 | 10.8 |
| MS2137–23 | 0.310 | 1 | June 2003 | 10.8 |
| Abell 68 | 0.255 | 1 | August 2004 | 5.4 |

Up to now, 6 clusters have been observed in low resolution mode ($R \sim 200$), with a useful spectral range from 4000 to 6800 Å(Table 1). Most clusters have a redshift in the range 0.2–0.3 which is optimal for our program. Data reduction was rather tricky, and is presented in detail in (6).

### 2.1 Abell 2667

Abell 2667 was observed with IFU/VIMOS in a single pointing, split in 4 exposures of 45 mn each (6) . It is a strong X-ray emitter and multi-color HST imaging revealed a very prominent and bright luminous arc corresponding to 3 merging images. Its spectrum was extracted and split in 3 for each sub-image. The redshift confirmation is obvious in this case (Fig. 2). A few very faint multiple image candidates were also identified. However, by exploring the final datacube, we were not able to determine more new arc redshifts, preventing the use of A2667 for cosmological purposes. But with the present data, we were able to determine the redshift of more than 20 cluster members, so a first value of the velocity dispersion could be used for the lens modeling. A detailed lens model was therefore built and the mass distribution compared with the mass deduced from X-ray data and the virial mass (6) (see also Covone et al., these proceedings).

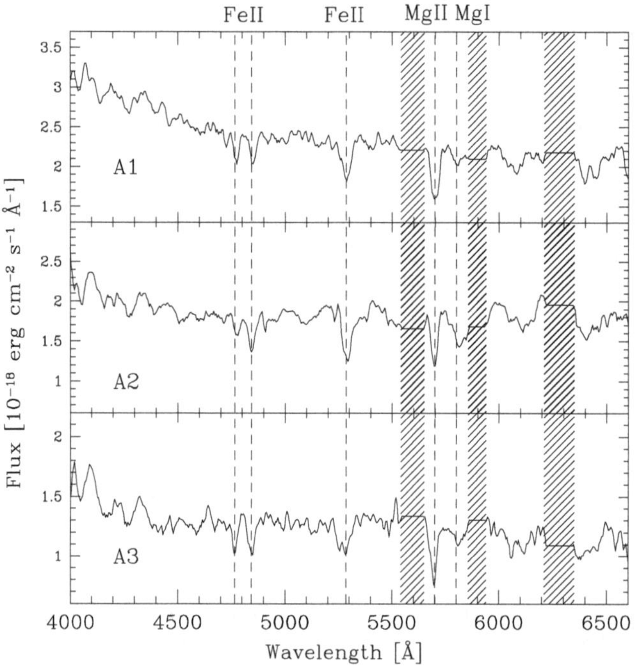


**Fig. 2.** Spectra of the three sub-images A1, A2 and A3 of the giant arc in A2667, with line identifications at redshift $z = 1.03$.

## 2.2 MS2137–23

MS2137–23 is a well known gravitational lens with two independent systems of lensed images: one tangential arc with its two counter-images and one radial arc with its counter-image candidate. The two main arcs have been observed spectroscopically with a redshift measurement at a redshift $z \simeq 1.5$ for both systems (7). Unfortunately these are not favorable for cosmography, although this lens has been studied in detail in order to constrain the properties of the mass distribution in the very center of the cluster (8). However, from the datacube built from our observations (2 exposures of 45mn each), we were able to extract the spectrum of the counter-image of the radial arc (A5), confirming its redshift at $z = 1.503$ (Fig. 3). This is the first redshift confirmation that A5 is the counter-image of the radial arc.

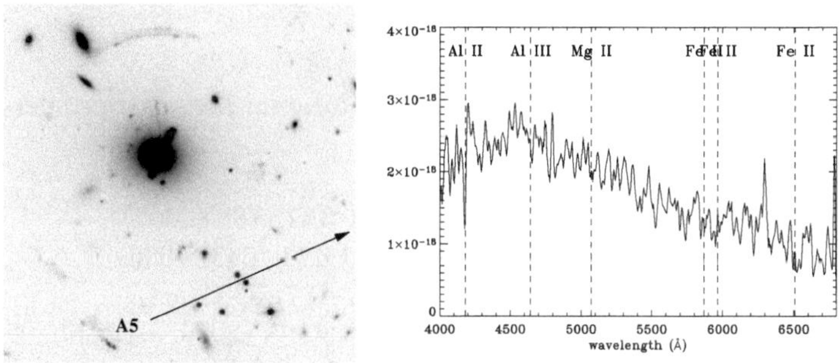


**Fig. 3.** Integrated spectrum of the counter-image of the radial arc in MS2137–23, and the line identification with redshift $z = 1.503$.

### 2.3 Abell 1689: the Ultimate Lens !

Fortunately, there exits presently at least one cluster-lens which deserves great attention because of its unprecedented number of arcs and arclets. From deep HST/ACS observations of Abell 1689, more than 30 systems of multiple images are identified in the center (9). The lens has a complex mass distribution and its lens modeling is not easy. In addition only a few multiple images presently have measured redshifts, because of their faintness. We have started an VIMOS/IFU survey of this lens in 2003, when the instrument had not completely stabilized. This prevented an accurate data reduction, although we still hope to extract some exciting results from the datacubes. However, several spectroscopic redshifts are already available, in particular from previous long slit observations and we should be able to provide a good lens model as well as some new constraints on cosmography. Note that a preliminary attempt (9) did not give interesting constraints, partly because only photometric redshifts were used for most of the arcs. The key point will be clearly to include additional constraints on the lens model, either from weak lensing measurements at large distance or from X-ray data.

## 3 Conclusions and Prospects

We have started an ambitious observing programme, well suited to 3D spectroscopy and aimed at studying in detail a sample of strong gravitational lenses. The main difficulty is the faintness of the arcs which requires to push the present spectrographs to their limits. The results are therefore very sensitive to the data quality at the output of the telescope and to the data reduction which must be done with great care. These fundamental steps are in progress and we have demonstrated that faint object 3D spectroscopy is feasible in rich environments like clusters of galaxies. Several scientific outputs from these data have been presented, and others are in Covone et al. (these proceedings), concerning the properties of the cluster galaxies.

## References

[1] P. Schneider, J. Ehlers, E. Falco: *Gravitational Lenses* (Springer-Verlag Berlin Heidelberg New York 1992)
[2] Y. Mellier: ARAA **37**, 127 (1992)
[3] G. Golse, J.P. Kneib, G. Soucail: A&A **387**, 788 (2002)
[4] J.P. Kneib, R. Ellis, I. Smail et al: ApJ **471**, 643 (1996)
[5] G. Soucail, J.P. Kneib, G. Golse: A&A **417**, 33 (2004)
[6] G. Covone, J.P. Kneib, G. Soucail et al: A&A (2006, astro-ph/0511332)
[7] D. Sand, T. Treu, R. Ellis: ApJ **574**, L29 (2002)
[8] R. Gavazzi: A&A **443**, 793 (2005)
[9] T. Broadhurst, N. Benitez, D. Coe et al: ApJ **621**, 53 (2005)

# First Results from the VIMOS-IFU Survey of Gravitationally Lensing Clusters at $z \sim 0.2$

G. Covone[1,2], J.-P. Kneib[2], G. Soucail[3], E. Jullo[4] and J. Richard[5]

[1] INAF – Osservatorio Astronomico di Capodimonte, Naples, Italy
covone@na.astro.it
[2] OAMP – Laboratoire d'Astrophysique de Marseille, France
[3] OMP – Laboratoire d'Astrophysique de Toulouse-Tarbes, France
[4] European Southern Observatory, Santiago, Chile
[5] California Institute of Technology, Pasadena, USA

**Summary.** We present the on-going observational program of a VIMOS Integral Field Unit survey of the central regions of massive, gravitational lensing galaxy clusters at redshift $z \simeq 0.2$. We have observed six clusters using the low-resolution blue grism ($R \simeq 200$), and the spectroscopic survey is complemented by a wealth of photometric data, including *Hubble Space Telescope* optical data and near infrared VLT data. The principal scientific aims of this project are: the study of the high-$z$ lensed galaxies, the transformation and evolution of galaxies in cluster cores and the use of multiple images to constrain cosmography. We briefly report here on the first results from this project on the clusters Abell 2667 and Abell 68.

## 1 Introduction

Because of their intense gravitational field, massive (i.e., $M > 10^{14} M_{\odot}$) clusters of galaxies act as Gravitational Telescopes (GTs) and are therefore an important tool to investigate the high-redshift Universe (see, e.g., (1)). In order to fully exploit the scientific potential of the GTs we have started an extensive integral field spectroscopy (IFS) survey of massive galaxy clusters. Targets have been selected among well-known gravitational lensing clusters between redshift $\sim 0.2$ and 0.3, for which complementary *Hubble Space Telescope* (HST) data are available. All the clusters are X-ray bright sources. Our sample partially overlaps the one analyzed in (12).

IFS is the ideal tool in order to obtain spatially complete spectroscopic information of compact sky regions such as cluster cores. Moreover, cluster cores are the regions where strong lensing phenomena are observed (i.e., giant arcs and multiply imaged sources): for clusters in our sample, strongly lensed galaxies are within $\theta \simeq 1$ arcmin from the cluster center.

VIMOS-IFU (8) is thus the natural choice for such observational program, since, at present, it provides the largest f.o.v. for among integral field spectrographs mounted on the 8-10m telescopes.

All the clusters in our sample have been observed using the low-resolution blue (LR-B) grism, with a spectral resolution $R \sim 200$. Taking into account the lower efficiency at the end of the spectra and the zero order contaminations, the final useful spectral range is limited between $\simeq$ 3900 Å and 6800 Å. This spectral range is suitable both for detecting high-redshift source (e.g., Ly$\alpha$ emitters in the redshift range $2.2 < z < 5.5$ or [OII] emitters out to $z \sim 0.8$) and to sample the rest-frame 4000 Å break for the cluster galaxy population. With a fiber size 0.66 arcsec, the IFU f.o.v. covers a contiguous region of $54 \times 54$ arcsec$^2$, sampled by 6400 fibers.

A subset of the clusters in the sample has also been observed using a higher resolution grism ($R \simeq 3000$) covering the $\lambda = 6300 - 8600$ Å range. These observations are useful to probe higher redshift Ly$\alpha$ emitters at $4.2 < z < 6.1$ or [OII] emitters at $0.7 < z < 1.3$.

Observations have been completed (see Table 1 in (13)), and data reduction is now in progress. The data reduction process has been described in (2), and (15) gives details about VIPGI, the VIMOS dedicated pipeline. Furthermore, we have developed a Sextractor-based tool to help in object detection and spectra extraction from the fully reduced data cube (5).

Altogether, our IFS survey covers a region of about 9 square arcmin in the central regions of six massive clusters.

Hereafter, we briefly report on the first results from this project: the mass distribution in Abell 2667 in Sect. 2, the properties of a magnified high-$z$ source in Abell 68 in Sect. 3 and an investigation of cluster galaxies in Sect. 4. We refer to the presentation by G. Soucail (13) for a detailed discussion of the cosmography aspect of the project.

## 2 Mass distribution model: A2667

Wide field IFS of the cluster central regions provides simultaneously spectroscopic redshifts of both the cluster members and the images of the gravitationally lensed sources, thus allowing a direct comparison of the strong lensing analysis with the dynamical one. Abell 2667 (hereafter, A2667) is a very remarkable galaxy cluster at $z = 0.233$: it is among the top 5% most luminous X-ray clusters (at its redshift), and shows one of the brightest gravitational arc in the sky (Fig. 1).

A2667 has been observed with VIMOS-IFU during two separate nights (June 2003), with a total of 4 pointings of 2400s each, centered on the cD galaxy, using the LR-B grism (see (2) for details). A small offset of about 2 arcsec was performed between the first and last two pointings. Therefore, as consequence of the small number of pointings, a not optimal dithering strategy and observations carried on two separate nights, the sky subtraction has not been optimal.

Nevertheless, we have obtained the spectroscopic measurements of the redshift for 34 sources in the central $54 \times 54$ arcsec$^2$ region of the cluster,

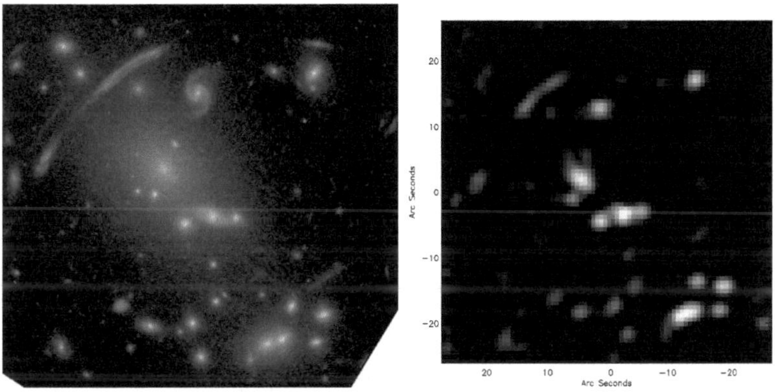


**Fig. 1.** Greyscale images of the inner arcmin square in the cluster A2667. Left panel: image created from the HST-WFPC2 data, filters $B_{450}$, $V_{606}$ and $I_{814}$. Right panel: from the VIMOS-IFU data cube, using slices centered at $\lambda = 4600, 5200$ and 6000 Å.

corresponding to a box of $200 \times 200\, h_{70}^{-2}\, \mathrm{kpc}^2$ at the cluster redshift[6]. It includes in particulars: 22 cluster members (i.e., all the cluster members brighter than $V_{606} = 23.2$, AB system) and the three separate images of the giant gravitational arc ($z = 1.0334$).

Using the spectroscopic redshift and the multiple images identified on the HST-WFPC2 image, we have built a strong lensing model and performed a dynamical analysis of the cluster core, both resulting in mass of $\simeq 7.2 \times 10^{13}\, M_\odot$ within the central $110\, h_{70}^{-1}$ kpc, and a velocity dispersion of $\simeq 950 \mathrm{kms}^{-1}$, close to the value derived from the X-ray temperature (assuming that the cluster follows the $\sigma - T$ relation).

Such agreement supports the idea that A2667 cluster core is in a relaxed dynamical state, as expected from its regular X-ray morphology (Rizza et al. 1998). Therefore, A2667 core appears to be dynamically evolved, in contrast with the large fraction ($70 \pm 20\,\%$) of unrelaxed clusters with similar X-ray luminosity at similar redshift (12).

## 3 Physical properties of a unusual lensed high$-z$ source

The combination of IFS and the large magnification provided from strong lensing gives a unique opportunity to study in detail (i.e., the spatially resolved) the spectral properties of intrinsically low-luminosity source in the high$-z$ Universe, (see, for instance, (14)).

VIMOS-IFU observation of the core of the galaxy cluster Abell 68 ($z = 0.255$) has revealed a surprisingly extended Ly$\alpha$ emission around a previously

[6]We use a cosmological model with $\Omega_\Lambda = 0.7$ and $\Omega_m = 0.3$

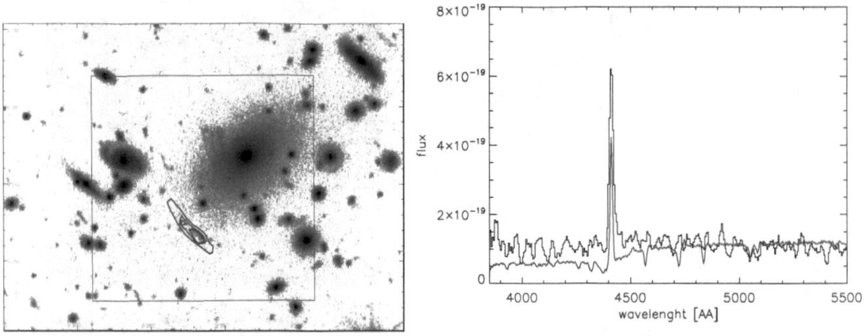


**Fig. 2.** Galaxy cluster A68. Left panel: HST $R_{707}$-band image of the cluster, with the VIMOS-IFU f.o.v. and the lensed source C4. Note its wider extension as detected by means of the IFS. Right panel: one-dimensional VIMOS spectrum of the arc (black solid line) compared with a template of a $z = 3$ Lyman-break galaxy from (11).

known gravitationally lensed source (3) (denoted as C4 in (12), $z = 2.625$). As shown in Fig. 2, the arc is seen to be $\sim 4$ arcsec in length on the HST-WFPC2 image (filter $R_{707}$, exposure time 7.5 ks). But in our shallow IFU pointing (4.8 ks), the arc is seen to be much more extended (3), reaching a maximal elongation of $\simeq 10.8$ arcsec. The emission line is not resolved in the IFU data, and its equivalent width is about 140 Å, and remains constant within the errors along the source length.

According to the strong lensing model, the source is single imaged and its magnification is $\mu \sim 35$: the intrinsic shape of the Lyman-$\alpha$ emitting region has a disk-like appearance (see (3)) with a maximum extension of $\simeq 10\,$kpc. It is therefore about 10 times smaller than Ly$\alpha$ blobs (see, for instance, (9)) but much bigger than a typical galaxy at $z \sim 2.5$ One possibility is that we are observing a $\sim L_*$ galaxy undergoing a strong star-formation event, with negligible dust-obscuration.

## 4 Cluster galaxies investigations

The central region of rich galaxy clusters at intermediate redshift is expected to host an old and red galaxy population, in great majority composed of early-type galaxies, see e.g. (7), which at a redshift of $z < 1$ is passively evolving, with very low levels of star-formation activity.

Recently, (4) have shown that *composite cluster spectra*, built from the light-weighted combination of all the cluster members long-slit spectra, are a useful tool to provide insights in the properties of the cluster population. In this respect, IFS offer a unique possibility to build cluster composite spectra in unbiased way, since, for a given field of view, all clusters members are

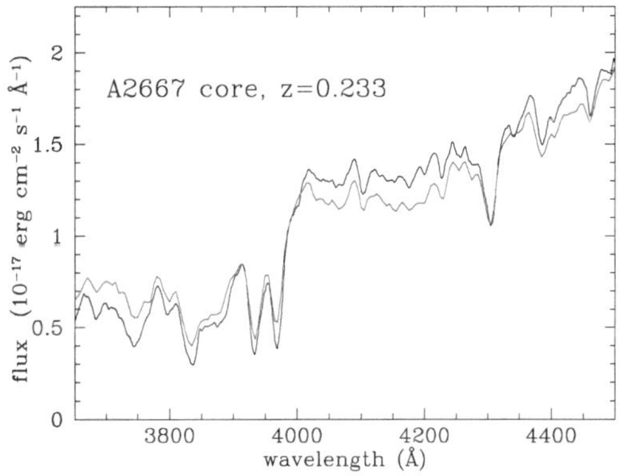


**Fig. 3.** Composite spectrum of the galaxy population in the core of the cluster A2667 (solid line), compared with the template spectrum of $z = 0$ early-type galaxies (dotted line) from (6). The contribution of the cD galaxy is excluded, since it appears to host an AGN.

observed (without any *a priori* selection) and, for each galaxy, all the light is collected (therefore avoiding the possibility that for larger galaxies some flux contribution might be missed due to the specific orientation of the slit). Preliminary work (see Fig. 3) appears to confirm that the galaxy population in the very central region of A2667 is dominated by an evolved and passively aging stellar population. We plan to build composite spectra for all the clusters, and to exploit the recent *Spitzer* mid-infrared observations to complement this result and provide quantitative upper limits on the hidden star-formation in the cluster cores.

## References

[1] Campusano, L.E., Pelló,R. Kneib, J.-P. et al. A&A 378, 394
[2] Covone, G., Kneib, J.-P., Soucail, G. et al. 2006a, A&A in press (astro-ph/0511332)
[3] Covone, G., Kneib, J.-P., Jullo, E. et al. 2006b, in preparation
[4] Dressler, A., Oemler, A., Poggianti, B.M. et al. 2004, ApJ 617, 867
[5] Jullo, E. 2006, Master Thesis , Université de Provence–Marseille
[6] Kinney, A.L., Calzetti, D. Bohlin, R.C. et al. 1996, ApJ 467, 38
[7] Kodama, T. & Bower, B. 2001, MNRAS 321, 18
[8] Le Fèvre, O. et al. 2003, The Messenger 111, 18
[9] Matsuda, Y., Yamada, T., Hayashino, T. et al. 2004, AJ 128, 569
[10] Richard, J. et al. 2006, in preparation
[11] Shapley, A.E. 2003, PhD Thesis, Caltech

[12] Smith, G.P., Kneib, J.-P., Smail, I. et al. 2005, MNRAS 359, 417
[13] Soucail, G., Covone, G., Kneib, J.-P. 2006, these proceedings
[14] Swimbank, A.M., Smith, J., Bower, R.G. et al. 2003, ApJ 598, 162
[15] Zanichelli, A., Garilli, B, Scodeggio, M. et al. 2005, PASP 117, 1271

# Integral Field Spectroscopy of the Core of Abell 2218

S. F. Sánchez[1], N. Cardiel[1,2], M. Verheijen[3] and N. Benitez[4]

[1] Centro Astronómico Hispano Alemán, Calar Alto, (CSIC MPI), C/Jesus Durban Remon 2-2, 04004-Almeria, Spain, sanchez@caha.es
[2] Departamento de Astrofísica, Facultad de Físicas, Universidad Complutense de Madrid, 28040 Madrid, Spain cardiel@caha.es
[3] Kapteyn Astronomical Institute, PO Box 800, 9700 AV Groningen, the Netherlands
[4] Instituto de Astrofísica de Andalucia CSIC, Camino Bajo de Huetor S/N, Granada, Spain

**Summary.** We report on integral field spectroscopy observations, performed with the PPAK module of the PMAS spectrograph, covering a field-of-view of ~74"×64" centered on the core of the galaxy cluster Abell 2218. A total of 43 objects were detected, 27 of them galaxies at the redshift of the cluster. We deblended and extracted the integrated spectra of each of the objects in the field using an adapted version of `galfit` for 3D spectroscopy (`galfit3d`). We use these spectra, in combination with morphological parameters derived from deep HST/ACS images, to study the stellar population and evolution of galaxies in the core of this cluster.

## 1 Introduction

Galaxy clusters have been used for decades to study the evolution of galaxies. Being tracers of the largest density enhancements in the universe, clusters are considered the locations where galaxies formed first. It is known that they are dominated by old and large elliptical galaxies, with colors that are consistent with a bulk formation at high redshift followed by a passive evolution (e.g., (18)). However, (5) have shown an increase in the fraction of blue galaxies in clusters from low to intermediate redshift, which disagrees with that simple scenario. Furthermore, a fast morphological evolution from late to early-type galaxies, claimed as a possible solution, does not predict the observed fractions of S0 galaxies at low redshift. Other processes like gas-stripping and gravitational harassment have to be considered. These notions predict that galaxies at the core of clusters must have globally older stellar populations than galaxies in the outskirts of the cluster. Consequently, we should detect deviations from passive evolution in the scaling relations for early-type galaxies (e.g., the Fundamental Plane). In order to test this hypothesis we started a complete spectroscopic survey of the core of Abell 2218.

Abell 2218 is one of the richest clusters in the Abell catalogue (2), with a richness class 4. It has a redshift of $z$ ~0.17, and a velocity dispersion

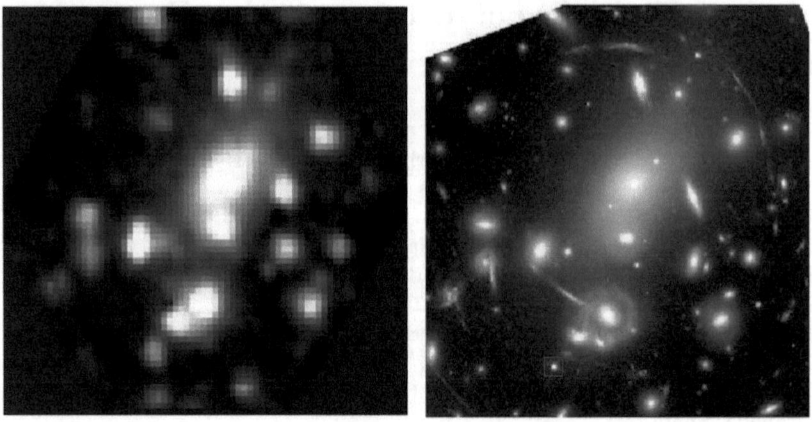

**Fig. 1.** *left-panel:* Three-color image created by coadding the flux of the final datacube through three broad-bands corresponding approximately to $V$,$R$ and $I$. *right-panel:* Similar image created using HST data (Fruchter et al.,(1)). The star used to derive the PSF is indicated with a square.

of 1370 km $s^{-1}$ (8; 9). Detailed high-resolution X-ray maps (11) and mass-concentration studies, based on the properties of the gravitationally lensed arcs, have shown that the cluster contains two density peaks, the strongest of them dominated by a large cD galaxy. Several observational programs have produced a large dataset of well deblended slit spectra for the galaxies in the outer parts of this cluster (18), and extensive multi-band, ground-based and HST imaging, e.g., (1). We focused our survey in the central arcmin region of the cluster, around the cD galaxy.

### 1.1 Observations and Data Reduction

Observations were carried out on 30/06/05 and 06/07/05 at the 3.5m telescope of the Calar Alto observatory with the PMAS (13) spectrograph and its PPAK module (7). The V300 grating was used, covering a wavelength range between 4687-8060 Å with a nominal resolution of ∼10 Å FWHM. The PPAK fiber bundle consists of 331 science fibers of 2.7" diameter, concentrated in a single hexagonal bundle covering a field-of-view of 72"×64". Following a dithered 3-pointing scheme, 3 hours of integration time was accumulated each night, 6 hours in total. In addition to our Integral Field Spectroscopy (IFS) data, we used a F850LP-band image of 11310s exposure time taken with the ACS camera on board the HST, obtained from the HST archive.

Data reduction was performed using R3D(16), in combination with IRAF packages and E3D(14). The reduction involved standard steps for fiber-based integral-field spectroscopy. First, science frames are bias subtracted. A continuum illuminated exposure, taken before the science exposures, is used to locate and trace the spectra on the CCD. Each spectrum is then extracted by coadding the flux within an aperture of 5 pixels. Wavelength and flux calibration are performed using an arc lamp and standard calibration star exposures, respectively. The three dithered exposures are then combined, and a datacube with 1"/pixel sampling is created for each night using E3D. Finally, the two datacubes are recentered and combined using IRAF tasks.

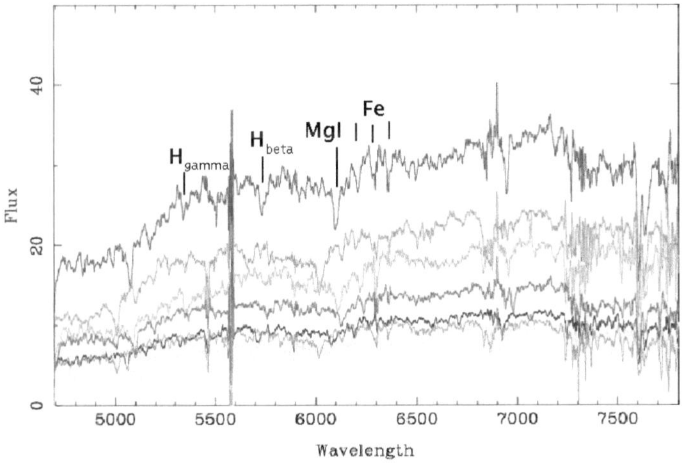


**Fig. 2.** Example of the dataset, showing the spectra of some of the brightest galaxies in the cluster. Some absorption features of interest are indicated.

Figure 1 shows a three-color image created by coadding the flux of the final datacube through three broad-bands: $V$,$R$ and $I$-band (left panel), together with a similar image created using HST observations (right panel)(1). It is interesting to note the similarities between the two images, despite the differences in sampling and resolution. As we quoted before, many galaxies are strongly blended. In particular, the central cD contaminates most of the galaxies in the field.

## 1.2 Galaxy Detection and Spectra Extraction

In order to deblend and extract the integrated spectra of each individual galaxy in the field, we used a technique developed by ourselves (15; 10). The technique is an extension to IFS of `galfit` (12), that we named `galfit3d`. It entails a deblending of the spectrum of each object in the datacube by fitting analytical models. IFS data can be understood as a set of adjacent narrow-band images, each with the width of a spectral pixel. For each narrow-band image it is possible to apply modelling techniques developed for 2D imaging, like `galfit`, and extract the morphological and flux information for each object in the field at each wavelength. The spectra of all the objects are extracted after repeating the procedure for each narrow-band image throughout the datacube. We have already shown that the use of additional information to constrain the morphological parameters increases the quality of the recovered spectra (10). For that purpose we have used the F850LP-band image obtained with the HST/ACS camera.

First, we use SExtractor (3) on the section of the F850LP-band image corresponding to the field-of-view of our IFS data. For each detected galaxy we recover its position, integrated magnitude, scale length, position angle and ellipticity. These parameters were used as an initial guess to fit each of the galaxies in the F850LP-band image with a 2D Sérsic profile model, convolved with a PSF, using `galfit`. The PSF was obtained using a stamp image of the star in the PPak field-of-view. The fit for each individual object was done in a sequential way, from the brightest to the faintest, masking all the remaining objects. After iterating over all the detected galaxies we obtain a final catalogue of their morphological parameters. Similar techniques are used to derive the morphological parameters of galaxies in different ACS imaging surveys (17; 6).

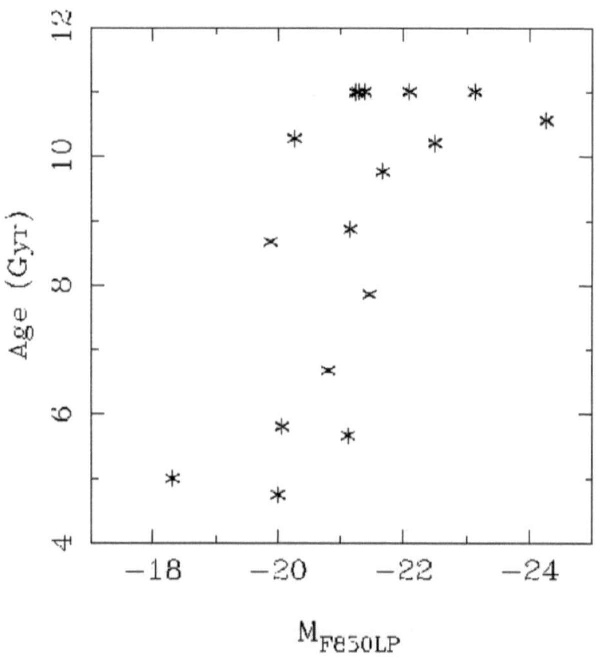


**Fig. 3.** Luminosity-weighted age of some galaxies in the cluster core is plotted against the absolute magnitude in the F850LP-band.

The catalogue of detected galaxies in the F850LP-band image was cross-checked visually with a 2D image derived from the datacube by coadding the flux over the entire wavelength range (4687-8060 Å). A final catalogue of 41 objects (40 galaxies and 1 star) was created, excluding two detected arc-lenses. Once we derived the morphological parameters for each galaxy, we extracted their integrated spectra by modelling each galaxy in the datacube with `galfit3d`. We use the same model used for the F850LP-band image, with the morphological parameters fixed, and fitting only the flux at each wavelength. The fit was performed again in a sequential way, from the brightest to the faintest, masking all the objects but the fitted one. In each iteration we used as input the residual datacube of the previous iteration, "cleaning" each object one-by-one. We finally got a spectrum for each of the 40 galaxies in the field-of-view.

### 1.3 Analysis and Results

Figure 2, shows a few examples of the extracted spectra. Different absorption features are detected in each spectrum, including some that are sensitive to age and metallicity (e.g., H$\beta$ and MgI). A few spectra show clear gaseous emission lines (e.g., H$\beta$, [OIII] and H$\alpha$). We derive redshifts for 28 of the 40 galaxies by comparing the observed wavelength of the absorption (or emission) features with the restframe values. Of those, 27 galaxies are at a redshift around $z \sim 0.17$, the nominal redshift of the cluster, and another one at $z = 0.104$. The signal-to-noise of the remaining 12 spectra is too low to unambiguously identify spectral features. We derived the age and metallicity of the galaxies by fitting each spectrum with single stellar population synthetic models, created using the GISSEL code (4).

Figure 3 shows a preliminary result obtained from the combination of parameters derived from the fitting procedure and the morphological analysis. It shows the luminosity-weighted age versus the F850LP-band absolute magnitude for the 17 galaxies for which we have completed the analysis. There is a large spread of ages in the stellar populations of the galaxies in the core of Abell 2218, contrary to the expectations from a single bulk formation and a passive evolution. Furthermore, the brightest (and more massive) galaxies are older than the fainter (and less massive) ones, which show a larger spread in ages. This may indicate that the smaller, less massive galaxies have formed later, being captured by the cluster, and/or they have enjoyed more recent periods of star formation. These results illustrate that the evolution of stellar populations in galaxies in clusters is far from passive, even in the central core.

## References

[1] Smail, I., Kuntschner, H., Kodama, T., et al. 2001, MNRAS, 323, 839
[2] Abell, G. O., Corwin, H. G., & Olowin, R. P. 1989, ApJS, 70, 1

[3] Bertin, E., & Arnouts, S., 1996, AAS, 117, 393
[4] Bruzual, G., & Charlot, S., 2003, MNRAS, 344, 1000
[5] Butcher, H., Oemler, A., 1984, ApJ, 285, 426
[6] Coe, D., Benitez, N., Sánchez, S.F., et al., 2006, ApJ, submitted
[7] Kelz, A., Verheijen, M.A.W., Roth, M.M. et al., 2006, PASP, in press
[8] Kristian, J., Sandage, A., & Westphal, J.A., 1978, ApJ, 221, 383
[9] Le Borgne, J.F., Pelló, R., Sanahuja, B., 1992, A&AS, 85, 87
[10] García-Lorenzo, B., Sánchez S.F., Mediavilla, E., et al., ApJ, 621, 146
[11] McHardy, I.M., Stewart, G.C., Edge, A.C., et al. 1990, MNRAS, 242, 215
[12] Peng, C.Y., Ho, L.C., Impey, C.D., Rix, H., 2002, AJ, 124, 266
[13] Roth, M.M., Kelz, A., Fechner, T., et al., 2005, PASP, 117, 620
[14] Sánchez S.F., 2004, AN, 325, 167
[15] Sánchez S.F., García-Lorenzo, B., Mediavilla, E., et al., 2004, ApJ, 615, 156
[16] Sánchez S.F. & N.Cardiel, Calar Alto Newsletter, num. 10
[17] Rix, H., Barden, M., Beckwith S.V.W., et al., 2006, ApJS, 152, 163
[18] Ziegler, B.L., Bower, R.G., Smail, I., et al., 2001, MNRAS, 325, 1571

# Part V

# Intermediate $z$ Galaxies

# Velocity Fields of z∼0.6 Galaxies with GIRAFFE and Perspectives for ELTs

M. Puech, F. Hammer and H. Flores

GEPI - Observatoire de Paris - 5, place Jules Janssen 92195 Meudon France

**Summary.** Using 15 deployable integral field units of FLAMES/GIRAFFE at VLT, we have recovered the velocity fields of 35 galaxies at intermediate redshift ($0.4 \leq z \leq 0.75$). This facility is able to recover the velocity fields of almost all the emission line galaxies with I(AB)$\leq$ 22.5. We find that less than 40% of intermediate redshift galaxies are indeed rotating disks, producing a Tully-Fischer relationship (stellar mass or M(K band) versus Vmax) which has apparently not evolved in slope, zero point and scatter, since z=0.6. The very large scatters found in previously reported Tully-Fischer relationships at moderate redshifts are apparently due to the difficulty to identify the nature of velocity fields with slits.

Indeed a majority of intermediate redshift starbursts are not rotating disks: 60% of galaxy velocity fields evidence the fact that they have not reached their dynamical equilibrium. Those galaxies include mergers, compact galaxies and/or inflow/outflows and their presence suggests a strong evolution in the dynamical properties of galaxies during the last 7 Gyrs. We also used the moderately high spectral resolution of FLAMES/GIRAFFE (R=10000) to derive 2D maps of electronic densities from the [OII] 3726,3729 line ratios. This allows us to identify an outflow and few giant HII regions with densities similar to Orion, but at z=0.6. It leads to a new technique for mapping extinctions, star formation rate densities, gas metal abundances in distant galaxies, which, combined with velocity fields, allows to investigate the details of galaxy physics at large lookback times.

Integral field spectroscopy is a mature technique which will be applied to Extremely Large Telescopes. We summarize some of the extragalactic science cases, and discuss their resulting requirements (image quality, field, multiplex & spectral resolution), which can be compared to adaptive optics techniques such as GLAO (Ground Layer Adaptive Optics) , FALCON (or Distributed Adaptive Optics) and Laser Guide Stars.

## 1 Velocity fields of z∼0.6 galaxies

In the frame of the Paris Observatory GTO (PI: F. Hammer), we observed the [OII] emission line of 35 galaxies selected both in the CFRS field and in the HDFS at $0.4 \geq z \geq 0.7$, using the multi-IFU mode of FLAMES/GIRAFFE. In this mode, GIRAFFE is able to observe simultaneously 15 galaxies with a 6x4 array of 0.52 arcsec micro-lenses. Although quite small in size, this sample is nevertheless representative of intermediate redshift galaxies with $EW_o$([OII])$\geq$ 15Å and $I_{AB} \leq$22.5 (4).

Because of the relatively large size of the GIRAFFE IFU pixel, the velocity width measured in a given pixel is expected to be the convolution of random motions with larger scale motions. In other words, in the case of a rotating disk, a peak at the dynamical center is expected in the $\sigma$-map: this peak corresponds to the rising part of the rotation curve. This purely spatial integration effect has been used as a basis of a kinematical classification. We split the sample into three distinct kinematical classes defined as follows (see Figure 1): the Rotating Disks (RD) class was composed of galaxies showing both a VF characteristic of a RD and a $\sigma$-map peaked in the dynamical center; Perturbed Rotations (PR) were those showing a still regular RD-like VF, but with a $\sigma$-map without any peak in the dynamical center, or with a peak significantly off centered; Finally, the Complex Kinematics (CK) class encompass all galaxies showing a VF without any apparent structure, or clearly incompatible with classical rotation (e.g. VF with a rotation axis perpendicular to the optical axis which are probably characteristics of outflows). This classification has been checked using numerical simulations which aims at reproducing the observed $\sigma$-map from the VF to ensure that the peak seen in the $\sigma$-map was really due to the rotation, as explained above. Finally, we simulated GIRAFFE/IFU observations of both RD and galaxies with more disturbed kinematics using 3D spectroscopy of local galaxies (see 5; 11): we checked that CK were real and not artifacts due to the low spatial resolution used.

This simple classification scheme leads us to the conclusion that *only 40% of z∼0.6 galaxies have reached a relaxed dynamical state*, corresponding to the RD class. Interestingly, this fraction seems to be lower in the subsample of the most compact galaxies (13). It will be important in the future, to understand whether PR and CK classes correspond to different physical processes, respectively minor and major mergers.

## 2 The Tully-Fisher relation at z∼0.6

We estimated $V_{max}$ as being half the maximal gradient found in the velocity fields, corrected from inclination. Hydrodynamical simulations from (3) were used to estimate a correction on $V_{max}$ for spiral galaxies to take into account the large spatial size of GIRAFFE pixels: we found that for galaxy sizes ranging between 1.5 and 3 arcsec, GIRAFFE underestimates the maximal rotational velocity by $\sim$ 20%. Absolute magnitudes in B and K bands were derived following (6; 9). We were then able to investigate the TF relation (Figure 3) at z∼0.6.

At first sight, our result in K band does not differ considerably from that of (2), except that they sample only the highest luminosity range (see Figure 3). As in (2), the TF relation shows a very large scatter, which overpasses by several magnitudes the scatter found in the local TF relation. Further examination taking into account the kinematical classification defined in the

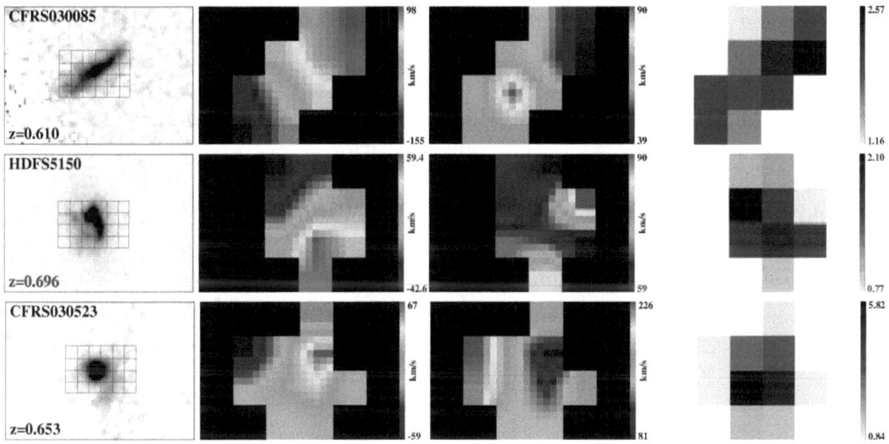


**Fig. 1.** Kinematical classification of z∼0.6 galaxies. From Left to right: I-band HST image with the IFU bundle superimposed; Velocity Field and $\sigma$-map (with a 5x5 interpolation for visualization ease); map of the mean spectral S/N in the [OII] doublet. Up: CFRS03.0085 classified as a RD; note the peak in the $\sigma$-map located at the dynamical center of the VF. Middle: HDFS5150 classified as a PR; The peak in the $\sigma$-map is off-centered relatively of the dynamical center of the VF; Down: CFRS03.00523 classified as CK; The VF is very distinct of the classical rotational motion as seen in the two previous examples.

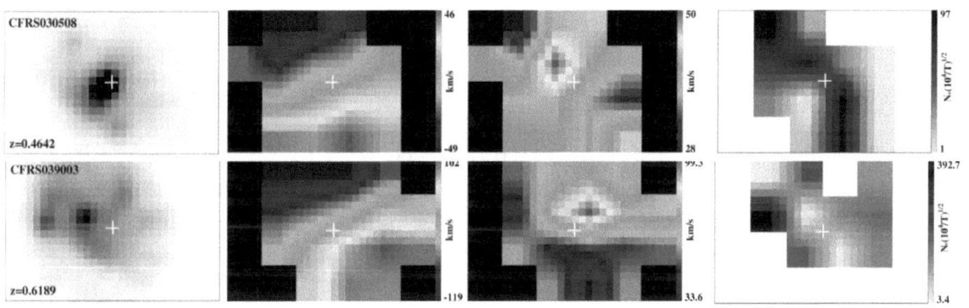


**Fig. 2.** Mapping of electron density in distant galaxies using the [OII] doublet. From left to right: I-band HST image corresponding to the GIRAFFE IFU FoV, GIRAFFE VF, $\sigma$-map and electron density map (with 5x5 interpolation, see Figure 1). Up: CFRS03.0508 (classified as CK). The VF and $\sigma$-map looks like a rotation, but note that the dynamical axis is almost perpendicular to the main optical axis. This probably reflects an outflow, which is confirmed by the electron density map: high densities are pointing to maximal velocities ends. Down: CFRS03.9003. This galaxy was classified as a rotating disk. Densities are characteristic of HII regions ($\sim 100\mathrm{cm}^{-3}$), except in the center which probably corresponds to the bulge identified by (16), and in the knot seen at the left of the HST image which corresponds to a relatively high electron density, comparable with that found in the outer areas of some nearby nebulae such as Orion. This know is probably responsible for the high SFR ($\sim$75 $M_\odot$/yr) of this LIRG.

last section shows that *all of the scatter is related with the galaxies which kinematics has been classified either complex or perturbed, i.e. galaxies which have not yet reached their dynamical equilibrium.* The same effect appears in the B band TF relation, a significant fraction of the dispersion being related to interlopers (see Figure 3). The only evolution seen correspond to a brightening of 1 magnitude for one third of the spirals in B band, which is probably linked with star formation. This important result will need more statistics to be firmly established. The ESO Large Program IMAGES (Intermediate Mass Galaxy Evolution Sequence, PI: F. Hammer) should answer this issue, raising the statistics to ~250 galaxies at 0.4≥z≥0.9 and $M_J \leq -20.5$.

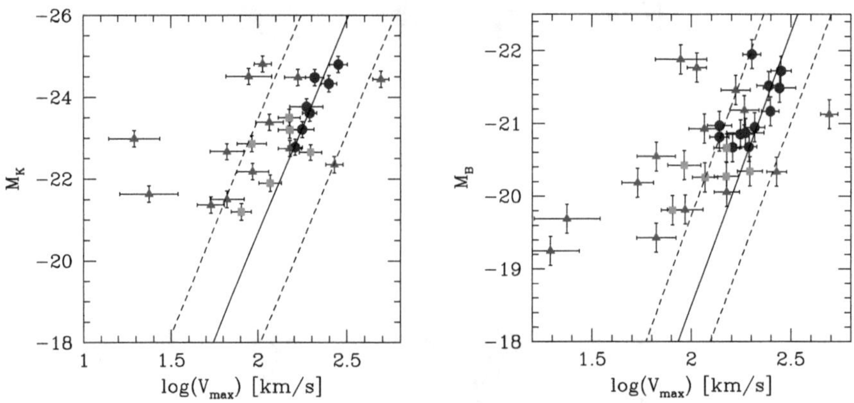


**Fig. 3.** Tully-Fisher relations at z~0.6 obtained with FLAMES/GIRAFFE. Left: TF relation in K band. Right: TF relation in B band. Magnitudes have been corrected following (15).

## 3 Mapping physical quantities in distant galaxies

Kinematics and dynamics are by far the most direct application of 3D spectroscopy. However, integral field spectroscopy can also be used to map physical quantities in distant galaxies. Thanks to its exceptional spectral resolution (R~10000) for an IFU mode, GIRAFFE is able to resolve the [OII] emission line doublet of z~6 galaxies. The [OII] line ratio can then be used as a tracer of electron density (see 10, for more details). We derived the very first mapping of electron density in a sub-sample of 6 galaxies, selected for their high S/N (14). The most impressive results are the confirmation of an outflow in CFRS03.0508 and the identification a giant HII region in CFRS03.9003 (see Figure 2).

With GIRAFFE and SINFONI at the VLT, we are now working to enlarge this approach to other emission lines such as H$\alpha$, H$\beta$ and [OIII]. This will allow us to establish extinction, star formation, metallicity and electron density maps of a sample of 15 galaxies at z∼1.2. These measurements will help us to investigate the physical processes that govern the increase in the high redshift star formation and the galaxy growth, when the stellar components of galaxies were growing very rapidly – much more rapidly than today. Notice that X-SHOOTER on the VLT (planed to be delivered in 2008) will be an exquisite tool for this kind of studies.

## 4 3D spectroscopy on ELTs

By many aspects, 3D spectroscopy on a 8 meter telescope reveals itself as a very powerful tool to understand the formation and evolution of galaxies. It now appears as an evidence that such an instrument will be very useful on an ELT to push the spectroscopy of galaxies up to $m_{AB}$ ∼26-27. This will open a new avenue in the understanding of galaxy physics and the coupling between visible and dark matters, especially near reionisation era, were the first proto-galaxies are also expected.

However, integral field spectroscopy of such distant sources will require a very high spatial sampling: the expected size of z∼6 sources is as low as 0.15 arcsec (1). Only Adaptive Optics will be able to provide enough S/N in a so small aperture. Note however that this does not account for the gas extension, but relies only on the continuum properties: GIRAFFE results suggest that gas is more extended than stars, especially in non-relaxed systems which are probably very common at z≥6. Even if this could relax the needs on the IFU spatial resolution, in any case, AO will be mandatory.

Another important requirement will be the size of the FoV. At least 10 arcmin in diameter is required to encompass the correlation lengths of cosmological structures at all redshifts, and then avoid cosmic variance (8). This is also required to reach a multiplex factor ranging from 10 to 100 objects and then mitigate the large exposure times required for so distant and faint objects.

We have undertaken a systematic comparison of AO systems for ELTs (12). From these preliminary results, it appears that MOAO (Multi-Objects AO) seems to be a very promising way to fulfill the scientific needs in both light concentration and FoV. MOAO consists in correcting only the region of interest selected on a larger FoV, instead of correcting the whole FoV like with GLAO or even MCAO. In this concept, several independent AO systems are spread over the focal plane. Each IFU has its own AO system, which uses atmospheric tomography techniques: three Wave-Front Sensors (WFS) per IFU measure the off-axis wavefront coming from stars located around the galaxy, and the on-axis wavefront from the galaxy is deduced from off-axis measurements and corrected thanks to an AO system within each IFU. The process of on-axis wavefront reconstruction from off-axis measurements is repeated as many times as there are spectroscopic IFUs. This concept was first proposed for the FALCON instrument project (7) in a Natural Guide Star (NGS) version. The concept has been recently transported for OWL with MOMFIS.

We now plan to study the relation between FoV, sky coverage, spatial resolution and light coupling in both cases of Natural and Laser Guide Stars.

## References


[1] Bouwens R.J., Illingworth G.D., Blakeslee J.P. et al., 2004, ApJ, 611, 1.
[2] Conselice C., Bundy K., Ellis R. et al., 2005, ApJ, 628, 160.
[3] Cox T.J., Primack J., Jonsson P. et al., ApJ, 2004, 607, 87.
[4] Flores H., Hammer F., Puech M. et al. 2006, A&A, submitted.
[5] Garrido O., Marcelin M., Amram P. et al., 2005, A&A 362, 127.
[6] Hammer F., Gruel N., Thuan T.X. et al., 2001, ApJ 550, 570.
[7] Hammer F., Sayède F., Gendron E. et al., 2002, Scientific drivers for ESO future VLT/VLTI instrumentation, Proceedings of the ESO workshop help in Garching, Germany, 139.
[8] Hammer F., Puech M., Assémat F. et al., 2004, SPIE Proc. 5382, 727.
[9] Hammer F., Flores H., Liang Y. et al., 2005, A&A 430, 115.
[10] Osterbrock D. & Ferland G., 2006, Astrophysics of Gaseous Nebulae.
[11] Östlin G., Amram P., Masegosa J. et al., 1999, A&AS, 137, 419.
[12] Neichel B., Fusco T., Puech M. et al., 2005, Proc. of the IAU Symp. 232, astro-ph/0512525.
[13] Puech M., Hammer F., Flores H. et al., 2006a, A&A, submitted.
[14] Puech M., Flores H. Hammer F. & M.D. Lehnert, 2006b, A&A, submitted.
[15] Tully R.B., Pierce M.J., Huang J.-S. et al., 1998, AJ 115, 2264.
[16] Zheng X.Z., Hammer F., Flores H. et al., 2004, A&A, 421, 847.

# News from the "Dentist's Chair": Observations of AM 1353-272 with the VIMOS IFU

P. M. Weilbacher[1] and P.-A. Duc[2]

[1] Astrophysikalisches Institut Potsdam, An der Sternwarte 16, D-14482 Potsdam, Germany, pweilbacher@aip.de
[2] Service d'astrophysique, CEA Saclay-Orme des Merisiers, Bat. 709, 91191 Gif sur Yvette cedex, France, paduc@cea.fr

**Summary.** The galaxy pair AM 1353-272 nicknamed "The Dentist's Chair" shows two ∼30 kpc long tidal tails. Previous observations using multi-slit masks showed that they host up to seven tidal dwarf galaxies. The kinematics of these tidal dwarfs appeared to be decoupled from the surrounding tidal material. New observations of the tip of the southern tidal tail with the VIMOS integral field unit confirm the results for two of these genuine tidal dwarfs but raise doubts whether the velocity gradient attributed to the outermost tidal dwarf candidate is real. We also discuss possible effects to explain the observational difference concerning the strongest velocity gradient seen in the slit data, which is undetected in the new integral field data, but arrive at no firm conclusion. Additionally, low-resolution data covering most of the two interacting partners show that the strongest line emitting regions of this system are the central parts.

## 1 Introduction

Following old ideas about the creation of dwarf galaxies during interaction of giant galaxies, and detailed investigations of several nearby examples of these Tidal Dwarf Galaxies (TDGs, 1; 2; 3; 4), we carried out a first small survey of interacting galaxies (6) with the aim of better understanding the star-formation history of TDGs and constraining the number that are built per interaction (5). During this survey, we studied a system cataloged as AM 1353-272, which we called "The Dentist's Chair" for its peculiar shape. Fig. 1 shows the two components of the system: 'A', a galaxy with ∼30 kpc long tidal tails; and 'B', a disturbed disk galaxy; both having a distance of $D \approx 160\,\mathrm{Mpc}$ ($H_0 = 75\,\mathrm{km\,s^{-1}\,Mpc^{-1}}$). Within the tails several obvious clumps with blue optical colors are visible. Using optical and near-infrared imaging, evolutionary models, and optical spectroscopy, *seven* of these clumps were classified as TDG candidates in formation (7, , marked 'a' to 'd' and 'k' to 'm'). The largest velocity gradient with an amplitude of $>300\,\mathrm{km\,s^{-1}}$ appeared in TDG candidate 'a', at the very end of the southern tidal tail. This raised the question how an object with relatively low luminosity could exhibit such fast "rotation". However, as these observations were done using the multi-slit technique and hence are spatially restricted due to the narrow slit, subsequent observations were planned using an integral field unit (IFU) to cover more of the tidal tails and view the velocity structure of the TDGs in two dimensions.

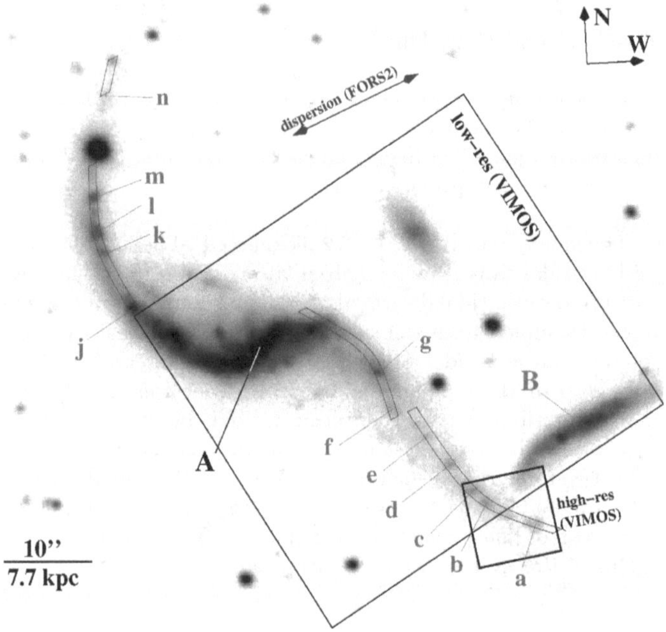


**Fig. 1.** The interacting system AM 1353-272. The two interacting galaxies (capital letters) and the relevant knots in the tidal tails are marked (lower case letters). Overlayed are the original FORS2 slits and the two VIMOS pointings in low and high resolution mode.

## 2 IFU Observations

Our new VIMOS data taken at the ESO VLT consists of 1.3 hours of exposure time in high resolution blue mode (field of view of $13'' \times 13''$) and good seeing conditions ($\sim 0\farcs7$), targeted at the tip of the southern tidal tail. This includes the TDG candidates ‘a’ to ‘c’. The center of galaxy ‘A’, much of the southern tidal tail, and the companion ‘B’ were targeted at low-resolution (blue grism, $54'' \times 54''$), and observed for 1 hour in mediocre conditions with $\sim 2\farcs0$ seeing. These two pointings are sketched in Fig. 1.

The data were reduced using the ESO pipeline for the VIMOS instrument. We made a small enhancement to the code that allowed us to interpolate the wavelength solution between adjacent spectra on the CCD. This was only used for the low-resolution data, where the errors introduced by this method

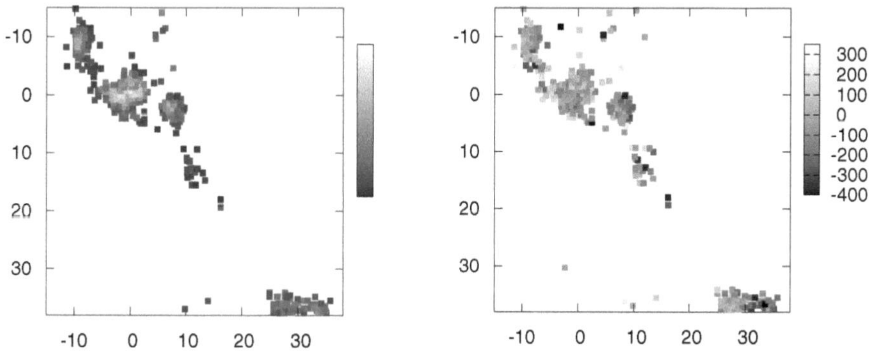


**Fig. 2.** Results of the low-resolution pointing, the center of AM 1353-272 A is at the coordinates (0,0); at the bottom right, near (30,32) companion 'B' is visible. Axes labels are in arcseconds. **Left**: relative H$\beta$ emission line flux (bright: high flux, dark: low flux). **Right**: velocity field derived from H$\beta$ emission line (the greyscale bar gives relative velocities in $\mathrm{km\,s^{-1}}$).

are smaller than the accuracy allowed for by the spectral resolution. In low-resolution mode, on the order of 15% of the spectra could not be wavelength calibrated due to overlapping spectral orders. From the final datacube of extracted and wavelength calibrated spectra, we measured the relative fluxes and velocities in each spectral element using Gaussian fits to the brightest usable emission line. In the low-resolution data [O III]5007 is the strongest emission line, but at this redshift ($z \approx 0.04$) it is strongly blended with a sky emission line, so H$\beta$ had to be used instead. In the high-resolution data, [O III]5007 is less affected by the sky-line and is the only line with sufficient S/N for the analysis in the low surface brightness region near the end of the tidal tail.

# 3 Results

Fig. 2 summarizes the results that can be derived from the low-resolution pointing using fits to the H$\beta$ emission line. The southern tidal tail in undetected in this exposure and the strongest line emission appears to be in the center of 'A' and in the two knots at the end of its bar-like central structure (designated 'g' and 'j' in 7). Galaxy 'B', despite being strongly reddened, also is a strong source of H$\beta$ line emission. As the velocity resolution is on the order of 100 $\mathrm{km\,s^{-1}}$, in this mode of VIMOS we cannot resolve the velocity structures in individual knots, but the bar-like structure in 'A' seems to rotate (the eastern end near knot 'j' is receding, the western end near knot 'g' is approaching). The same is true for the companion 'B'.

To verify our original FORS2 multi-slit observations, we try to "reconstruct" them from the VIMOS datacube. To that end we average over the

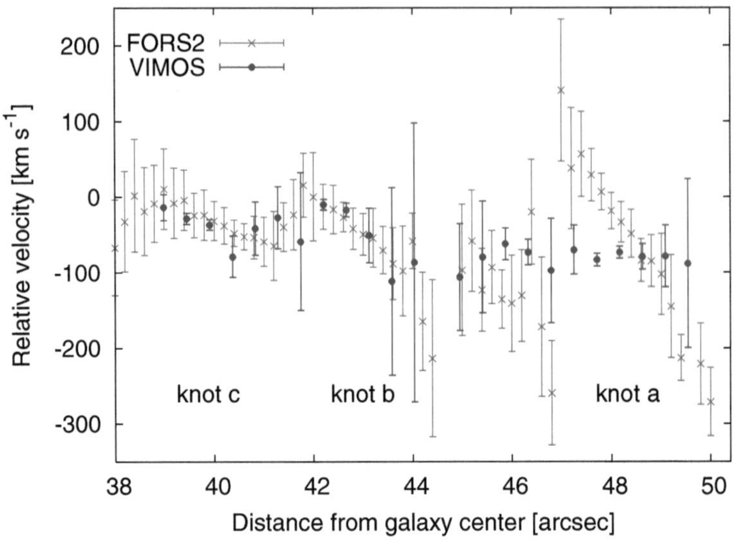


**Fig. 3.** Velocity field as derived from FORS2 observations (black points) and reconstruction of this velocity field from the VIMOS IFU datacube (grey pluses).

spaxels and derive an average redshift over the slit width. Fig. 3 shows the resulting velocity field along this artificial slit and compares it with the velocity profile of the FORS2 data. From this plot it can be seen that the velocity gradients of knots 'b' and 'c' are very well recovered, while the steep slope within knot 'a' with an amplitude of $\sim$350 km s$^{-1}$ in the FORS2 data appears almost flat in the reconstructed VIMOS data. We tested several alternative slit positions and curvatures, varied the effective slit width, and also tried to add slit effects (velocity offsets due to non-centered emission within the slit) to our reconstruction. None of these changes improved the match for knot 'a'. In fact, slit effects significantly worsened the agreement for all three knots. The FORS2 observations were done on these extended objects in 1$''$.0 seeing with a 1$''$.2 slit, so that it appears unlikely that slit effects would have a strong contribution to the observed velocity gradient. Other problems, like instrumental flexures should have been removed by the data reduction procedure as detailed in Weilbacher et al. (7). We are therefore confident that slit effects do not play an important role in the FORS2 data. On the other hand, if we assume that the original slit-based data give the correct results, it is unclear how the VIMOS data could be flawed so as to hide this one velocity gradient. The wavelength calibration works well for the high-resolution mode as confirmed by checks with sky emission lines.

## 4 Summary and Outlook

We presented a few tentative results for the interacting system called "The Dentist's Chair" from new observations with the VIMOS integral field mode: line emission seems to be concentrated within the centers of the interacting partners, while the tidal tails themselves are not detected H$\beta$ narrowband slices. Three knots, previously identified as TDG candidates near the end of the southern tidal tail, are detected in [O III]5007 emission. For two of them the velocity profiles were confirmed. However, the strongest velocity gradient in the outermost TDG candidate (knot 'a') as measured on FORS2 data is not confirmed by the VIMOS datacube. The reason for this discrepancy is unknown.

To solve this mystery and find more clues to the origin of the velocity fields seen in this interacting system, further, deeper IFU observations, taken with appropriate dither offsets to facilitate more accurate sky subtraction, are required. With other instruments like e. g. the GMOS IFU would be possible to cover both the Ca-triplet and H$\alpha$ in the same exposure and directly compare the stellar velocity field with ionized gas dynamics. Since good S/N is required to detect the absorption lines, this can only be done in the brighter northern tidal tail.

*Acknowledgement.* PMW received financial support through the D3Dnet project from the German Verbundforschung of BMBF (grant 05AV5BAA). We are grateful to Ana Monreal-Ibero and Lise Christensen for practical hints on IFU data handling. The data we discuss was taken in service mode at Paranal (ESO Program 074.B-0629).

## References

[1] Duc, P.-A., Brinks, E., Springel, V., et al., 2000, AJ 120, 1238
[2] Duc, P.-A., Brinks, E., Wink, J.E., & Mirabel, I.F., 1997, A&A 326, 537
[3] Duc, P.-A. & Mirabel, I.F., 1998, A&A 333, 813
[4] Hibbard, J.E., Guhathakurta, P., van Gorkom, J.H., & Schweizer, F., 1994, AJ 107, 67
[5] Weilbacher, P. M., Duc, P.-A., & Fritze-von Alvensleben, U., 2003, A&A 397, 545
[6] Weilbacher, P. M., Duc, P.-A., Fritze-von Alvensleben, U., et al., 2000, A&A 358, 819
[7] Weilbacher, P. M., Fritze-von Alvensleben, U., Duc, P.-A., & Fricke, K. J., 2002, ApJ 579, L79

# 3D Spectroscopy of Low-$z$ (Ultra)Luminous Infrared Galaxies

M. García-Marín[1], L. Colina[2], S. Arribas[3], A. Monreal-Ibero[4], A. Alonso-Herrero[5] and E. Mediavilla[6]

[1] DAMIR/IEM/CSIC maca@damir.iem.csic.es
[2] DAMIR/IEM/CSIC colina@damir.iem.csic.es
[3] STScI/IAC arribas@stsci.edu
[4] AIP amonreal@aip.de
[5] DAMIR/IEM/CSIC aalonso@damir.iem.csic.es
[6] IAC emg@iac.es

**Summary.** We are using integral field spectroscopy (IFS) data together with archival *Hubble Space Telescope* (HST) images, for carrying out a program aimed at studying a representative sample of local luminous and ultraluminous infrared galaxies (LIRGs and ULIRGs; $11 \leq \log(L_{IR}/L_{\odot}) \leq 13$). Our goals are to characterize the stellar and ionized gas structure, the two dimensional extinction and star formation, the kinematical properties and the dynamical mass tracers. Here we will review some of the most interesting results obtained from our sample, including detailed studies for some individual galaxies. The implications for deriving the star formation rate (SFR) and mass when studying high-redshift galaxies will be briefly discussed too.

## 1 Introduction

### 1.1 The sample of galaxies

We have selected a representative sample of about 25 local (U)LIRGs in an effort for deriving global results for this galaxy type. The main characteristics of the sample are: (i) redshift <0.2, (ii) $11.3 \leq \log(L_{IR}/L_{\odot}) \leq 12.6$, (iii) activity level covering H II, LINER and Seyfert and (iv) merging phase from early to late. In Fig. 1 we show HST images for some galaxies of the sample.

### 1.2 Instrumentation

To achieve the goals of our program, we have mainly used IFS data obtained with INTEGRAL (2), a fiber-fed system connected to WYFFOS (Wide-Field Fiber Optic Spectrograph) (5) and mounted on the 4.2 m William Herschel Telescope (La Palma). INTEGRAL has three different configurations that we have detailed in Table 1. In all cases the fibers are forming a rectangular area covering the object; in addition, external fibers forming a circle are used for measuring the sky simultaneously. Although we have used the three configurations for studying the galaxy sample, the intermediate bundle SB2

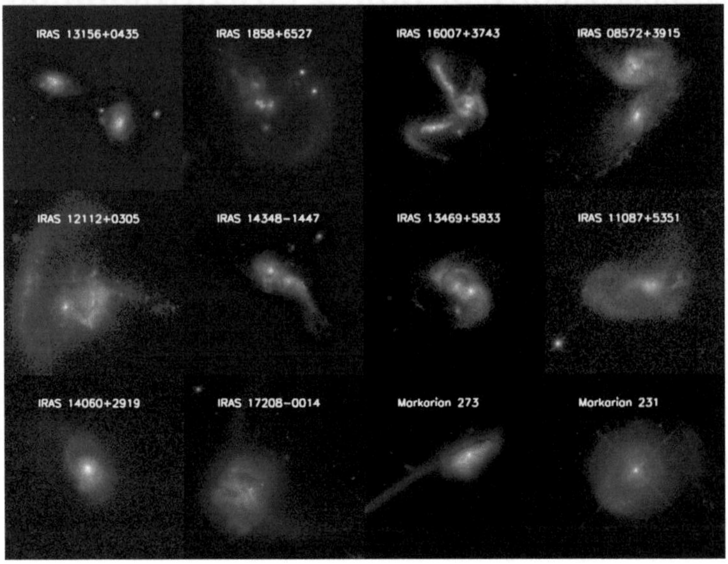


**Fig. 1.** Archival HST I-band images obtained for some galaxies of our sample, covering different stages of interaction.

is the most common in our analysis. These data combined with the high resolution HST multi-wavelength images, are ideal and complementary to study the complex structure of the ULIRGs.

**Table 1.** Main characteristics of the INTEGRAL bundles

| Mode | FoV(arcsec$^2$) | ⊘(arcsec) | Fibers (main+sky) |
|---|---|---|---|
| SB1 | 7.80×6.40 | 0.45 | (175+30) |
| SB2 | 16.0×12.3 | 0.90 | (189+30) |
| SB3 | 33.6×29.4 | 2.70 | (115+20) |

## 2 Morphology of the stellar component and ionized gas. Extinction effects

Using INTEGRAL data and HST images, it is possible to compare the morphology of the stellar and the ionized gas components of the galaxy under study. As an example, we present the results obtained for IRAS 12112+0305 (6), a ULIRG ($L_{IR} \sim 12.34\ L_{\odot}$, z=0.0723) where strong evidences for decoupled stellar and ionized gas components exist. As we can see in Fig. 2, the stellar main body is concentrated in three regions ($N_n$, $N_S$ and R2), separated by about 2-3 arcsec. The global morphology of the HST image corresponds well to the INTEGRAL continua, taking into account the different spatial resolution. When comparing the continuum and the emission line maps, the morphology changes dramatically being the peak of emission associated with R1 in all three cases. These morphological changes are caused by the internal extinction due to the dust, and by changes on the ionization conditions. In this particular case, the extinction derived from the H$\alpha$ H$\beta$ ratio ranges from $A_V$=0.9 in R1 to $A_V$=7.8 in $N_S$. Comparing the observed H$\alpha$ flux ratio F($N_S$)/F(R1)=0.4 with the extinction corrected luminosity L($N_S$)/L(R1)=85, it is clear the strong effect of the dust and the importance of its correction. It is remarkable that this morphological variations and displacements of the intensity peaks when comparing the stellar continuum and the ionized gas, are common on the (U)LIRGs that we have studied.

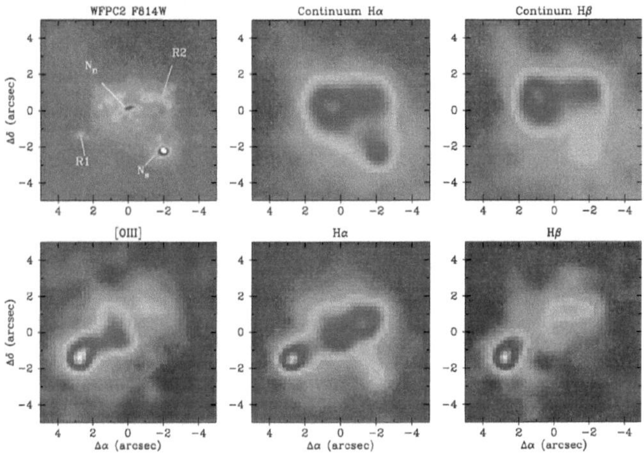


**Fig. 2.** INTEGRAL images of the ionized gas and stellar light distribution in the central regions of IRAS 12112+0305 as traced by different emission lines (H$\beta$, [O III]$\lambda$5007 and H$\alpha$). The high resolution HST I-band image is also shown.

## 3 Overall star formation and Star Forming Regions

In this section we will present some of the results that we have obtained for Arp 299 (IC 694+NGC 3690) (9), a nearby LIRG ($L_{IR} \sim 11.8\ L_\odot$, z=0.01) in an early interaction stage. For the study of this galaxy we used two INTEGRAL configurations: the SB2 bundle for covering the individual galaxies and the SB3 for covering the interface between (Fig. 3 *Left*). Complementary HST multi-wavelength images both in continuum and emission lines are also available (1). Once the emission line maps are studied, it is possible to derive the extinction due to the dust using the H$\alpha$ H$\beta$ ratio (Fig.3 *Right*); in this case NGC 3690 presents a wide range of extinction (0.5<$A_V$< 4), while IC 694 hosts the most extinct region A, with $A_V$~6. In NGC 3690 the most dimmed area is associated to the nuclear region B1 where, as we will see, there is evidence for a buried AGN. Once the extinction is corrected, the NGC 3690 H$\alpha$ map is very similar to the Pa$\alpha$ image (Fig.3 *Right*), with the peaks located in the same regions: C1 (a very active star forming region) and B1, and the same overall structure. The H$\alpha$ luminosity is about 1.5 times larger in C1 than in B1, and 2.5 larger in A (nucleus of IC 694). We have analyzed the individual

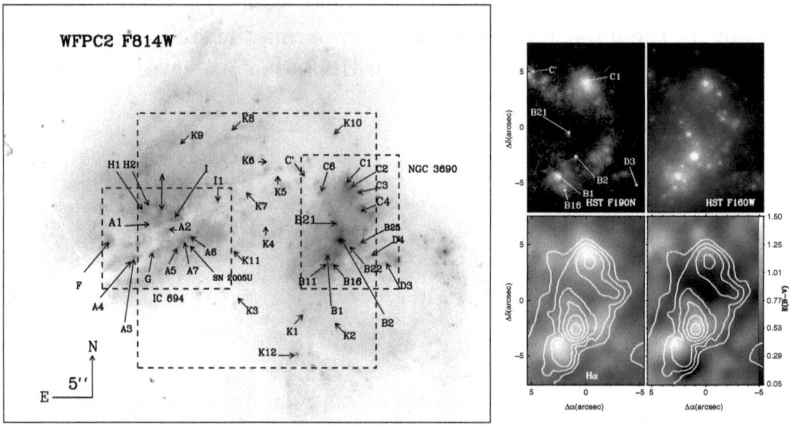


**Fig. 3.** *Left:* HST I-band image of the interacting galaxy Arp 299. The rectangular overlays indicate the three INTEGRAL pointings (SB2 and SB3 bundles). We have marked the position of the regions under study. *Right:* NGC 3690 maps. On the top we show the HST-NICMOS Pa$\alpha$ (left) and H-band (right). On the bottom we show the H$\alpha$ extinction corrected map (left) and the extinction map (right).

star forming regions, comparing the INTEGRAL H$\alpha$ equivalent width (EW) values and the HST colours with Starburst99 instantaneous models (11) (13). The results obtained are showed on Fig. 4; the ages derived for the regions under study are ranging between 3.4 and 7.5$\times10^6$ years, and the stellar mass content ranges between 6 and 650$\times10^6 M_\odot$. The nuclear regions of NGC 3690 (B1, B2) present differences among them: B2 has an EW(H$\alpha$) 10 times lower than B1 and is about 2$\times10^6$ years older. The $A_V$ derived for several regions is compatible with those of the models, taking into account the errors. In spite of that, in another regions like A (nucleus of IC 694), B1 (nuclear region of NGC 3690) and C1 (young star forming region) and they are compatible with an additional reddening due to the presence of a hot dust component (B1: 85% of emission from dust at 800 K).

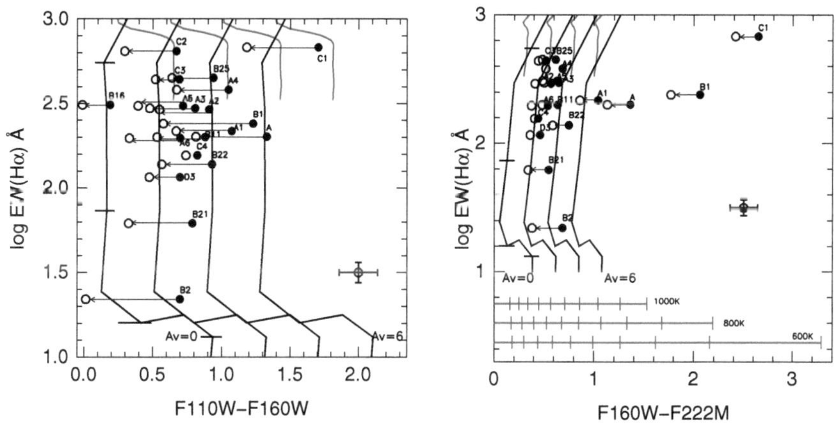


**Fig. 4.** Theoretical models (Starburst99) compared to measured values. The x-axis indicate HST colors similar to J-H (left) and H-K (right). The black tracks indicate instantaneous bursts, and the grey continuum ones, with different extinction values. The filled points are observed colours for the regions under study, while the hollow ones are extinction corrected. The horizontal lines represent mixing curves of hot dust (from 0% to 100%, increasing 10% on each step) and a normal stellar population of 6 Myr old, at different temperatures, normalized to the K-band. In both cases the black horizontal marks are indicating, from top to bottom, 4, 6, 8 and $10 \times 10^6$ years.

# 4 Two dimensional Ionization Structure

IFS is the ideal technique for mapping the 2D ionization state of complex systems like ULIRGs. In this section we will present results regarding this topic for Arp 299 using the INTEGRAL (SB3 bundle) data. The ionized state has been derived using standard optical emission lines diagrams (4). As showed in Fig. 5 three types of ionization are present: H II, LINER and Seyfert, with a spatial distribution associated to the spatial location of the different ionization sources. We have found a Seyfert-like nebula ([O III]/H$\beta$ $> 0.5$) with conical morphology, and the apex oriented towards B1, which is located at a projected distance of 1.5 kpc from its center. As shown on the I-band HST image (Fig. 3) this nebula is located in a region of very low stellar density. This is the first optical evidence of a buried AGN in B1, supported too by the facts that B1 presents a hot dust component contribution, it is a high extinct region in NGC 3690, and it has a high velocity dispersion ($\sigma$). The presence of a buried AGN in B1 is in agreement with previous X-ray studies (8). Apart from the off-nuclear Seyfert region and the H II-like ionization of the star forming regions, the LINER-type ionization is dominant (see (12)).

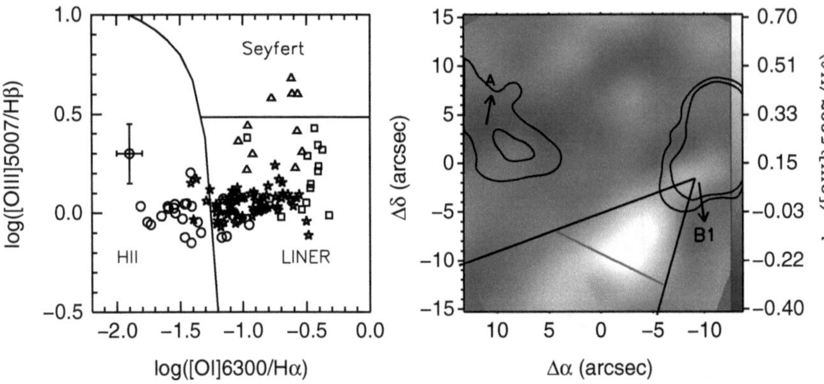


**Fig. 5.** Classic diagnostic diagram and 2D diagnostic diagram for Arp 299 observed with the INTEGRAL SB3 bundle. The triangles represent values within the inner part of the cone marked on the map. The squares represent the zones in the outer part of the cone. The stars represent the interface region values, and the circles are associated to the galaxies (isocontours marked on the map), except for B1, which is part of the cone.

## 5 Kinematical properties and dynamical mass tracers

In this section we will review the conclusions derived from INTEGRAL IFS kinematics data for 11 ULIRGs, compared to data obtained from CO (molecular gas) and IR (stellar) spectroscopy. In Fig. 6 we present the results for four ULIRGs using INTEGRAL data, including the H$\alpha$ velocity dispersion and velocity field, together with HST-NICMOS data. Again the morphological differences between the continuum and emission line maps are relevant, unveiling that the stellar and the gaseous components are decoupled (see details in (7)). The main conclusions regarding this kinematical study are:

- Peak-to-peak velocity amplitudes of up to 600 km s$^{-1}$ are detected in tidally-induced structures: tails and extra-nuclear star-forming regions.
- The peak of the velocity dispersion in 60% of the ULIRGs studied does not coincide in position with the stellar nucleus.
- The comparison of the stellar and gas kinematics tracers suggests that the cold molecular gas does not share the same velocity field as the stars, and ionized gas could be more rotationally supported.
- The observed velocity amplitudes on scales of several kpc do not trace rotations in general, and are therefore not reliable mass estimators of these interacting/merging systems (Fig. 7 *Left*).
- Central velocity dispersion measured using high equivalent width from optical emission lines should be considered, in general, the most reliable tracer of the dynamical mass (Fig. 7 *Right*).

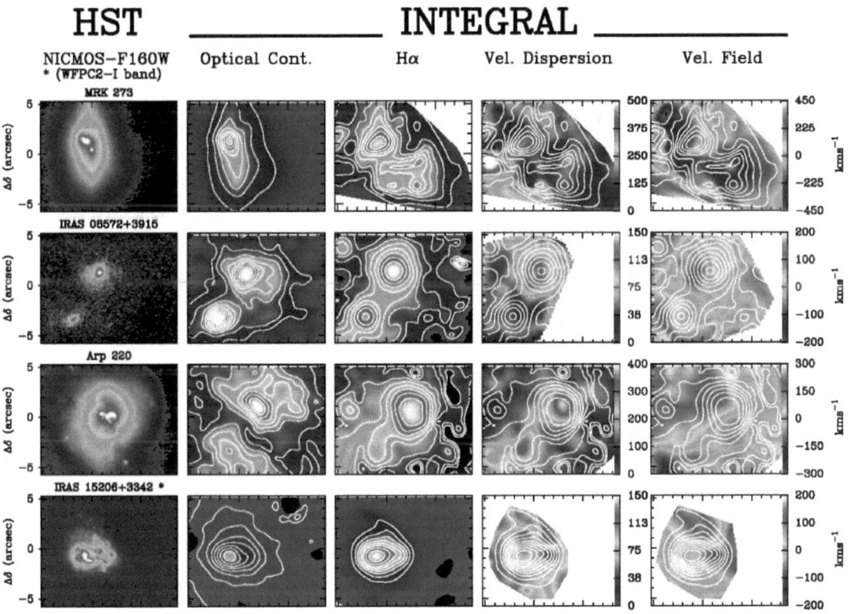


**Fig. 6.** Results for four ULIRGs with available IFS. The kinematics properties of the ionized gas are represented by the velocity dispersion map and the velocity field. The Hα contours are superposed on these maps.

- For the studied sample, the mass derived from the central ionized gas velocity dispersion gives an average value of $0.4m_{*} \pm 0.3m_{*}$, confirming that ULIRGs are moderate mass ($<m_{*}$) systems.

## 6 Implications for high-redshift galaxies: mass and SFR estimates

Over the last years, the study of galaxies at high-redshift has become a very active field. It is known that LIRGs and ULIRGs carry a large fraction of the stellar formation at z>1, and that one of the tracers of the star formation rate (SFR) and the dynamical mass will be Hα emission line. This line is shifted into the near and mid-IR for z>1, and due to the own nature of these galaxies (differential extinction effects, decoupled ionized and stellar structures, very complex velocity fields due to the interacting process) the use of multi-slit spectrographs can lead to the underestimation in a factor 2 or 3 of the Hα based SFR and masses in some high-z luminous dust-enshrouded starbursts. The natural option for studying these systems is the IFS. Finally, the central velocity dispersion from optical emission lines should be the more reliable tracer for the dynamical mass. The velocity amplitudes should not be used to estimate mass of high-s systems, if their stellar structure shows irregular morphological features like the low-z systems that we have study. For analyze these high-z galaxies, the future integral field units for the James Webb Space Telescope, NIRSpec (3) and MIRI (10), will be key instruments.

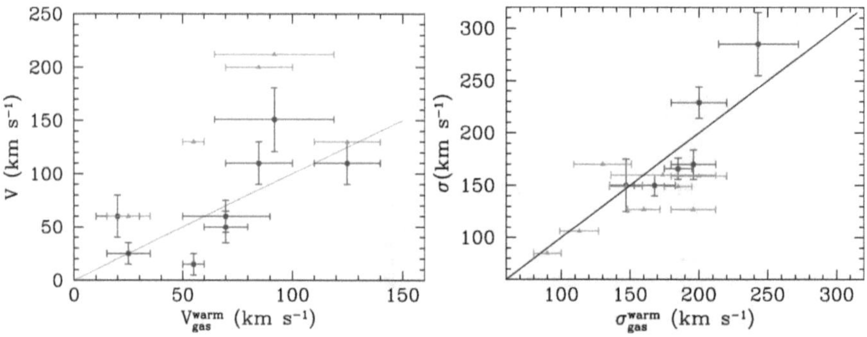


**Fig. 7.** *Left:* Central stellar (filled circles) and cold molecular gas (filled triangles) velocity amplitude vs. warm ionized gas values obtained from our IFS data. *Right:* Central stellar (filled circles) and cold molecular gas filled triangles) velocity dispersion vs. the warm ionized gas values obtained from our IFS data. In both plots the line represents a slope of 1.

## References

[1] Alonso-Herrero A. et al. 2000, ApJ, 532, 845
[2] Arribas et al. 1998, SPIE, 3355, 821
[3] Arribas et al. 2006, this proceeding
[4] Baldwin, J. A., Phillips, M. M. & Terlevich, R. 1981,PASP, 93, 5
[5] Bingham, R. G. et al. 1994, SPIE, 2198, 56
[6] Colina, L., Arribas, S., Borne, K. D. & Monreal, A. 2000, ApJ, 533, 9
[7] Colina, L., Arribas, S. & Monreal-Ibero, A. 2005, ApJ, 621, 725
[8] Della Ceca et al. 2002, ApJ, 581, 9
[9] García-Marín et al. 2006, in prep.
[10] García-Marín et al. 2006, this proceeding
[11] Leitherer, C. et al. 1999, ApJS, 123, 3
[12] Monreal et al. 2006, this proceeding
[13] Vázquez G. A. & Leitherer, C. 2005, ApJ, 621, 695

# Optical Spectra in the Non-Nuclear Regions of ULIRGs: Evidence of Ionization by Shocks

A. Monreal-Ibero[1,2], L. Colina[3] and S. Arribas[4,2]

[1] Astrophysicalisches Institut Potsdam amonreal@aip.de
[2] Instituto de Astrofísica de Canarias
[3] Instituto de Estructura de la Materia (CSIC) colina@isis.iem.csic.es
[4] Space Telescope Science Institute arribas@stsci.edu

**Summary.** In this contribution we present the analysis of the ionization structure of a sample of six low-$z$ Ultraluminous Infrared Galaxies using Integral Field Spectroscopy. We discuss the possible ionizing mechanisms responsible, focussing especially on the external regions ($d \sim 2 - 10$ kpc). The observed line ratios are better explained by the presence of shocks with velocities of 150 to 500 km s$^{-1}$. This is supported by the existence of a positive correlation between the ionization state and the velocity dispersion, especially when using the [O I]$\lambda$6300/H$\alpha$ ratio. We discuss the origin of these shocks. We find no evidence for signatures of superwinds in the extranuclear regions. We propose as an alternative explanation the existence of tidally induced large-scale gas flows caused by the merging process itself as the mechanism causing these shocks. Finally, there is one galaxy where the line ratios in the external regions cannot be explained by shocks.

## 1 Introduction

Ultraluminous Infrared Galaxies (ULIRGs) were discovered by IRAS at the end of the 80's (15). They are defined as those objects with a luminosity similar to that of optically selected quasars ($L_{bol} \approx L_{IR} \gtrsim 10^{12} L_{\odot}$) and they are locally twice as numerous. All of them present some properties in common: they are systems rich in gas and dust, they present emission lines in the optical and most (if not all) show signs of mergers and interactions (see 16, for a review). It has been suggested that ULIRGs could be the progenitors of some elliptical galaxies (6; 18) and maybe of quasars (15). Also, they seem to be the local counterparts of the so-called submillimeter galaxies (17).

Ionization in these systems has been mainly studied in the nuclear regions via traditional spectroscopy with long-slit (e.g. 9). On the other hand, evidence of superwinds has already been reported in several systems on the basis of the properties of the emission (e.g. 10) or absorption (e.g.(11)) lines, although their importance in the extranuclear regions is unclear.

Here, we present the analysis of the ionization mechanisms in the external regions ($\geq 2$ kpc) of six ULIRGs using Integral Field Spectroscopy (IFS). A full description of the ionization structure, that includes also the more internal regions, can be found in (13).

## 2 Sample, Observations, Data Reduction and Analysis

We have analyzed six low-$z$ ULIRGs covering a relative wide range of dynamical states of the merging process. In Table 1 we show their relevant properties for this work. They are at a typical redshift of about 0.07 that corresponds to a linear scale of 1.5 kpc arcsec$^{-1}$. The last column shows the optical classification found in the literature. We see that there are several cases where this is not clearly determined.

**Table 1.** Some relevant properties of the analyzed systems

| Galaxy | $z$ | Scale (kpc arcsec$^{-1}$) | log($L_{IR}/L_{\odot}$) | Optical classification |
|---|---|---|---|---|
| IRAS 08572+3915 | 0.058[2] | 1.20 | 12.15 | Sy 2 / LINER |
| IRAS 12112+0305 | 0.073[2] | 1.52 | 12.30 | LINER |
| IRAS 14348−1447 | 0.083[2] | 1.72 | 12.31 | LINER / Sy 2 |
| IRAS 15206+3342 | 0.125[2] | 2.60 | 12.18 | Sy 2 / H II |
| IRAS 15250+3609 | 0.054[3] | 1.12 | 12.03 | H II / LINER |
| IRAS 17208−0014 | 0.043[3] | 0.89 | 12.40 | LINER / H II |

References: [2](9), [3](8)

Data were obtained with the INTEGRAL system (1) plus the WYFFOS spectrograph (3). Details about the configuration as well as the reduction process can be found elsewhere (13). For the present work we have used the [O I]λ6300, Hα, [N II]λλ6548,6584 and [S II]λλ6717,6730 emission lines. We fitted each emission line profile to a single Gaussian function using the DIPSO package inside the STARLINK environment. For each spectrum, we calculated the [O I]λ6300/Hα, [N II]λ6584/Hα, and [S II]λλ6717,6730/Hα line ratios. Also, we derived the ionized gas velocity dispersion from the Hα line width (after subtracting the instrumental profile in quadrature).

## 3 Results and Discussion

We show in Fig. 1 the [N II]λ6584/Hα vs. the [S II]λλ6717,6730/Hα line ratios for all the spectra where we have this information ([S II]λλ6717,6730/Hα instead of [O I]λ6300/Hα has been selected in order to cover a larger area). Many more spectra are classified as LINER according to the [S II]λλ6717,6730/ Hα ratio than according to [N II]λ6584/Hα. In addition, most of the data points are located either in the LINER-like region according to both line ratios, or in the region where the [S II]λλ6717,6731/Hα ratio is typical of a LINER but [N II]λ6584/Hα is typical of H II regions.

For the sake of the following discussion we define circumnuclear regions as those confined within the central ∼ 3 arcsec and extranuclear regions as

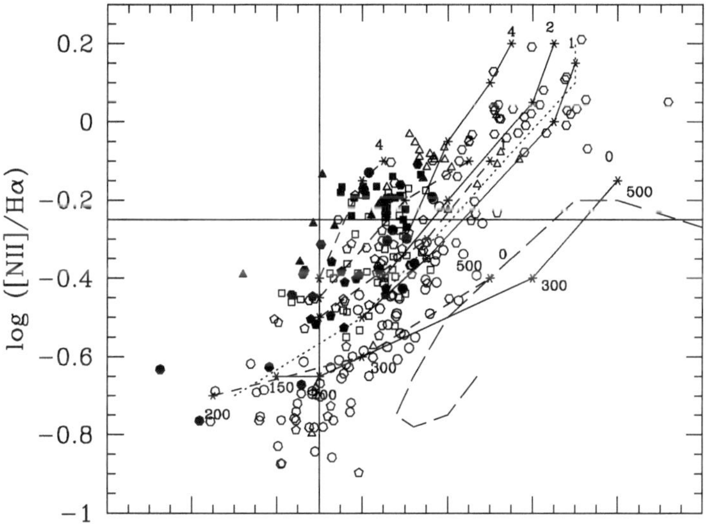


**Fig. 1.** [N II]λ6584/Hα vs. [S II]λλ6717,6731/Hα ratios. The left bottom corner represents the region occupied by typical H II regions while the top right corner of is the locus for typical LINER-like spectrum. Shape code is the same as in Figure 2. *Solid symbols:* circumnuclear regions; *hollow symbols:* extranuclear regions. Models of (5) for shocks without precursor (continuous lines) and with precursor (dashed lines) have been superposed. At the beginning of each line, it is shown the magnetic parameter $B/n^{1/2}$ ($\mu$G cm$^{3/2}$). Shocks velocity range from 150 to 500 km s$^{-1}$ for models without precursor and from 200 to 500 km s$^{-1}$ for models with precursor. The long-dashed line indicates the values predicted for photoionization by a power-law model for a dusty cloud at $n_e = 10^3$ cm$^{-3}$ and $Z = Z_\odot$ (7). The dotted line indicates the locus for an instantaneous burst model of 4 Myr, $Z = Z_\odot$, IMF power-law slope of −2.35 and $M_{up} = 100$ M$_\odot$; dust effects have been included and $n_e = 10^3$ cm$^{-3}$ (2).

those which typically extend for several kpc outwards of this region. The so-defined *nuclear regions* corresponds roughly to the areas previously studied via long-slit spectroscopy and will not be discussed in this contribution.

The line ratios predicted by different mechanisms have been overplotted to our data in Fig. 1. The AGN models are not, in general, likely to be representative of these low-density ($n_e < 10^3$ cm$^{-3}$) regions, although IRAS 17206−0014 may represent an exception in this context. An alternative mechanism could be ionization by young stars. The model that best fits our data is plotted as a dotted line. However, the conditions for this models are very specific and it seems unlikely to be representative of the extranuclear regions of all these ULIRGs, especially taking into account that for clusters younger than 3 Myr, or older than 6 Myr, and for models with a constant star formation rate, the predicted line ratios are more typical of H II regions. The most likely mechanism to explain the observed ionization in the extended, extranuclear regions according to Fig. 1 is the presence of large scale shocks.

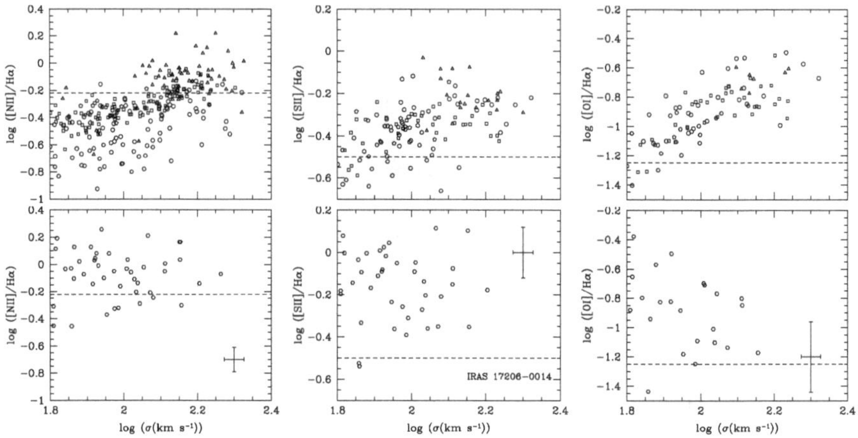


**Fig. 2.** Relation between the velocity dispersion and [N II]λ6584/Hα (left), [S II]λλ6717,6731/Hα (middle) and [O I]λ6300/Hα (right) for the five systems that follow the correlation (first row) and IRAS 17208−0014 (second row). Dashed horizontal lines mark the frontiers between H II region and LINER-type ionization. The symbol code is as follows: IRAS 08572+3915, pentagon; IRAS 12112+0305, circle; IRAS 14348−1447 square; IRAS 15206+3342, heptagon; IRAS 15250+3609, triangle; IRAS 17208−0014, hexagon.

A positive correlation between the velocity dispersion and ionization has been claimed in the literature as further evidence supporting the presence of shocks (e.g. 10). In Fig. 2 we present the relation between the line ratios and velocity dispersion for our systems. The correlation is clearly visible in five of our systems, especially when using the [O I]λ6300/Hα ratio (the best indicator to disentangle between ionization by star or by shocks (5)). This reinforces the idea that shocks are also the dominant ionization source at large scales (> 2-3 kpc). What is causing these shocks? One possibility is that they are due to superwinds. However, the distances where we are measuring the shocks are much bigger than the typical size of a superwind ($\lesssim$2 kpc). In addition, no evidence for distinct components in the emission lines has been detected in the external regions. This can be interpreted as if the high velocity outflows associated with the nuclear superwinds were not present at these distances or as if they were of much lower amplitude.

An alternative is that shocks are due to the merging process itself. This is supported by the large velocity gradients with peak-to-peak velocities of upto several hundreds km s$^{-1}$, associated with tidal tails and extranuclear regions, and the existence of extranuclear regions with large values of velocity dispersion (up to 200 km s$^{-1}$). These velocities imply therefore the presence of an extended, highly turbulent medium on kpc-size scales (4). Also, specific models for the nearest ULIRG, Arp220, show that the existence of flows with velocity gradients much greater than the impact velocity of the collision is a natural consequence of the merging process itself (12).

However, there are exceptions: IRAS 17208−0014 does not follow the mean behavior observed in the other systems, showing a wider range of line ratio values that cannot be explained by shocks. One hint that could explain the observed ratios is the detection of an extended ($\sim$ 4 kpc) hard X-ray nebula in this galaxy (14) which would produce a spectrum more similar to that of an AGN.

AMI acknowledges support from the Euro3D Research Training Network, funded by the EC (HPRN-CT-2002-00305). Financial support was provided by the Spanish Ministry for Education and Science through grant AYA2002-01055. Work based on observations with the William Herschel Telescope operated on the island of La Palma by the ING in the Spanish Observatorio del Roque de los Muchachos of the Instituto de Astrofísica de Canarias.

## References

[1] Arribas, S. et al. 1998 Proc. SPIE, **3355**, 821
[2] Barth, A. J. & Shields, J. C. 2000, PASP, **112**, 753
[3] Bingham, R. G. et al., 1994, Proc. SPIE, **2198**, 56
[4] Colina, L., Arribas, S., & Monreal-Ibero, A. 2005, ApJ, **621**, 725
[5] Dopita, M. A. & Sutherland, R. S. 1995, ApJ, **455**, 468
[6] Genzel, R. et al., 2001, ApJ, **563**, 527
[7] Groves, B. A., Dopita, M. A., & Sutherland, R. S. 2004, ApJS, **153**, 9
[8] Kim, D.-C. et al., 1995, ApJS, **98**, 129
[9] Kim, D.-C. & Sanders, D. B. 1998, ApJS, **119**, 41
[10] Lehnert, M. D. & Heckman, T. M. 1996, ApJ, **462**, 651
[11] Martin, C. L. 2005, ApJ, **621**, 227
[12] McDowell, J. C., et al. 2003, ApJ, 591, 154
[13] Monreal-Ibero, A., Arribas, S. and Colina, L. 2005, ApJ, *in press*, (astro-ph/0509681)
[14] Ptak, A. et al. 2003, ApJ, 592, 782
[15] Sanders, D. B. et al., 1988, ApJ, **325**, 74
[16] Sanders, D. B., and Mirabel, I. F. 1996, ARA&A, **34**, 749
[17] Smail, I., Ivison, R. J., & Blain, A. W. 1997, ApJL, **490**, L5
[18] Tacconi, L. J. et al., 2002, ApJ, **580**, 73

# Part VI

# Centres of Galaxies

# Stellar Populations in the Galactic Center with SINFONI

F. Martins[1], R. Genzel[1,2], T. Paumard[1], F. Eisenhauer[1], T. Ott[1], S. Trippe[1], R. Abuter[1], S. Gillessen[1] and H. Maness[2]

[1] Max Planck Institüt für Extraterrestrische Physik, Postfach 1312, D-85741 Garching bei München, Germany
[2] Department of Physics, University of California, CA 94720, Berkeley, USA

**Summary.** We present results of the analysis of the population of massive stars recently discovered in the central cluster of the Galaxy. A spectral classification of all stars is made based on their K band spectra. Population synthesis models and the HR diagram are used to show that the population was born in a burst of star formation $\sim$ 6 Myrs ago. Finally, special emphasis is given to two stars with special properties: GCIRS34W and GCIRS16SW.

## 1 Star Clusters in the Galactic Center

Due to improvements in the infrared and high energy observational capabilities in the last years, the central regions of the Galaxy have become more and more accessible. This has triggered a number of studies in various fields of astrophysics, going from the properties of supermassive black holes to accretion phenomena and physics of compact objects. In terms of stellar populations, the major breakthrough was the discovery of three massive clusters in the central few tens of parsecs, namely the Arches (2), the Quintuplet (8; 14) and the central cluster (10). They all contain a large number of massive young stars in an advanced evolutionary status (1; 7; 18; 4; 6) which led authors to the conclusion that a recent episode of star formation took place in the Galactic Center in the last ten million years (11).

To date, the most detailed studies of the Quintuplet and Arches clusters are those of Figer et al. (4), Figer et al. (5), Cotera et al (2) and Figer et al. (6). Several tens of Wolf-Rayet and other evolved massive stars such as Luminous Blue Variables (LBV) and transition "slash" stars (Ofpe/WN) have been spectroscopically classified, allowing a quantitative estimate of the age of the clusters: 4 Myrs for the Quintuplet cluster, and 2.5 Myrs for the Arches cluster. In the case of the former, the IMF was derived and turned out to be top heavy, with a slope $\Gamma \approx -0.65$ (compared to $-1.35$ for the Salpeter IMF). From this result and the number of observed stars, the total mass of the cluster was found to be a few $10^4$ $M_{\odot}$. A similar value was derived for the Quintuplet cluster assuming a Salpeter IMF, showing that both clusters are among the most massive ones known in the Galaxy.

Concerning the central cluster, evolved massive stars were discovered by Allen et al. (1), Forrest et al. (7) and Krabbe et al. (10). In particular, the so-called "AF" star showed very strong emission lines from H and He in the K band, revealing a post main sequence status. Other such stars were soon discovered (10) and were nicknamed "Helium stars" since at that time stars with such strong emission lines were observed only in the Galactic Center and could not be related to other objects. A detailed analysis with atmosphere models by Najarro et el. (15) and Najarro et al. (16) confirmed that these objects are evolved massive stars related to Wolf-Rayet and LBV stars. From the global properties of this population of massive stars, Krabbe et al. (11) showed that the hydrogen ionizing flux emitted could explain the ionization of the local interstellar medium, but certainly not the helium ionization. It was then inferred that a population of normal OB stars with much harder spectral energy distributions should be present, but was too faint to be observed. It is this population that we have just discovered (17) and on which we concentrate now.

## 2 Spectroscopy of the OB stars in the Central Cluster

Spectroscopic imaging of the central cluster was performed in August 2004 with the new integral field spectrograph SINFONI on ESO Very Large Telescope. The excellent sensitivity of the instrument coupled to the use of adaptive optics allowed the discovery of several tens of new massive stars. Whereas all massive stars known so far in this region showed strong emission lines, most of the new ones display absorption in Br$\gamma$ and He I 2.112 $\mu m$. The presence of such absorption lines reveals wind densities smaller than in the previously known "Helium" stars and is typical of OB stars.

In order to better classify these new stars, we have used the recent K band atlas of intermediate resolution spectra of OB stars provided by Hanson et al. (9). Comparing our SINFONI spectra to templates from this atlas we were able to make a detailed spectral classification with an accuracy of $\sim$ one sub spectral type (see Fig. 1). The luminosity class was also derived from spectral morphologies, but additional use of the estimated absolute K band magnitudes was made to refine the classification. In the end, we found that most of the new massive stars are late O – early B supergiants, although a number of giants and even dwarfs are also detected.

## 3 Star Formation History of the Central Cluster

With the discovery of the population of OB stars in the central cluster, we have now probably seen most of its massive star content (although OB dwarfs/giants may still be missing). A study of the global properties of this population is then possible. In particular, the star formation history

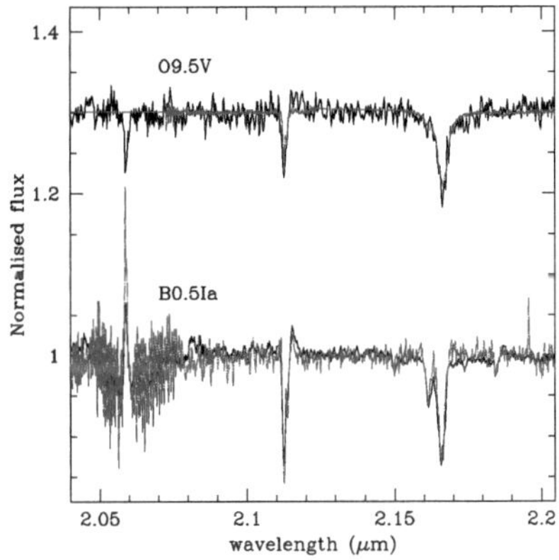


**Fig. 1.** Comparison between the average spectrum (black solid lines) of luminosity class III−V (top) and I (bottom) OB stars recently discovered in the central cluster and the templates from Hanson et al. (9) giving the best match (dot−dashed line). Such a comparison applied to individual spectra allows an accurate spectral classification.

can be constrained from the inspection of the ratios of different types of massive stars. Indeed, massive stars initially on the main sequence evolve to the Wolf-Rayet state and this evolution is faster for the most massive stars. Hence, the ratio of Wolf-Rayet stars to OB stars is an age indicator (e.g. 13). We have thus computed synthesis population models for different assumptions regarding the star formation history and the IMF. We have compared the observed ratios of different types of massive stars to the predicted numbers and results are shown in Fig. 2. A burst of star formation $\approx$ 6 Myrs ago best explains the observations.

An independent estimate of the age can be made from the position of the OB stars in the HR diagram. This is shown in Fig. 2 where the effective temperatures have been derived from the spectral types using the $T_{eff}$ scales of Martins et al. (12) and Schmidt-Kaler (19), while the K band absolute magnitudes were estimated from the observed photometry, the distance to the Galactic Center of Eisenhauer et al. (3) and the extinction of Scoville et al. (20). The bulk of OB stars lies between the 4 and 8 Myr isochrones, confirming the above result.

It is now clear that the center of the Galaxy experienced a star formation event recently. Preliminary results of a detailed study of a "deep field" observed with SINFONI $\sim 20''$ north of SgrA* seems to indicate an older ($\approx$ 500 Myr) population of stars, possibly revealing a previous episode of star formation.

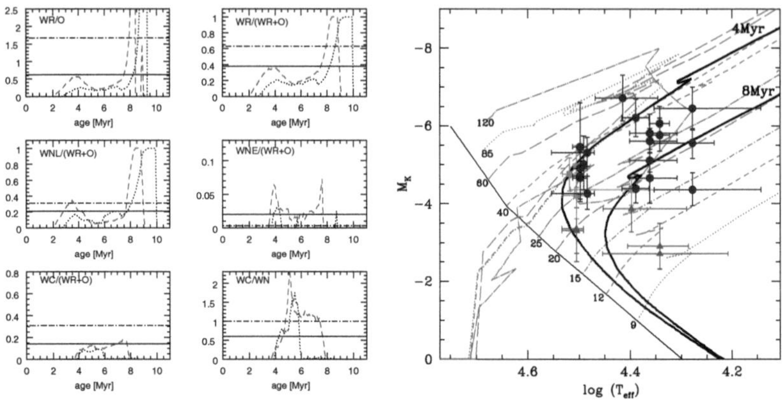


**Fig. 2.** *Left:*Ratios of different types of massive stars as a function of age for a single stellar population at solar metallicity (dotted line) and twice solar metallicity (dashed line). The two horizontal lines show the values observed in the two stellar disks of the central parsec of the Galaxy. On average, the observed population is consistent with an age of ≈ 6 Myrs. *Right:* Qualitative HR diagram of the population of OB stars recently discovered by SINFONI. An age of 4-8 Myrs can be estimated, as revealed by the corresponding isochrones.

# 4 Two Particular Stars: GCIRS16SW and GCIRS34W

New observations of the Galactic Center reveal little by little the population of the central cluster, but they also improve our knowledge of already known stars, especially the so-called "Helium stars". Among them, a few are thought to be related to LBVs. GCIRS34W is an example and was monitored photometrically and spectroscopically for the last few years. Irregular variations in photometry have been observed, but spectroscopy showed almost no variation in the shape and strength of lines observed in K band. This is surprising since LBV-like objects should show both photometric and spectroscopic variability. The most likely explanation for this behaviour is the formation of dust, probably from material previously ejected by the star itself (21). The presence of dust is indeed another characteristic of LBVs. Overall, GCIRS34W is thus probably a transition object, possibly observed after it experienced an LBV type ejection of matter.

Another star also sometimes classified as an LBV candidate is GCIRS16SW. This star is known to be a short period photometric variable (in K band), which can be explained by either a binary or by pulsations. Recent spectroscopic monitoring with SINFONI has shown that radial velocities were variable with a period twice as large as the photometric period. Such a behaviour can only be explained if the star is a massive binary with two components of similar K band magnitude. Work is under way to derive the properties of both components and will likely put constraints on the evolutionary status of these LBV candidates.

*Acknowledgments:*
FM acknowledges support from the Alexander von Humboldt foundation.

## References

[1] Allen, D. A., Hyland, A. R. & Hillier, D. J.: MNRAS **244**, 706 (1990)
[2] Cotera, A. S, Erickson, E. F., Colgan et al: ApJ **461**, 750 (1996)
[3] Eisenhauer, F., Genzel, R., Alexander, T. et al: ApJ **628**, 246 (2005)
[4] Figer, D. F., McLean, I. S. & Morris, M.: ApJ **514**, 202 (1999a)
[5] Figer, D. F., Kim, S. S., Morris, M. et al: ApJ **525**, 750 (1999b)
[6] Figer, D. F., Najarro, F., Gilmore, D. et al: ApJ **581**, 258 (2002)
[7] Forrest, W. J., Shure, M. A., Pipher, J. L. et al: Brackett Alpha Images. In: *The Galactic Center*, ed by D.C. Backer (AIR COnf. Proc. 155, 1987) p. 153
[8] Glass, I. S., Catchpole, R. M. & Whitelock, P. A.: MNRAS **227**, 373 (1987)
[9] Hanson, M.M., Kudritzki, R.P., Kenworthy, M.A. et al: ApJS **161**, 154 (2005)
[10] Krabbe, A., Genzel, R., Drapatz, S. et al: ApJ **382**, 19 (1991)
[11] Krabbe, A., Genzel, R., Eckart, A. et al: ApJL **447**, L55 (1995)
[12] Martins, F., Schaerer, D., Hillier, D.J.: A&A **436**, 1049 (2005)
[13] Mas-Hesse, J.M. & Kunth, D.: A&A Suppl. **88**, 399 (1991)
[14] Nagata, T., Woodward, C. E., Shure, M. et al: ApJ **351**, 83 (1990)
[15] Najarro, F., Hillier, D. J., Kudritzki, R. P. et al: A&A **285**, 573 (1994)
[16] Najarro, F., Krabbe, A., Genzel, R. et al: A&A **325**, 700 (1997)
[17] Paumard, T., Genzel, R., Martins, F. et al: ApJ submitted (2006)
[18] Okuda, H., Shibai, H., Nakagawa, T. et al: ApJ **351**, 89 (1990)
[19] Schmidt-Kaler, T. In Landolt-Börnstein, New Series Group, VI, vol. 2, ed by K. Schaifers & H.H. Voigt (Berlin: Springer-Verlag, 1982) p. 1
[20] Scoville, N.Z., Stolovy, S.R., Rieke, M. et al: ApJ **594**, 294 (2003)
[21] Trippe, S., Martins, F., Ott., T. et al.: A&A, in press (2006)

# Stellar Populations in the Centers of Nearby Disk Galaxies

O. K. Sil'chenko

Sternberg Astronomical Institute, Moscow, Russia olga@sai.msu.su

**Summary.** The main results on the stellar population properties of nearby disk galaxies obtained with the Multi-Pupil Fiber/Field Spectrograph (MPFS) of the 6m telescope during the last ten years are reviewed.

The results which I present are obtained with the Multi-Pupil Fiber/Field Spectrograph (MPFS) of the 6m telescope. The integral-field unit of the 6m telescope of the Special Astrophysical Observatory of RAS, the MPFS, had started its work in 1989 (1), thus being one of the first integral-field spectrographs in Europe, together with TIGER (4) and HEXAFLEX (7). During these 16 years several modifications of the MPFS have been made. Now we have a field of view of 16 arcsecond and spectral range of 2000 elements, and of 1500 – 6000 Å with various gratings. Three main topics concern the following results obtained with the MPFS during the last ten years:

- chemically decoupled galactic stellar nuclei, or cores;
- statistics of the mean stellar ages and metallicities over the sample of nearby lenticular galaxies;
- inner polar gas in lenticular and spiral galaxies – in normal spiral galaxies it was found by us for the first time (9).

## 1 Chemically Decoupled Nuclei

Chemically decoupled stellar nuclei, or cores, present a sharp increase of the metal-line equivalent widths in the very centers of galaxies (8). They are seen best of all when mapping the MgbI$\lambda$5175 equivalent width which is usually expressed by the corresponding Lick index Mgb (15). They can be star-like compact nuclei or spatially resolved elongated structures. Often they are also distinguished kinematically. The chemically decoupled nuclei have nothing in common with the well-known metallicity gradients in stellar spheroids which are very shallow, not more than 0.2–0.3 dex per radius dex, whereas in the case of chemically decoupled unresolved nuclei, the metallicity rises by a factor two and more over a radial distance equal to one spatial element extension.

Extended chemically decoupled cores may represent quite various spatial configurations. They may be circumnuclear stellar disks. One of the best examples of the chemically decoupled circumnuclear disks seen edge-on is NGC 3623 (3). Here the Magnesium index shows an elongated maximum, and the iron indices have compact peaks in the center. The fast rotation along the major axis and the minimum of the stellar velocity dispersion are the kinematical evidence for the disk nature of the decoupled core. But the resolved elongated chemically decoupled cores are not always disks. They may also be compact triaxial structures. We can mention NGC 1023 as an example of the triaxial chemically decoupled cores (11). In this galaxy we see also an elongated area of enhanced Magnesium index and a compact iron-index peak. But this time isovelocities turn in the very center by some $30^\circ - 40^\circ$, and the stellar velocity dispersion demonstrates a resolved maximum aligned with the kinematical major axis. We have compared the orientations of the photometric and kinematical major axes in NGC 1023. The latter is obtained by applying a tilted-ring method to the stellar velocity field of NGC 1023 derived from the SAURON data. In the center, the photometric and kinematical major axes deviate in opposite senses with respect to the line of nodes, implying rotation within a triaxial potential.

## 2 Stellar Populations in the Centers of S0s

Nearby lenticular galaxies are studied by us by using a rather representative sample. We have retrieved a list of bright, northern, nearby lenticulars from the HYPERLEDA. It includes 122 objects without strong AGN and star formation bursts, among them 40 are Virgo members. We have taken half of the rest, have added 8 Virgo members and a few more faint or more distant, so more luminous, galaxies. A total of 58 nearby lenticulars uniformly distributed over four types of environments is analysed (12). We consider separately the star-like unresolved nuclei and the bulges taken as the rings between $4''$ and $7''$ from the centers. Due to the benefits of 3D spectroscopy, the S/N ratios of the nuclear and off-nuclear spectra are comparable despite the difference in surface brightness. The Lick indices H$\beta$, Mgb, Fe5270, and Fe5335 are calculated for the nuclei and for the bulges. By comparing the measurements for the nuclei with the classic aperture data of the Lick team (13) for 27 common galaxies, we have ensured that our index system is properly calibrated onto the Lick standard one, so we can determine the stellar population properties by confronting our data to evolutionary synthesis models. The Mg/Fe ratios of the bulges are mostly between 0 and +0.3. One interesting thing is an obvious difference between the mean Mg/Fe ratio of the brightest group galaxies and of the second-rank group members. This difference still persists if we take galaxies with the same masses/stellar velocity dispersion. According to their abundance ratio distribution, the central group galaxies resemble cluster galaxies and the second-rank group members resemble field galaxies.

In the diagram H$\beta$ versus [MgFe] we separate age and metallicity effects and determine both parameters. The nuclei look younger than the bulges, and the lenticulars in clusters and group centers are older than those in the field and group non-central areas. This impression is confirmed by the cumulative age distributions. We have united the cluster galaxies and the group central galaxies into the subsample of a dense environment, and the field galaxies and the second-ranked group members into the subsample of a sparse environment. There exists a well-known correlation between the age and stellar velocity dispersion for early-type galaxies (5). Since our dense environment subsample contain the galaxies more massive on average than our sparse environment subsample, we check if the age difference reported above is a true environment effect. The low-mass lenticulars appear to show identical age vs $\sigma_*$ trends in any type of environments; but above $\sigma_* = 170$ km/s the dense and sparse environment galaxies separate, with the dense environment galaxies looking older.

## 3 Inner Gas Polar Rings/Disks

A spectacular phenomenon which we study with the MPFS is inner gas polar rings/disks in lenticular and spiral galaxies. As for the lenticulars, after a few occasional findings we have selected targets where the inner polar rings are seen 'by eye' in optical-band HST images as dust lanes perpendicular to the isophote major axes. Gas has to be coupled to dust. We have obtained the gas and star velocity fields with the MPFS and indeed, the kinematical major axes of the ionized gas and of the stars seem to be nearly orthogonal (10). By searching for common and perhaps unusual properties of the lenticulars with the inner polar gas, we have found that

1. many of them have large bars;
2. the majority are the central group galaxies or are in clusters, so belonging to the 'dense environment' subsample;
3. practically all of them are detected in the HI 21 cm line which is rather unusual for lenticular galaxies.

The two latter properties seem to favour a hypothesis of polar gas accretion from some other galaxy. However, when the large-scale maps of neutral hydrogen distribution and velocities are available, they demonstrate decoupling between the inner ionized gas and outer neutral gas. The outer gas is usually confined to the planes close to the main symmetry planes of the global stellar disks. Taking into account the regular rotation of the outer gas in the disk galaxies with inner polar rings, we can propose two possible scenarii. We may be dealing with a consequence of *exactly* vertical central infall of gas, which seems improbable due to the large number of galaxies with inner polar rings. Or we must admit that outer planar gas is pushed onto polar orbits when drifting to the center by some dynamical mechanism. There exist theoretical considerations for both these scenarii. Van Albada et al. (14) considered accretion of external gas, with an arbitrary orientation of the spin, onto a tumbling triaxial potential. In the center of the galaxy the accreted gas settled into the plane orthogonal to the biggest axis of the ellipsoid, so forming a polar ring; but the outer parts of the gas distribution warped by up to 90° so that the outer gas appeared to counter-rotate with the stellar triaxial body. On the contrary Friedli and Benz (6) started from the presence of counter-rotating gas in the global disk; the disk being unstable formed a bar, the gas then lost its momentum in the bar and drifted to the center. But near the center the counter-rotating gas left the plane of the global disk and settled on highly inclined orbits, demonstrating stable rotation in the nearly polar plane. Both theories require the presence of counter-rotating gas beyond the radius of polar rings. Recently I (11) have found some observational signatures satisfying the theoretical expectations. We reported the inner polar ring in NGC 7280 in 2000 (2) by having observed it with the MPFS in 1998. Later this galaxy has been observed with the SAURON. By appealing to the larger field of view of SAURON, we have seen that the gas counter-rotates compared to the stars at radii larger than $7''$, whereas inside $4''$ it is polar. The same situation is observed in NGC 7332. Both galaxies have bars.

## 4 Summary

Let us consider once more the age distribution of the nuclei of the lenticular galaxies, this time without separation according to the environment type (Fig. 1). I have taken chemically decoupled nuclei that are more metal rich than the bulges by two times and more. They involve 42% of the sample. Whereas all S0 nuclei have rather a flat age distribution up to 8 Gyr, the chemically decoupled nuclei demonstrate a clear maximum at 3 Gyr, so they are on average younger than the other nuclei. The nuclei of the lenticulars with the inner polar rings have two maxima of the age distribution. The 'younger' peak includes all 5 galaxies with the inner polar gas *and* the chemically decoupled nuclei. The nuclei of the lenticulars with the inner polar gas which *are not* chemically decoupled are all very old. This is consistent with the prediction of dynamical models that polar gas orbits which are perpendicular to the major axis of a triaxial potential are very stable, so the polar gas has little chance to fall into the very center.

I am grateful to the astronomers of the Special Astrophysical Observatory of RAS V.L. Afanasiev, A.N. Burenkov, V.V.Vlasyuk, S.N. Dodonov, and A.V. Moiseev for supporting the MPFS observations at the 6m telescope. The 6m telescope is operated under the financial support of Science Ministry of Russia (registration number 01-43); I thank also the Programme Committee of the 6m telescope for allocating the observational time. During the data analysis we have used the Lyon-Meudon Extragalactic Database (LEDA) supplied by the LEDA team at the CRAL-Observatoire de Lyon (France). This research is partly based on data taken from the ING Archive of the UK Astronomy Data Centre and on observations made with the NASA/ESA Hubble Space Telescope, obtained from the data archive at the Space Telescope Science Institute, which is operated by the Association of Universities for Research in Astronomy, Inc., under NASA contract NAS 5-26555. The study of galaxies in the nearby galaxy groups is supported by the grant of the Russian Foundation for Basic Researches 04-02-16087.

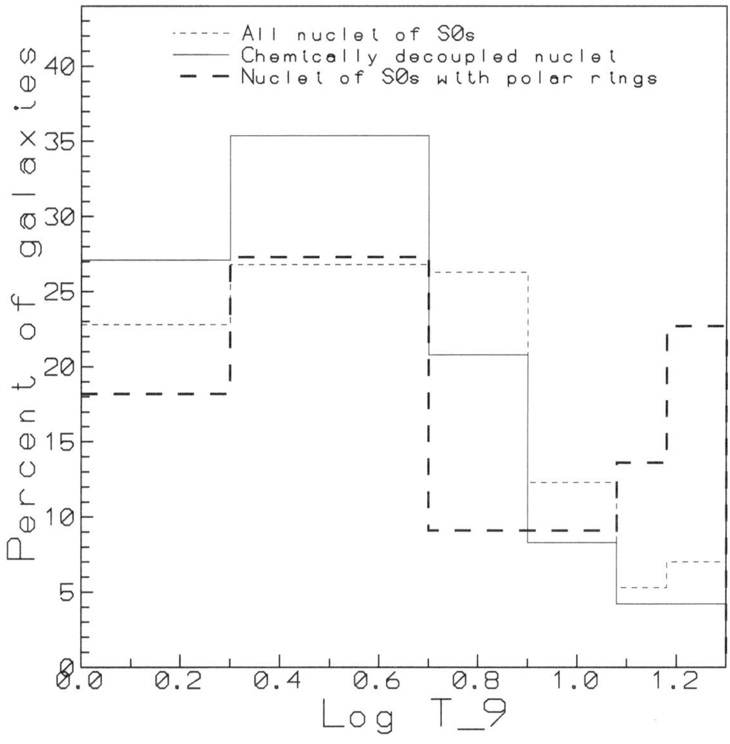


**Fig. 1.** The stellar population age distribution for the nuclei of the nearby lenticulars is shown for all galaxies, for galaxies with chemically decoupled nuclei, and for galaxies with inner gas polar rings.

## References

[1] V.L. Afanasiev, V.V. Vlasyuk, S.N. Dodonov, O.K. Sil'chenko: Preprint SAO RAS **N54**, 1 (1990)
[2] V.L. Afanasiev, O.K. Sil'chenko: Astron. J. **119**, 126 (2000)
[3] V.L. Afanasiev, O.K. Sil'chenko: Astron. Astrophys. **429**, 825 (2005)
[4] R. Bacon, G. Adam, A. Baranne, G. Courtes, D.Bubet et al: Astron. Astrophys. Suppl. Ser. **113**, 347 (1995)
[5] N. Caldwell, J.A. Rose, K.D. Concannon: Astron. J. **125**, 2891 (2003)
[6] D. Friedli, W. Benz: Astron. Astrophys. **268**, 65 (1993)
[7] J.L. Rasilla, S. Arribas, E. Mediavilla, J.L. Sebastian: Astrophys. and Space Sci. **171**, 301 (1990)
[8] O.K. Sil'chenko, V.L. Afanasiev, V.V. Vlasyuk: Astron. Zh. **69**, 1121 (1992)
[9] O.K. Sil'chenko, V.V. Vlasyuk, A.N. Burenkov: Astron. Astrophys. **326**, 941 (1997)
[10] O.K. Sil'chenko, V.L. Afanasiev: Astron. J. **127**, 2641 (2004)
[11] O.K. Sil'chenko: Astronomy Letters **31**, 227 (2005)
[12] O.K. Sil'chenko: Astrophys. J. **641**, in press (2006)
[13] S.C. Trager, G. Worthey, S.M. Faber, D. Burstein, J.J. González: Astrophys. J. Suppl. Ser. **116**, 1 (1998)
[14] T.S. Van Albada, C.G. Kotanyi, M. Schwarzschild: MNRAS **198**, 303 (1982)
[15] G. Worthey, D.L. Ottaviani: Astrophys. J. Suppl. Ser. **111**, 377 (1997)

# AGN Shocks in 3D

G. Cccil

University of North Carolina, Chapel Hill NC 27599-3255, USA

**Summary.** I review observable manifestations of various AGN shocks, discuss some key past 3D results, and outline some recent improvements in the computational framework within which spectra are interpreted. New models of dusty clouds broaden the scope of photoionization by providing a self-consistent basis for the successful matter-bounded/ionization-bounded model sequences, and can undermine the case for shocks in some AGN. However, we are only beginning to explore computationally the time evolution of shocks in 3D. The cyclical nature of high-velocity shock compression may provide a natural time scale for some AGN phenomena. It would be premature to dismiss a rôle for shocks in datasets that encompass only a few strong emission-lines. We need comprehensive IFU spectra.

## 1 Detecting Shocks

Taking an X-ray to I-band perspective, I focus on the smallest radius within an AGN where high velocities can be resolved spatially — the narrow-line region (NLR). NLR gas often moves virially within the kpc-scale bulge (28), but is it in hydrostatic equilibrium? A coincident radio jet often traces the outflow, and occasionally ionized gas is associated that is moving far faster than the virial value. Is this gas shock excited or has it simply been concentrated for AGN photoionization? Even a weak AGN can illuminate its host (38, for example). Reference (23) notes that shocks carry negligible energy in a luminous AGN: non-gravitational release would need a huge mass flow. But are shocks viable for fainter structures (e.g. M51 (5)) more than an arcsecond from the AGN?

Shocks were first posited to solve the [O III] "temperature problem" in AGN: plasma temperatures derived from various emission-line diagnostics exceed significantly the ionization level of the gas inferred from the same spectra. Two solutions were posited: a distribution of photoionized clouds differing in pressure (at different radii from the AGN), and hot, cooling gas behind shock fronts. Only the rare optical spectrum at pre-HST resolution could distinguish between the two. Studies have used few of the possible panchromatic spectral diagnostics because the flow emits from diverse regions of uncertain geometry with numerous shocks/counter-shocks at different velocity and uncertain pressure, each of which may be running into non-stationary gas of uncertain density that most instruments can't resolve spatially!

NLR gas dynamics are poorly constrained because we have few limits on relaxation time scales. It *is* a high-pressure zone with several gas phases from diffuse to denser clouds that seek pressure equilibrium. Early computation of fast astrophysical shocks (3) highlighted the flat, power-law spectrum of their post-shock recombination zone (Fig. 1); detailed micro-physics of gas cooling is included in these single cloud, 1D models.

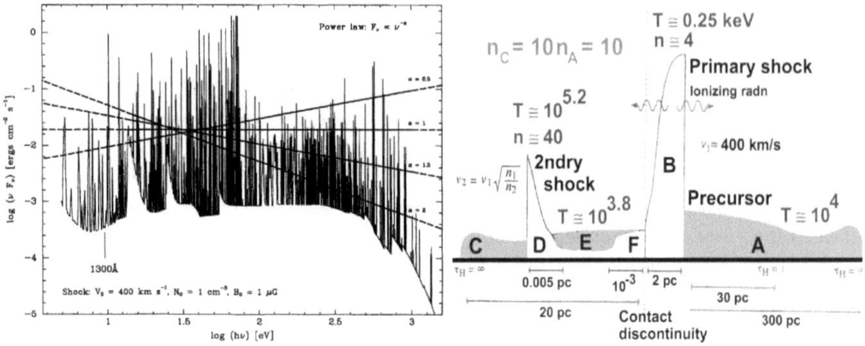


**Fig. 1.** (**a**) 400 km s$^{-1}$ shock compared to flat power-laws popular in AGN (26); (**b**) Typical parameters for a high-velocity cloud/cloud collision (3)

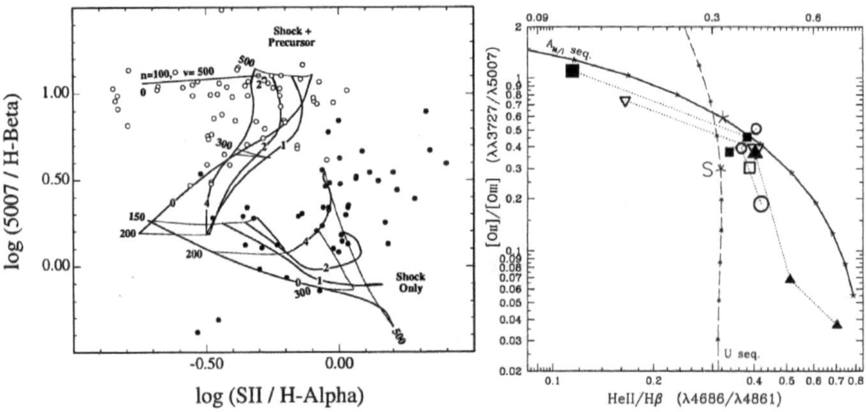


**Fig. 2.** Visible-band emission-line ratios from (**a**) shock grids, with/without radiative precursor (15), and (**b**) from the photoionized $A_{M/I}$ sequence (solid lines) and optically thick clouds (dashed) (4)

With more computer power, sophisticated physics is retained while adding a snapshot of the cloud's hydrodynamical evolution. Models with detailed local cooling omit hydrodynamics entirely, merely "painting" the fluxes of planar shocks of various velocities onto a static curved surface that matches the de-projected, observed shape (1.5D treatment). These models reproduced the spectra of Herbig-Haro bow-shocks and, in a couple of cases, of extragalactic bow shocks (11). Models generate grids of line ratios from the FIR to X-ray (14; 1) that diagnose gaseous ionization and temperature. But, such steady flows omit much time-dependent phenomena, and are unphysical. They are parameterized by abundances (without dust destruction or re-emission), and the presence/absence of a precursor that is irradiated upstream of the shock by hot, post-shock downstream gas (Figs. 2 & 1). 3D simulations are underway (40), and the complexities evident in recent 2D slab simulations (34) emphasize the need for high resolution to follow the extremely dynamic flow.

Although UV emission-line fluxes are most sensitive to shock conditions (1), dust absorption is a binary switch. So, few NLR's were mapped with UV spectra before HST/STIS died, and our only potential UV tool — HST's Cosmic Origins Spectrograph — can work only on unresolved sources.

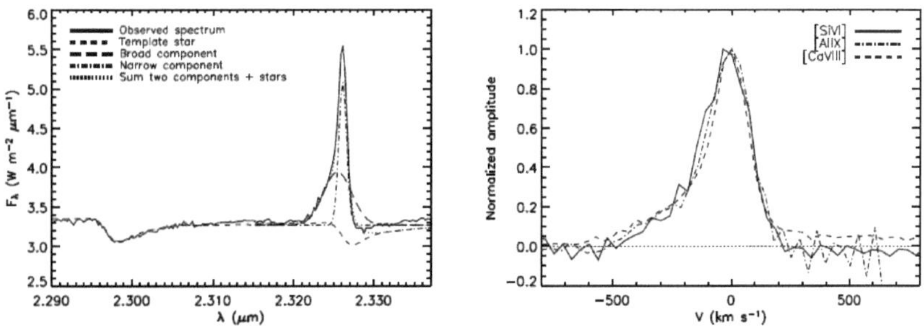


**Fig. 3.** SINFONI nuclear spectrum of Circinus, showing various high ionization lines with strongly asymmetric blue wings (27)

Photoionization models were refined with nested clouds that differ principally in pressure. Later, reference (4) found that their "$A_{M/I}$" combinations of matter and ionization bounded clouds take a different track on diagnostic diagrams from that of a single cloud sequence of different ionization parameters ($U$), e.g. Fig. 2b. Stimulated by the presence of high-excitation "coronal" lines in optical/NIR spectra of many NLR's (e.g. (27) Circinus, Fig. 3, and in NGC 1068) and especially by our (12) STIS spectra of NGC 1068 (Fig. 7), references (13; 19) found that dusty clouds explain both the observed narrow range of $U$ and elevated [O III] temperature of NLR's. Fig. 4a shows a dusty cloud surrounded by diffuse ablata that emits the coronal lines. The *apparent low U* comes from the isobaric cloud core. Dust also enhances gas/radiation coupling, allowing acceleration to the high observed velocities.

Most studies have been of gas-rich galaxies, where we can trace putative shocks by their X-ray to IR line emission. Shocks have been studied most effectively in the visible with both Fabry-Perot (FPS) and integral-field

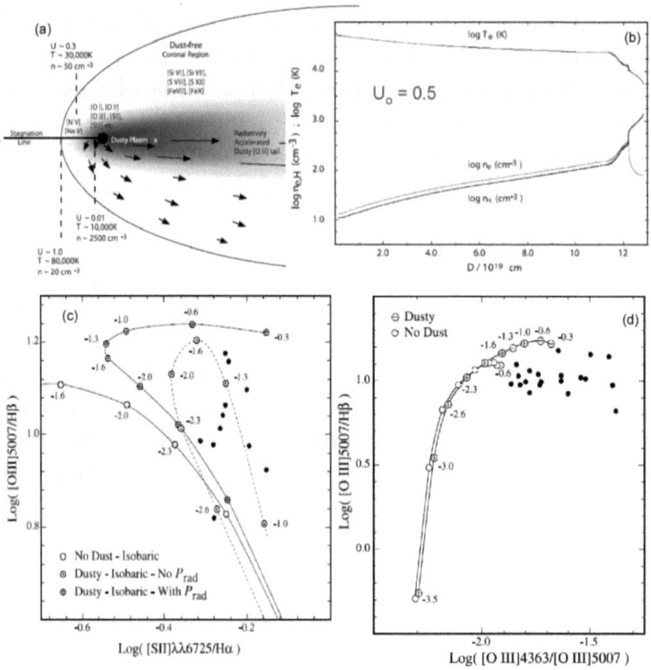

**Fig. 4.** (**a**) Cartoon of a photo ablating, dusty cloud. The average gas $U$, temperature, and density are shown; (**b**) the internal structure of the emitting cloud; (**c**) photoionization models with/without dust. Reference (13) shows how line ratios vary for different $U$ on the cloud. The dots plot spatially integrated spectra of various AGN. (**d**) The X-axis is an emission-line flux ratio that is sensitive to gas temperature. Note how the dust-free cloud stagnates along the X-axis as $U$ increases. In contrast, the dusty model continues to increase its $O^{2+}$ temperature with $U$

(IFUS) 3D spectrometers. An IFUS has no positional slit bias, maps much faster and more reliably than stepped slits, and can map more spectral diagnostics than an FPS. The downside of IFU's — their small FOV compared to FPS/tunable filters — is more than compensated when several diagnostic lines are studied. In practice, separation of composite regions within the target along the sight line needs a large spectral-R, requiring multiple grating angles, atmospheric dispersion correction, and photometric/PSF stability to merge "postage stamp" data cubes. Such studies were not undertaken with early, small-format IFU's.

Shocks >150 km s$^{-1}$ emit soft X-rays ($T_s \sim 10^5 v_s/100$ km s$^{-1}$ K). Chandra has provided a pivotal result on the NLR of NGC 1068: reference (20) found no warm gas. The X-ray "plume" in this NLR coincides with little high-velocity optical/UV gas (12). Evidently, this NLR is photoionized, so shocks *do not* accelerate its gas clouds.

In contrast, in M51 (Fig. 5a) Chandra ACIS spectra (37) show a composite AGN+thermal spectrum on the AGN but only warm $\sim 6 \times 10^6$ K gas in the extra-nuclear N ring and S cloud several arcseconds outside. STIS optical spectra (5) (Fig. 5b) imply $\sim (675, 450)$ km s$^{-1}$ shocks on these, respectively. Moreover, velocities at the edge-brightened cloud from my ancient FP data cube (8) de-project to $\sim 500$ km s$^{-1}$ with respect to to galaxy systemic. The higher X ray temperature may indicate heating by cosmic rays (8), boosting [N II] fluxes that would otherwise require (5) enhanced nitrogen abundance. Motions and X-ray spectra are fairly consistent despite the crude de-projection geometry and use of a single temperature MEKAL model.

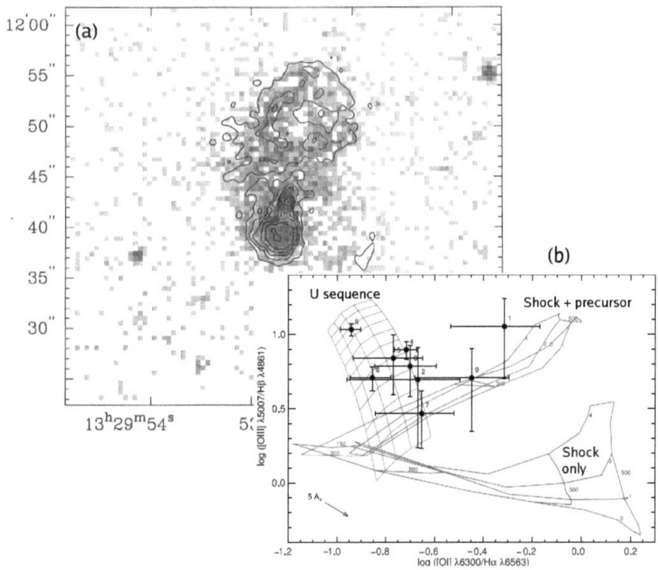


**Fig. 5.** (**a**) A VLA 20 cm map (17) contoured atop (37)'s Chandra image in the 0.5-8 keV band. The XNC is the bright region $\sim$ 3 arc-seconds S of the nucleus. The X-ray spectrum of the nucleus is fit with a power-law+thermal composite, but the XNC has only a thermal spectrum; (**b**) STIS spectra of the northern ring/XNC (points 9/1) and the nucleus (4-6), compared to photoionization and shock grids (5)

Radio synchrotron traces a flow even without entrained line emitting gas, but with the usual ambiguity of electron density/B-strength and topology that can sometimes be disentangled with polarization maps. Only in Galactic objects are synchrotron time scales short enough to be interesting. But, in AGN, such regions are often depolarized by thermal gas shells along our sight line, undermining polarization constraints. NGC 1068 is an interesting counter-example (44), where gas densities are low enough this far from the nucleus for polarization vectors to trace compression of the magnetized ambient ISM, confirming a bow shock.

## 2 Shocks From Gravitational Pressure Gradients

Shocked gas could be a iniquitous, diffuse mist of emitting filaments. Surveys (28, for example) find [O III] line profiles with broad, blue shifted wings on a narrow core. This is an ambiguous signature; is this gas even shocked?

Seyferts are more barred than non-Seyferts at the $2.5\sigma$ level (21). Bar shocks could extend to small radii in the "bars within bars" L-dumping scenario, even if they do not reach the fueling scale (16). Nuclear bars are not yet resolvable nearby but will be targets for high spatial-resolution optical spectra from laser-AO, IFU equipped 4m telescopes (WHT OASIS, SOAR SAM+IFU).

Large-scale shocks clump enough to be mapped by Chandra/XMM (32; 31; 43). IFU-scale examples are discussed in this volume. Potential confusions are photoionization from localized star formation, and spectral discontinuities from projection of tidal arms/filaments along our sight line.

## 3 Collimated Jet/ISM Collisions

These are spectacular in powerful radio galaxies (see e.g. van Bruegel's contribution). In an early study, reference (25) used TIGER/CFHT to map 3C 171. They found two components $2.''7$ from the nucleus that span $3''$ with $\Delta v \sim 600$ km s$^{-1}$ and with large [O III]/H$\beta$ flux ratios. Reference (43) used ARGUS/CFHT to find 2 nuclear components in hyper-luminous IRAS F20460+1925 separated by 1000 km s$^{-1}$. Here and in IRAS F23060+0505 they infer very fast shocks from [O III]/H$\beta$ vs [O II]/[O III] ratios.

Recent larger IFU's increase wavelength coverage: reference (42) finds interactions $6''$ outside the cores of Coma A, 3C171, and 3C265. Interactions are discovered in deep tunable filter images, then detailed over $\lambda$300 nm coverage at $\lambda$0.5 nm resolution with WHT's INTEGRAL IFUS. Broadest lines extend perpendicular to the radio axis, suggesting a cocoon. Jet/ISM interactions are feasible energetically, but the velocity field is undisturbed. Interactions are found in other distant systems: PKS 2250 (41), three CSS sources (22), and a SCUBA galaxy at $z = 2.385$ (35) are good examples.

In nearby galaxies, the X-ray, radio, and optical emitting filaments of jet/ISM interactions sometimes do not coincide. Displacements may correlate with kinematical, flux, and excitation discontinuities, and with bends and spectral steepening in the radio jet (18). Post-shock cooling may be responsible. A 3D spectrometer with high spectral R can separate and map any unrelated material that is projected along our sight line.

Clear examples are NGC 1068 (30; 2; 18), Mkn 3 (7) (see Fig. 6), and the jet termini of NGC 4258 (11) and M51 (see previous section and Fig. 5). Oosterloo and Morganti (29) find an ionized trail from an H I cloud at least projected along Cen A's jet. Here the kinematic evidence of a jet/ISM interaction in H I is only a 100 km $s^{-1}$ gradient over 1 kpc. But, X-rays mapped by Chandra also coincide with [O III] emission. These authors find very large velocity gradients in optical emission and H I absorption along the jet in IC 5063.

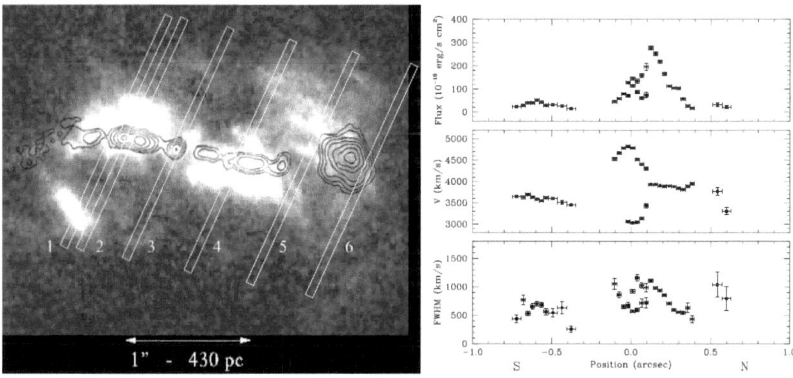


**Fig. 6.** (**a**) FOS long-slit positions across the NLR of Mkn 3 (6; 7); (**b**) split [O III] profiles across the jet cocoon at POS 2

The cocoons of 3C sources are poorly resolved (several kpc per arcsecond). Seyferts get us close to the action of a less powerful jet and, usually, lower gas velocities. An expanding cocoon may trigger a low-z analog of the jet induced star formation seen in radio galaxies: 8 and 15 kpc NE of the nucleus along the radio/X-ray jet of Cen A are young (15 Myr), hot stars. Jet bending and associated split emission-profiles show a larger surrounding shell (200 vs 15 pc jet-width) within the NLRs of NGC 1068 and Mkn 3 (2; 6; 7). From bubble size and observed velocities, these authors estimate an upper age of the radio sources of $< 0.15$ Myr and a lower limit on jet power. The KE of the fast gas agrees with the jet power integrated over its lifetime. There are complications:

- The vertical pressure gradient guides break-out on the "top" side, while the flow stagnates on the "lower disk" side (6; 7). This leads to different shock velocities and (re)entry of the expanding gas into the ionization cones.
- Overpressure can evaporate the cloudier ISM, allowing the nuclear ionization cone to torch remaining gas.

# 4 Cloud Ablation Shocks

Wind/jet acceleration of denser gas depends on an ISM cloud's "damage control" after it wanders into the outflow. Early simulations smacked the cloud with a SNR/ISM blast-wave; the cloud is crushed into catastrophic cooling that soon shreds it. However, if the cloud is exposed to the hard radiation field of the AGN's wide-angle ionization cone before entering the wind, the pressure discontinuity can ameliorate. As the cloud orbits into the wind, a stand-off bow shock builds upstream in a region whose density may already be enhanced by photo-ablated gas. The gas post-shock is initially subsonic, perhaps becoming transonic as more gas sheds by hydrodynamical instabilities that build along the cloud surface and penetrate to the sonic distance. Reference (33) shows that this layer then detaches in a distinctive "fire polish" event that dumps the cloud's entropy into the flow, resetting the instability clock. Dusty ablata continue to couple to the AGN ionizing radiation and any nuclear wind. Because the ablata are so light and are less self-shielded than the bulk cloud, they can be pushed to high velocities. Thus we explain the gas "knots" in NGC 1068's NLR; Fig. 7 shows that in $< 15$ pc they are boosted by 1500 km $s^{-1}$, we argue by radiation pressure not shocks. The knots resemble in their kinematics (12) and ionization (19) the "associated absorbers" seen in some quasar spectra.

Early axis-symmetric simulations spuriously stabilized a cloud along the outflow axis, producing e.g. the magnetized "nose-cone" of early MHD jet models. 3D simulations at high resolution are underway on dedicated computer clusters. But, 2D slab simulations are important, e.g. reference (24) extends (33) to show how thermal conduction damps gas instabilities to pre-

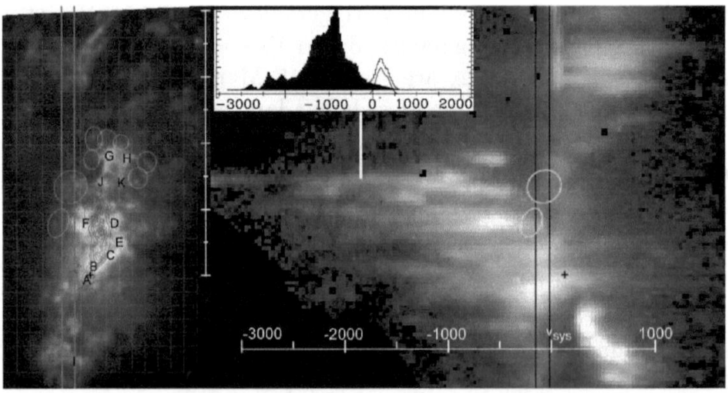


**Fig. 7.** (**a**) HST/FOC [O III] image of NGC 1068's NLR (12); vertical ticks every 0.″5; (**b**) An example STIS [O III] profile extracted along the vertical slice shown in (**a**); horizontal velocities with respect to galaxy systemic (12). The insert plots the profile of a very high-velocity knot

serve a cloud during acceleration. O VI and soft X-ray images are predicted, and the latter may be feasible in long Chandra exposures.

## 5 Shock Systems Associated With Wider-Angle Winds

1. The accretion shock near the galaxy disk plane. Simulations (36) (Fig. 8) show that this shock is needed to maintain an open throat for gas to fuel a steady wind. Do we see it in NGC 3079? (9). Perhaps. But it is subtle among circumnuclear star-formation in the plane, and IFU spectra are required to distinguish it from ionization fronts around hot stars.

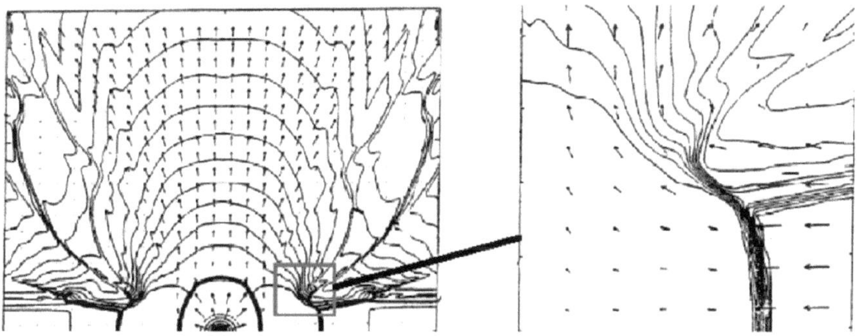

**Fig. 8.** (**a**) (Top row) This 2D axis-symmetric model of galactic wind outflow (36) shows pressure contours and velocity vectors in its steady-state flow. The closeup at right details the "ring" accretion shock near the galaxy disk plane

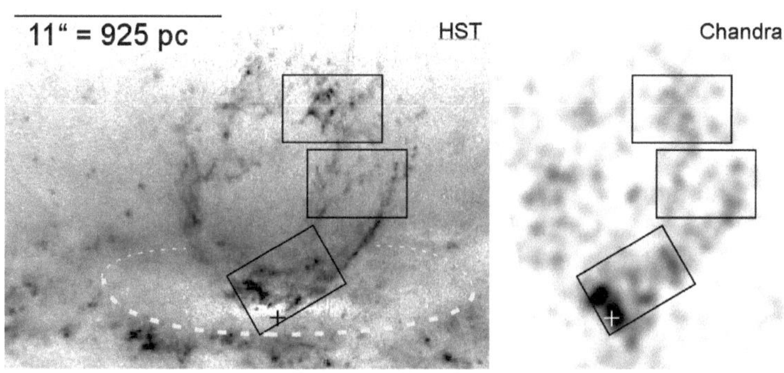


**Fig. 9.** The nuclear bubble in NGC 3079 in (**a**) [N II]+H$\alpha$ w/ HST, and (**b**) (10) w/ Chandra. The accretion ring shock may lie along the dashed line. What is likely a Rayleigh-Taylor instability has formed the disruptive vortex at the top of the bubble. Gas concentrates into 4 vertical towers in the walls. These are often unresolved by HST, tightening the constraint on the important gaseous filling factor (39)

2. The outer shock/contact discontinuity between unshocked ISM and shocked wind. This can be very narrow (9, for example) and faint because rarefied gas above the galaxy disk inhibits precursor emission.
3. The inner shock between the free wind and the bubble of shocked wind can be compact and dust enshrouded, but may be separable with an NIR IFU such as NIFS.

## References


[1] M.G. Allen, M.A. Dopita, Z.I. Tsvetanov: ApJ **493**, 571 (1998)
[2] D.J. Axon et al: ApJ **496**, 75 (1998)
[3] L. Binette,M.A. Dopita, I.R. Tuohy: ApJ **297**, 476 (1985)
[4] L. Binette, A.S. Wilson,T. Storchi-Bergmann: A&A **312**, 365 (1996)
[5] L.D. Bradley, M.E. Kaiser, W.A. Baan: ApJ **603**, 463 (2004)
[6] A. Capetti, D.J. Axon,F.D. Macchetto: ApJ **487**, 560 (1997)
[7] A. Capetti et al: ApJ **516**, 187 (1999)
[8] G. Cecil: ApJ **329**, 38 (1988)
[9] G. Cecil, J.Bland-Hawthorn,S. Veilleux, A.V.R. Filippenko: ApJ **555**, 338 (2001)
[10] G. Cecil, J. Bland-Hawthorn, S., Veilleux: ApJ **576**, 745 (2002)
[11] G. Cecil et al: ApJ **536**, 675 (2000)
[12] G. Cecil et al: ApJ **568**, 627 (2002)
[13] M.A. Dopita, B.A. Groves, R.S.Sutherland, L. Binette, G. Cecil: ApJ **572**, 753 (2002)
[14] M.A. Dopita et al: ApJ **490**, 202 (1997)
[15] M.A. Dopita, R.S. Sutherland: ApJ **455**, 468 (1995)
[16] E. Emsellem: In: *The Interplay among Black Holes, Stars and ISM in Galactic Nuclei, Proceedings of IAU Symposium*, vol 222, ed by T. Storchi-Bergmann, L.C. Ho, H.R. Schmitt. (Cambridge University Press, Cambridge 2004) pp 419–422
[17] H.C. Ford et al: ApJ **293**, 132 (1985)
[18] J. Gallimore et al: ApJ **464**, 198 (1996)
[19] B. Groves, G. Cecil, P. Ferruit, M.A. Dopita: ApJ **611**, 786 (2004)
[20] A. Kinkhabwala et al: ApJ **575**, 732 (2002)
[21] J.H. Knapen et al: In: *Active Galactic Nuclei: from Central Engine to Host Galaxy*, ed by S. Collin, F. Combes, I. Shlosman. Astronomical Society of the Pacific Conference Series, vol 290, p. 419 (2003)
[22] A. Labiano et al: A&A **436**, 493L (2005)
[23] A. Laor: ApJ **496**, L1 (1998)
[24] A. Marcolini et al: MNRAS, **362**, 626 (2005)
[25] I. Márquez, E. Pécontal, F. Durret, P. Petitjean: A&A **361**, 5 (2000)
[26] J.A. Morse, J.C. Raymond, A.S. Wilson: PASP **108**, 426 (1996)
[27] F. Mueller Sánchez, R.I. Davies, F. Eisenhauer, L.J. Tacconi, R. Genzel: astro-ph/0508424 (2005)

[28] C.H. Nelson, M. Whittle: ApJ **465**, 96 (1996)
[29] T.A. Oosterloo, R. Morganti: A&A **429**, 469 (2005)
[30] E. Pécontal, P. Ferruit, L. Binette, A.S. Wilson: Astrophys. & Space Science, **248**, 167 (1997)
[31] S.F. Sánchez et al: astro-ph/0407128 (2004)
[32] S.F. Sánchez et al: astro-ph/0310293 (2003)
[33] A.V.R. Schiano, W.A. Christiansen, J.M. Knerr: ApJ **430**, 237 (1995)
[34] R.S. Sutherland, G.V. Bicknell, M.A. Dopita: ApJ **591**, 238 (2003)
[35] A.M. Swinbank et al: astro-ph/0502096 (2005)
[36] G. Tenorio-Tagle, C. Munoz-Tunon: ApJ **478**, 134 (1997)
[37] Y. Terashima, A.S. Wilson: ApJ, **560**, 139 (2001)
[38] S. Veilleux et al: AJ **126**, 2185 (2003)
[39] S. Veilleux et al: ApJ **433**, 48 (1994)
[40] S. Veilleux,G. Cecil, J. Bland-Hawthorn: Galactic Winds. In: *Annual Review of Astronomy and Astrophysics*, vol 43, ed by R. Blandford, G. Burbidge, J. Kormendy, E. van Dishoeck (Annual Reviews, Palo Alto 2005) pp 769–826
[41] M. Villar-Martin et al: MNRAS **307**, 24 (1999)
[42] M. Villar-Martin, M., C. Tadhunter: New Astronomy Reviews **47**, 249 (2003)
[43] R.J. Wilman, C.S. Crawford, R.G. Abraham: MNRAS, **309**, 299
[44] A.S. Wilson, J.S. Ulvestad: ApJ **319**, 105 (1987)

# Young Kinematically Decoupled Components in Early-Type Galaxies

R. M. McDermid[1], E. Emsellem[2], K. L. Shapiro[3], R. Bacon[2], M. Bureau[4], M. Cappellari[1], R. L. Davies[4], P. T. de Zeeuw[1], J. Falcón-Barroso[1], D. Krajnović[4], H. Kuntschner[5], R. F. Peletier[6] and M. Sarzi[7]

[1] Leiden Observatory, Postbus 9513, 2300 RA Leiden, The Netherlands
mcdermid@strw.leidenuniv.nl
[2] CRAL-Observatoire, 9 Avenue Charles-André, 69230 Saint-Genis-Laval, France
[3] UC Berkeley Department of Astronomy, Berkeley, CA 94720, USA
[4] Denys Wilkinson Building, University of Oxford, Keble Road, Oxford, UK
[5] STECF/ESO, Garching, Germany
[6] Kapteyn Institute, Postbus 800, 9700 AV Groningen, The Netherlands
[7] Centre for Astrophysics Research, University of Hertfordshire, Hatfield, UK

**Summary.** We present results from a series of follow-up observations of a subsample of the representative SAURON survey elliptical (E) and lenticular (S0) galaxies using the OASIS integral-field spectrograph. These observations focus on the central $10'' \times 10''$, with roughly double the spatial resolution of the SAURON observations. This increased spatial resolution reveals a number of interesting and previously unresolved features in the measured stellar kinematics and absorption-line strengths. We find that galaxies exhibiting the youngest *global* stellar populations (as measured with SAURON) often contain a distinctly young *central* region (on scales of a few hundred parsec or less) compared to the rest of the galaxy. Moreover, these compact, young components are found to be mostly counter-rotating with respect to the rest of the galaxy. Given that there is no well-established reason for such young components to 'prefer' counter- over co-rotation, this finding raises the following questions: How common are these small KDCs as a function of age? Why are there more young than old compact KDCs? Where are the equivalent co-rotating components? We explore these questions using simple simulated velocity fields and stellar population models, and find that the fading of the young component as it evolves, coupled with the fact that counter-rotating components are more easily detected in the velocity field, may help explain the observed trends.

## 1 Young Kinematically Decoupled Components

Since the first applications of absorption line indices as a diagnostic tool for studying stellar populations in early-type galaxies, it was found that some of these evolved and dynamically relaxed objects contain a non-negligible population of young stars (e.g. (7)). The distribution of this 'frosting' of young stars within a galaxy was then largely uncertain, given the difficulties in measuring spatially-resolved absorption-line strengths. With the advent of integral-field spectroscopy, it is however now possible to obtain

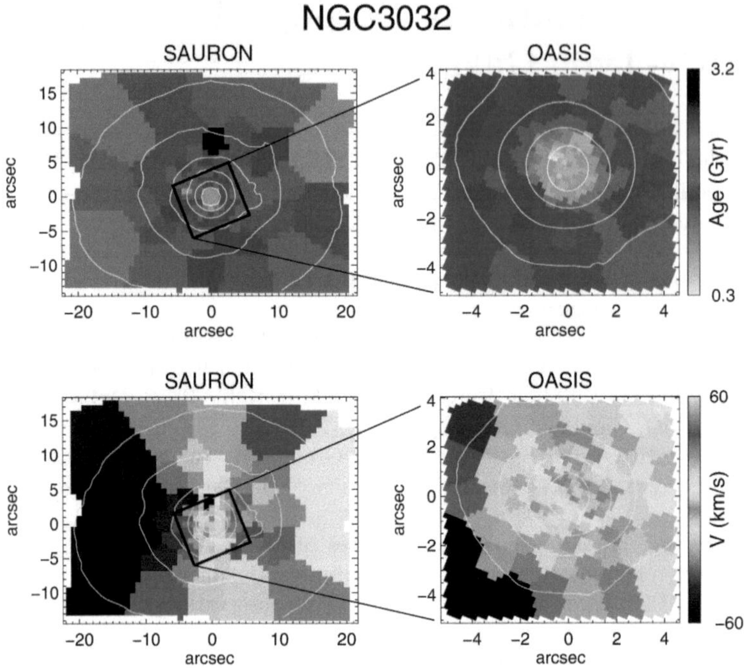


**Fig. 1.** *Top row:* Maps of mean stellar age, derived from line-strength measurements. *Bottom row:* Maps of mean stellar rotation velocity. The maps are derived from observations using SAURON (left) and OASIS (right), where the box indicates the OASIS field.

high-quality 'maps' of absorption-line strength distributions within galaxies (e.g. see Kuntschner et al., these proceedings), and in turn, maps of luminosity-weighted age, metallicity and abundance ratio by applying modern stellar population models.

From such stellar population maps, it is evident that many galaxies showing globally young ages also tend to show centrally concentrated young components. Figure 1 shows a clear example of this in the dusty S0 galaxy NGC 3032. The top row shows the mean stellar age, derived from maps of line-strength indices (H$\beta$, Fe5015, Mg$b$, and Fe5270) observed with SAURON (left) and OASIS (right), using the single-burst stellar population (SSP) models of (6), where the SSP which best reproduced the multiple observed indices was found at each position. The galaxy shows rather young ages across the whole field, but shows a distinct decrease in age in the central $1''$ radius. The bottom panels of Figure 1 show the corresponding velocity maps for this galaxy from the same two instruments. The increased spatial resolution of the OASIS observations reveals a small ($3.5'' \equiv 370$ pc diameter) kinematically decoupled component (KDC), which coincides with the location of the central young population.

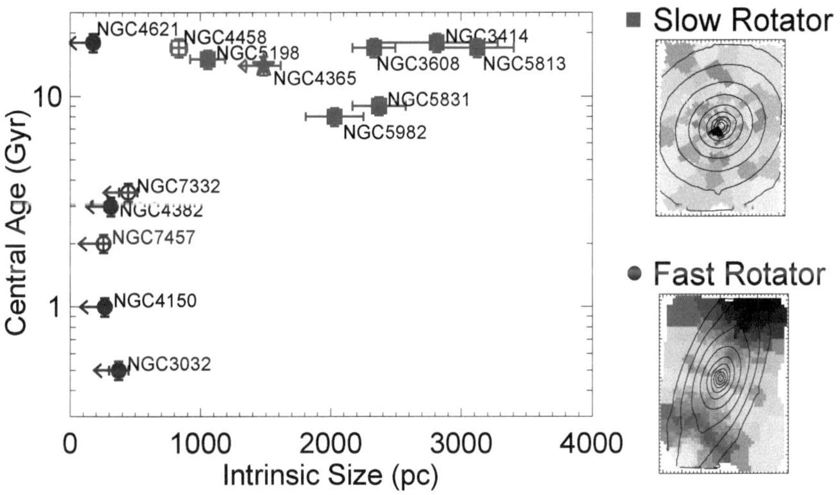


**Fig. 2.** Plot of central stellar age (where the KDC is assumed to dominate) against intrinsic KDC diameter (measured from velocity maps). Symbols indicate galaxies with both OASIS and SAURON data (filled) or only SAURON (open); and whether the galaxy is a slow (square) or fast (circle) rotator (illustrated by the right-hand maps). Arrows indicate upper limits on the KDC diameter estimate. NGC 4365 is included (star symbol) from (4).

Combining the available SAURON and OASIS data, we find a number of galaxies within our sample which contain similar small (< few hundred parsec), young (< 5 Gyr) KDCs, several of which are only resolved with OASIS. Figure 2 shows the distribution of KDC size (estimated from the velocity maps) against the mean luminosity-weighted age measured within the central arcsecond of the galaxy (where light from the KDC is assumed to dominate) for all E/S0 galaxies in the SAURON sample which show a clear KDC (i.e. neglecting co-rotating components). From this figure, we see that intrinsically large (> 1kpc) KDCs tend to be rather old, and the compact KDCs tend to be young, although they cover a range in age. Moreover, the large KDCs are found exclusively in galaxies showing low global angular momentum, which we term 'slow rotators'; the small KDCs on the other hand inhabit 'fast rotators', which show significant net rotation. The young KDCs are also generally counter-rotating systems, sharing almost the same rotation axis as the outer parts (within $\sim 10°$ in most cases).

## 2 Interpretation

Why are the small KDCs mostly young? Figure 3 shows that, of the fast rotating galaxies in the SAURON sample, five of the seven youngest objects

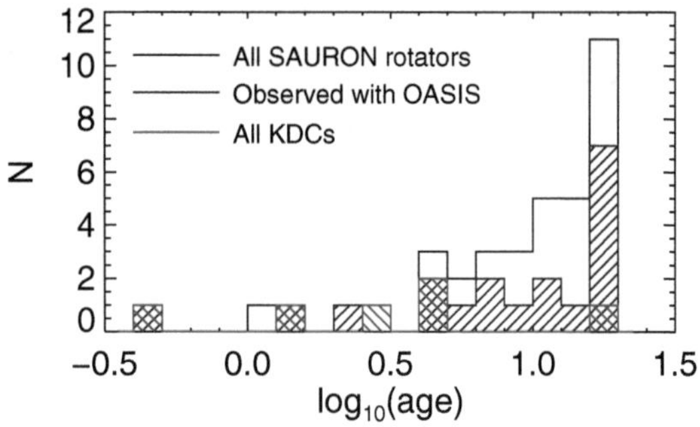


**Fig. 3.** Histogram of ages for the fast rotator galaxies from the SAURON survey. From the complete sample (no hatching), we indicate those observed so far with OASIS (hatching direction lower left to upper right), and those with a clear KDC (hatching direction upper left to lower right).

(< 5 Gyr) show a counter-rotating core, and one of these has yet to be observed with high spatial resolution, and so may yet reveal a KDC. Only one of the remaining thirteen objects older than 5 Gyr shows a detectable KDC.

Is the counter-rotation significant in producing the young stars? There are also galaxies showing young/intermediate-aged global and central populations, but which do not show strong substructure in their velocity fields. Are these 'single component' systems fundamentally different?

Figure 4 shows a possible answer to these questions. Using the measured central line-strengths from one of our young KDC galaxies, we constrain the amount of mass of young stars that can be added to a background 'base' population (assumed to be that of the main body of the galaxy) within the central aperture. Taking the example of NGC 4150, we add 8% *by mass* of a 0.5 Gyr population on a 5 Gyr base population, to give a combined H$\beta$ absorption strength of $\sim$ 4 Å. We simulate a two-component velocity field using a Fourier expansion technique similar to the 'kinemetry' method of (5). We assign these velocities at each position to SSP model spectra of (1), assigning the young population to the KDC component, and make the mass-weighted combination of spectra. The simulated data cube is then spatially binned using a Voronoi tessellation (2), and the kinematics were extracted using the penalised pixel fitting technique of (3).

The result is a realistic-looking KDC velocity field. We then hold the mass fraction fixed, and 'evolve' the populations in step. As the KDC population ages, its mass-to-light ratio increases, resulting in a dimming of the KDC stars. The effect is to 'fade' the KDC into the background rotation field, as the luminosity-weighted contribution becomes less significant. After 5 Gyr, the KDC is barely visible. This helps explain the apparent lack of intermediate and old aged small KDCs.

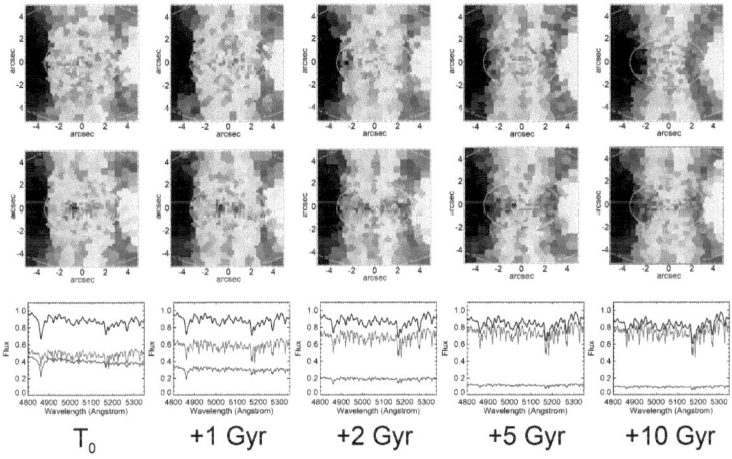


**Fig. 4.** Model velocity fields for an evolving counter-rotating (top row) and co-rotating (middle row) young subcomponent. The spectrum of the young decoupled component (lower line) and older main component (middle line) are shown on the bottom row, indicating their luminosity-weighted contribution to the observed spectrum (upper line).

The importance of counter-rotation is harder to address. The middle row of Figure 4 uses the same kinematic components and populations as the top row, but in this case the subcomponent is co-rotating. The impact of the co-rotating component on the total observed velocity field is clearly more subtle, and the component becomes difficult to separate from the background field after only $\sim$ 2 Gyr. For this reason, the intrinsic distribution of co- and counter-rotating central young components could be rather similar, but since the co-rotating cases can only be clearly identified at young ages, our sample of young galaxies is currently too small to answer this question satisfactorily.

## References

[1] G. Bruzual, S. Charlot, MNRAS, **344**, 1000 (2003)
[2] M. Cappellari, Y. Copin, MNRAS, **342**, 345 (2003)
[3] M. Cappellari, E. Emsellem, PASP, **116**, 138 (2004)
[4] R. L. Davies, et al., ApJL, **548**, L33 (2001)
[5] D. Krajnović, M. Cappellari, P. T. de Zeeuw & Y. Copin, MNRAS, *in press* (astro-ph/0512200)
[6] D. Thomas, C. Maraston, R. Bender, MNRAS, **339**, 897 (2003)
[7] G. Worthey, S. M. Faber, J. J. Gonzalez, D. Burstein, ApJS, **94**, 687 (1994)

# SINFONI's take on Star Formation, Molecular Gas, and Black Hole Masses in AGN

R. Davies, R. Genzel, L. Tacconi, F. Müller Sanchez, J. Thomas and S. Friedrich

Max Planck Institut für extraterrestrische Physik, Postfach 1312, 85741, Garching, Germany

**Summary.** We present some preliminary (half-way) results on our adaptive optics spectroscopic survey of AGN at spatial scales down to $0.085''$. Most of the data were obtained with SINFONI which provides integral field capability at a spectral resolution of $R \sim 4000$. The themes on which we focus in this contribution are: star formation around the AGN, the properties of the molecular gas and its relation to the torus, and the mass of the black hole.

## 1 The AGN Sample

The primary criteria for selecting AGN were that (1) the nucleus should be bright enough for adaptive optics correction, (2) the galaxy should be close enough that small spatial scales can be resolved, and (3) the galaxies should be "well known" so that complementary data can be found in the literature. These criteria were not applied strictly, since some targets were also of particular interest for other reasons. The resulting sample of 9 AGN is listed in Table 1. The observations of these are now completed, and while the data for some objects has been fully analysed, others are still in a preliminary stage. Additional AGN will likely be added once the Laser Guide Star Facility is commissioned.

One immediate result, which has a bearing on the classifications in the table, is the frequent detection of broad Br$\gamma$ – i.e. with FWHM at least $1000\,\mathrm{km\,s^{-1}}$. An example of this is given in Fig. 1. In only 3 galaxies was no broad Br$\gamma$ detected: Circinus, NGC 1068, and NGC 1097 (in which even the narrow Br$\gamma$ is so weak that it is almost lost in the stellar absorption features).

## 2 Star Formation

The topics we address here are the spatial scales on which stars exist around the AGN, the age and star formation history of these stars, and their contribution to the bolometric luminosity with respect to that of the AGN itself.

The stellar K-band (or equivalently H-band) continuum can be distinguished from the non-stellar continuum via the depth of stellar absorption

**Table 1.** AGN sample

| Target | Classification | Dist. (Mpc) | Observations Date | Instrument |
|---|---|---|---|---|
| Mkn 231[1] | ULIRG / Sy 1 / QSO | 170 | May '02 | Keck / NIRC2 |
| NGC 7469[2] | Sy 1 | 66 | Nov '02 | Keck / NIRSPAO |
| IRAS 05189-2524 | ULIRG / Sy 1 | 170 | Dec '02 | VLT / NACO |
| Circinus[3] | Sy 2 | 4 | Jul '04 | VLT / SINFONI |
| NGC 3227[4] | Sy 1 | 17 | Dec '04 | VLT / SINFONI |
| NGC 3783 | Sy 1 | 42 | Mar '05 | VLT / SINFONI |
| NGC 2992 | Sy 1 | 33 | Mar '05 | VLT / SINFONI |
| NGC 1068 | Sy 2 | 14 | Oct '05 | VLT / SINFONI |
| NGC 1097 | LINER / Sy 1 | 18 | Oct '05 | VLT / SINFONI |

[1] Davies et al. 2004a (2); [2] Davies et al. 2004b (3); [3] Müller Sanchez et al. 2006 (6); [4] Davies et al. 2006 (4);

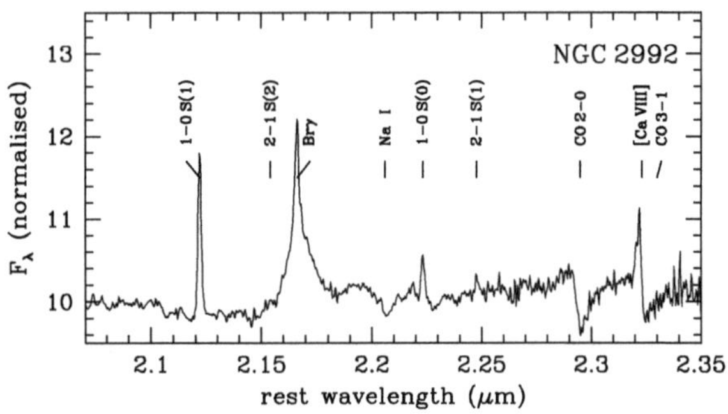


**Fig. 1.** Spectrum of the central 0.5″ of NGC 2992, taken with SINFONI at a spatial resolution of 0.3″ and a spectral resolution of $R \sim 3400$. The stellar absorption features are clear, as is the coronal [Ca VIII], the $H_2$ 1-0 S(1), and both narrow and broad Brγ.

features such as the CO band heads, because for any ensemble of stars the intrinsic depth will not vary much once late-type stars appear (see Davies et al. 2006 (4) for a more detailed discussion of this). Doing so immediately allows one to assess the physical size scale of the stellar population close to the AGN (see Fig. 2). In addition it permits a lower limit to be put on the bolometric luminosity originating in stars. This is because, while a stellar population which is still forming stars will have $L_{\rm bol}/L_{\rm K} \sim 50$ (or even higher if it is very young), even an old passively evolving population has $L_{\rm bol}/L_{\rm K} \sim 20$. In most cases we are able to apply tighter constraints than this by considering other diagnostics. For example, from the morphology and kinematics one can estimate the fractions of the narrow Brγ flux that are associated with stars and with the AGN's narrow line region. Similarly, it is often possible to estimate the fractions of the radio continuum associated with the AGN and stars: the former will be unresolved and have very high brightness temperatures (see Condon et al. 1991 (1)). The ratio of either of these to the stellar K-band continuum can provide strong constraints on the star formation time scales and hence the bolometric luminosity from stars close around the AGN.

Our preliminary results are:

- In all 9 cases we have resolved a stellar population around the AGN on the scales we have achieved (0.08–0.3″); and the stellar luminosity increases as one approaches the AGN.
- In the 5 cases we have analysed in detail so far (Mkn 231, NGC 7469, IRAS 05189-2524, Circinus, NGC 3227), the stellar population is young: the range of ages we find is 40–120 Myr
- The (young) stellar luminosity is comparable to that of the AGN on scales of 1 kpc (Mkn 231, IRAS 05189-2524); is 10–50% of the AGN on scales of 50–100 pc (NGC 7469, NGC 3227); and is a few percent of the AGN on scales of 10–20 pc (Circinus).

# 3 Molecular Gas

The $H_2$ morphologies traced by the 1-0 S(1) line show a much greater diversity than the stellar distributions, as typified in Fig. 2. This might be expected since it is known that distribution of gas is strongly influenced by dynamical resonances and outflows. However, when analysing the morphologies on ~10 pc scales, one needs to remember that the 1-0 S(1) line traces only hot (typically 1000–2000 K) gas, and hence the very local environment will have an important impact on the observed luminosity distribution: for example, is there a particularly massive star cluster nearby or has there been a recent supernova? With this caveat in mind, our preliminary results are:

- the 1-0 S(1) emission is stronger closer to the AGN (with the exception of NGC 1068) indicating the gas distribution is also concentrated towards the nucleus on scales of 10–50 pc.
- the kinematics show ordered rotation (again excepting NGC 1068) but also remarkably high velocity dispersion – in the range $\sigma = 70$–$140\,\mathrm{km\,s^{-1}}$, giving $V_{\rm rot}/\sigma \sim 1$. This means that the gas must be rather turbulent, most likely due to heating from the AGN and/or star formation, and as a result is probably geometrically thick.
- Given the size scales on which models predict the molecular torus around AGN should exist (10–100 pc, e.g. most recently Schartmann et al. 2005 (8)), and the fact that the torus must have a large enough scale height to collimate ionisation cones, it is reasonable to propose that the gas we have seen in these data is associated with the torus.

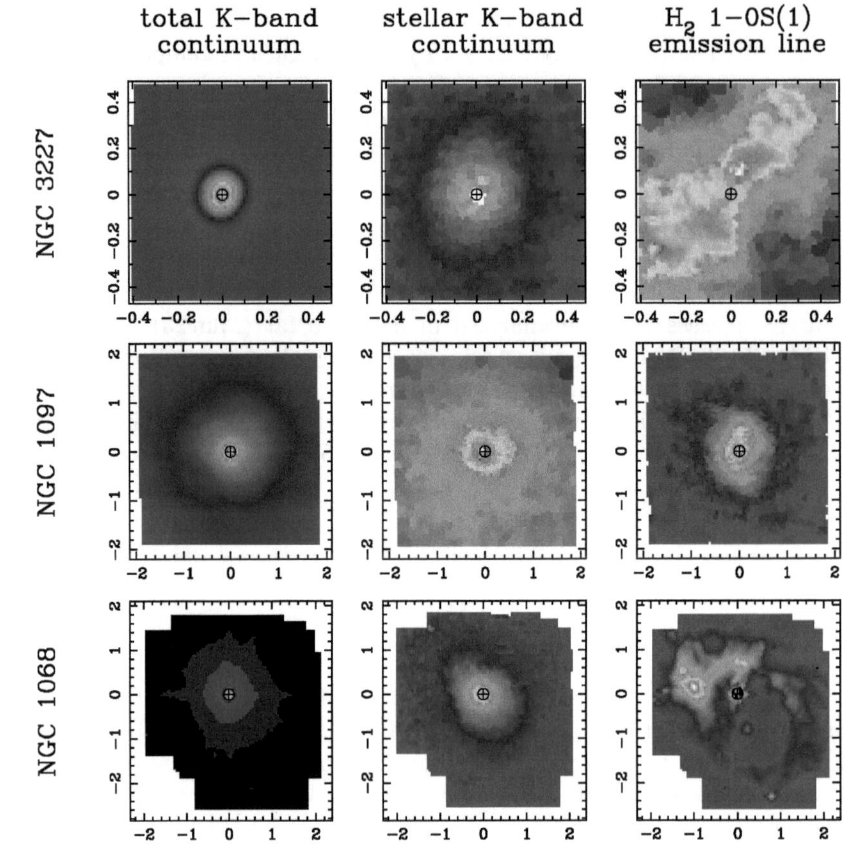


**Fig. 2.** Images from SINFONI of the 3 AGN NGC 3227, NGC 1097, and NGC 1068, which are at approximately the same distance so that 1″ ∼ 70–80 pc. The left panels show the full continuum at 2.1$\mu$m; the centre panels the stellar K-band continuum (derived from the CO band heads); and the right panels the $H_2$ 1-0 S(1) line emission.

## 4 Black Hole Masses

Since it was first discovered, the relation between the mass of the supermassive black hole $M_{\rm BH}$ and the velocity dispersion $\sigma_*$ of the surrounding spheroid has become a cornerstone of galaxy evolution and black hole growth in the cosmological context. However, almost without exception the 'reliable' black hole masses (typically based on stellar kinematics and resolving the black hole's radius of influence) have been derived only for nearby bulge dominated E/S0 quiescent galaxies (see the review by Ferrarese & Ford 2005 (5)). While extremely challenging, it is therefore crucial to determine stellar dynamical black hole masses in AGN – not only to verify that the $M_{\rm BH}$–$\sigma_*$ relation holds for galaxies which are by definition active, but to assess its scatter for these galaxies, and to provide a comparison to reverberation masses which might then allow one to constrain the geometry of the broad line region.

The high spatial resolution and integral field capability of SINFONI provide an ideal combination to do this, and we have successfully derived $M_{\rm BH}$ in NGC 3227 from stellar kinematics – the first time for a Seyfert 1 – using Schwarzschild orbit superposition techniques. Details of the specific code, which is based on that used by the Nuker team, are given in Thomas et al. (2004) (10). While the inclination and mass-to-light ratios are often uncertain parameters, for NGC 3227 they are relatively well constrained. Nevertheless, we have explored the range of values which the modelling would permit and find it to be consistent with those expected, giving us confidence that the results are physically meaningful and reasonably robust. The resulting range of permissible black hole masses is $M_{\rm BH} = 5 \times 10^6$–$2 \times 10^7\,M_\odot$.

The range is a result of the degeneracy between the black hole mass and the 'effective' mass-to-light ratio of the stellar population, which includes the contribution of the gas mass. If the gas is significantly less concentrated than the stars, then the higher $M_{\rm BH}$ is possible; on the other hand if the gas is strongly centrally concentrated in a similar way to the stars, then $M_{\rm BH}$ must be correspondingly lower.

That the mass we find is within a factor of 2–3 of the masses found by other methods suggests that all are satisfactory to this level of accuracy. However, the fact that the mass is also likely to be a factor of a few below that implied by the $M_{\rm BH} - \sigma_*$ relation, while in contrast the stellar dynamical mass of Cen A (Silge et al. 2005, (9)) is a factor of several greater, may indicate that for AGN the scatter around this relation could be very considerable.

## References


[1] Condon J., Huang Z.-P., Yin Q., Thuan T., 1991, ApJ, 378, 65
[2] Davies R., Tacconi L., Genzel R., 2004a, ApJ, 602, 148
[3] Davies R., Tacconi L., Genzel R., 2004b, ApJ, 613, 781
[4] Davies R., et al., 2006 ApJ, submitted
[5] Ferrarese L., Ford H., 2005, SSRv, 116, 523
[6] Müller Sanchez F., et al., 2006 A&A, submitted
[7] Peterson B., et al., 2004, ApJ, 613, 682
[8] Schartmann M., Meisenheimer K., Camenzind M., Wolf S., Henning Th., 2005, A&A, 437, 861
[9] Silge J., Gebhardt K., Bergmann M., Richstone D., 2005 AJ, 130, 406
[10] Thomas J., et al., 2004 MNRAS, 353, 391

# The Unified Model in Nearby Seyfert Galaxies Through Integral Field Spectroscopy

F. Di Mille[1], V. L. Afanasiev[2], S. Ciroi[1], S. N. Dodonov[2], A. V. Moiseev[2], P. Rafanelli[1] and A. A. Smirnova[2]

[1] Dipartimento di Astronomia, Università di Padova, vicolo dell'Osservatorio 2, 35122 Padova
[2] Special Astrophysical Observatory, Nizhnij Arkhyz, 369167, Russia

**Summary.** Active Galactic Nuclei (AGNs) are among the most intriguing extragalactic sources. Notwithstanding they have been studied since several decades, their internal structure is still a matter of debate. Indeed, large efforts have been made in testing the AGNs Unified Model, since several indications seem to run against the Type 1 - Type 2 dichotomy of AGNs caused by a pure inclination effect. Other mechanisms are likely to play important roles in switching on and fueling the central engine. An effective way to investigate the presence and effects of such additional features in the frame of the Unified Model, is to spectroscopically study the stellar and gaseous nuclear environment in nearby AGNs, in particular Seyfert 1, Seyfert 2 and intermediate type Seyfert. The integral field spectroscopy technique offers a powerful tool to perform these aims. During the last decade several authors, including us, have demonstrated the great capabilities and advantages of using 3D data in exploring the physical properties of extended nuclear and extranuclear galactic regions. In this contribution we present new integral field data of three nearby Seyfert galaxies : Mrk 3, Mrk 1157 and Ark 564. Data have been collected at the SAO 6m telescope (SAO-RAS, Russia) with the Multi-Pupil Fiber Spectrograph.

## Observations

We have performed spectrophotometric observations of a sample of low redshift Seyfert galaxies ($z < 0.03$) with the Multi Pupil Fiber Spectrograph (MPFS; Afanasiev et al. 2001) mounted at the 6-m telescope of the Special Astrophysical Observatory (SAO-RAS, Russia). A 600 $mm^{-1}$ grating was generally used to cover two overlapping spectral ranges: 3700-6300 Å and 4800-7400 Å , with a resolution of ~7 Å and a dispersion of ~ 2.6 Å $px^{-1}$. The field-of-view (FoV) of the MPFS was 16x15 arcsec with a sampling of 1 arcsec per spatial element.

## Mkn 3

Mrk 3 is a Seyfert 2 galaxy at z=0.013 with a hidden Broad Line Region (Miller & Goodrich 1990). High resolution HST imaging and spectroscopy had revealed an extended bi-conical Narrow Line Region with complex kinematics (Capetti et al. 1995, Ruiz et al. 2001). MPFS data (0.26 kpc per spatial element) reveals a high ionization emission region up to 2 kpc extended in the direction perpendicular to the major axis of the V-band continuum light reconstructed map. This region is well visible in all emission line maps and much better in the [O III]$\lambda$5007/H$\beta$ ratio, where a clear S-shape appears. Values of these ratios within the high ionization region are around 3-4. On the contrary the [S II]$\lambda\lambda$6716+6731/H$\alpha$ ratio map and the FWHM of the [O III]$\lambda$5007 show their highest values distributed along the continuum major axis. The velocity field of gas obtained from [O III] line positions is highly disturbed with strong deviation from circular motion in the center. In particular, we observe the kinematical minor axis oriented along the continuum major axis, and an apparent nuclear counter-rotation. We report some spectra extracted from the MPFS data cube corresponding at different regions: they show different slopes of their continua. The spectra of the East/South-East side of the galaxy have a bluer continuum than the spectra of the North/North-West side. A similar result was obtained by Kotilainen & Ward (1997) using B-I broad band colors.

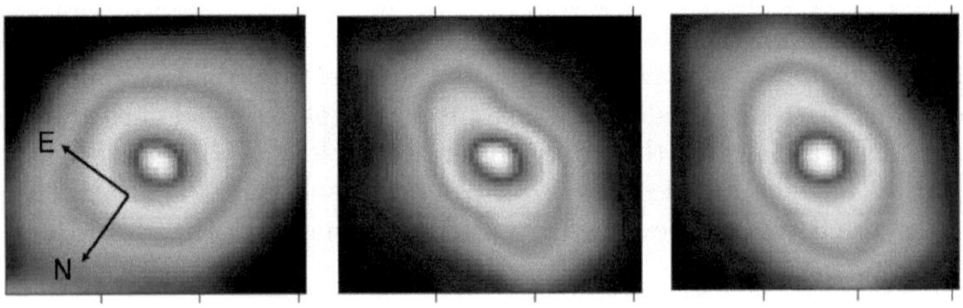


**Fig. 1.** Mkn 3 flux maps. *Left*: Continuum; *center:* [O III]$\lambda$5007; *right*: H$\alpha$

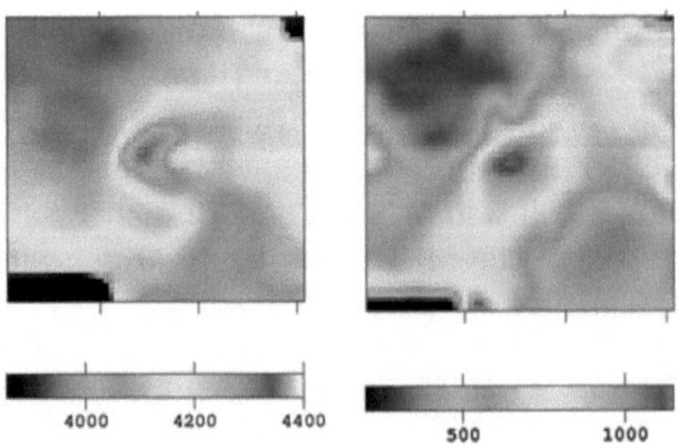


**Fig. 2.** Mkn 3 [O III]$\lambda$5007 velocity field (*left*) and FWHM (*right*).

## Mkn 1157

Mkn1157 (NGC591) is a Seyfert 2 galaxy at z=0.015 with SB(R')SB0/a morphological type. This galaxy has a large bar extending in the East-West direction and an arc-like feature in the north side. Our reconstructed V-band map shows an elongated shape in the direction of the bar. The H$\alpha$ emission is concentrated at the nucleus with two peaks at the end of the bar. The [O III]$\lambda$5007 emission is distributed in a region with an aperture of 10 arcsec, corresponding to 3 kpc (0.29 kpc per spatial element) slightly elongated toward the North-South direction. The flux ratio map of [O III]/$H\beta$ lines, after the subtraction of the underlying stellar absorption features (as in Ciroi et al. 2005), clearly show a high degree of ionization in this region, which suggests the possibility of an ionization cone in this galaxy. A more detailed analysis is required to confirm these findings. The velocity field of gas reconstructed by means of the [N II]$\lambda$ 6583 emission line shows deviation from circular motion probably influenced from the presence of the large bar.

## Ark 564

This is a Narrow Line Seyfert 1 at z=0.0247 hosted by a galaxy with morphological type SB. The spatial distribution of the [O III]$\lambda$5007 emission line fluxes (0.48 kpc per spatial element) has a halo morphology extending up to a radius of 5 arcsec, $\sim$ 2.4 kpc, larger than the seeing disk ($\sim$2 arcsec) during the observations. Within this region the [O III] $\lambda5007/H\beta_{narrow}$ ratio, calculated after the multi-Gaussian fitting of the broad+narrow H$\beta$ emission line, assumes values which are typical of the Narrow Line Region. Moreover, the velocity field of the highly ionized gas shows a weak but measurable circular rotation ($V_{max}$ $\sim$100 km/s), due to low inclination of the galaxy. These results are in agreement with the Unified Model, since the bi-conical shape frequently observed in Seyfert 2 galaxies is in principle expected to appear with a halo shape in Seyfert 1 galaxies.

## References


[1] Afanasiev,V. L., Dodonov,S. N., Moiseev,A. V. 2001, in Proc. Int. Conf. Stellar Dynamics.
[2] Antonucci, R. 1993, ARA&A 31, 473
[3] Capetti, A., Macchetto, F., Axon, D. J., et al. 1995, ApJ, 448, 600
[4] Ciroi,S., Afanasiev,V.L., Moiseev,A.V., et al. 2005, MNRAS, 360, 253
[5] Kotilainen,J. K. & Ward, M. K. 1997, A&AS, 121, 77
[6] Miller, J. S., Goodrich,R.W. 1990, ApJ, 355, 456
[7] Ruiz, J. R., Crenshaw D.M., Kraemer S.B., et al. 2001 AJ, 122, 2961
[8] Wilson, A. S. & Tsvetanov, Z. I., 1994, AJ, 107, 1227

# A 3D View of the Central Kiloparsec of the Seyfert Galaxy NGC 1358

G. Dumas[1,2], E. Emsellem[1] and P. Ferruit[1]

[1] CRAL Observatoire de Lyon, 9 avenue Charles André, 69561 Saint Genis Laval Cedex, France gdumas@obs.univ-lyon1.fr

[2] ARI Liverpool John Moores University, Twelve Quays House, Egerton Wharf, Birkenhead CH41 1LD, UK.

**Summary.** We present a study of the stars and ionised gas in the central kpc of the Seyfert 2 galaxy NGC 1358, using three-dimensional imaging spectroscopy obtained with OASIS (CFHT). We have derived both stellar and gaseous kinematical maps, including higher order Gauss-Hermite stellar velocity moments. Emission line (H$\alpha$, H$\beta$, [OIII], [NII]) and emission line ratios ([OIII]/H$\beta$ and [SII]/H$\alpha$) maps have also been derived. These data reveal an inner gaseous spiral structure (size $<$ 1 kpc) which clearly shows up as a perturbation in the ionised gas kinematic maps, but is not seen in the stellar velocity maps. The presence of an inner bar could explain this feature. Simple dynamical models are built to constrain the mass-to-light ratio of the galaxy and the properties of the presumed inner bar.

## 1 Introduction

Seyfert galaxies being relatively nearby, they are ideal objects to study the AGN phenomenon. At the kiloparsec scale, we often observe structures as bar or spiral perturbations which might be able to transport gas toward the nucleus (1), (4). Such structures are spatially well resolved with ground-based facilities, and can be mapped in detail using 3D spectroscopy. In this context, we have conducted an observational project using the OASIS 3D spectrograph at the CFHT to probe the central kiloparsec of a few Seyfert galaxies. Here we present the results obtained on one target : the barred Seyfert 2 galaxy NGC 1358. The inclination of this galaxy (i = 38$^\circ$, (3)) allow us to have a privileged view of the central morphological and dynamical features.

## 2 Observations and Results

The distance to NGC 1358 is 53.6 Mpc ((3)) : 1$''$ corresponds to 260 pc. The galaxy was observed in November 2000 with OASIS (CFHT) in two spectral configurations, MR1 and MR2. The first one includes the [OIII], H$\beta$ and [NI] emission lines , with a field of view of 10$''\!.$4x8$''\!.$3, and the second one includes [OI], [NII], H$\alpha$ and [SII], with a field of view of 16$''\!.$1x13$''\!.$7 [3]. In figure 1,

[3] The spectral domain also includes absorption lines which have been analyzed. We will focus here on the gaseous components.

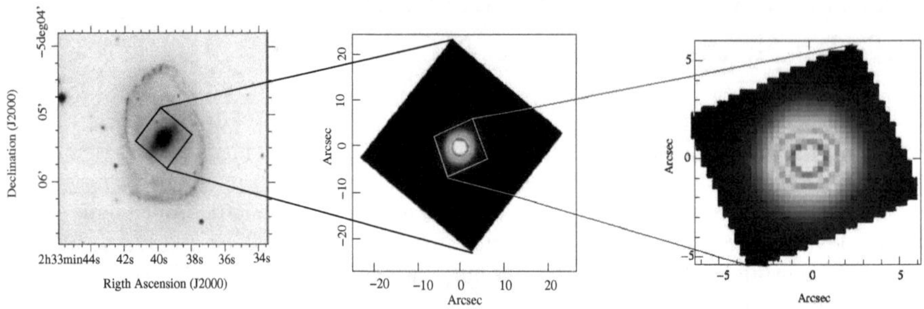


**Fig. 1.** NGC 1358. Left : DSS red image; middle : HST/WFPC2 image (F606W); right : OASIS MR1 continuum reconstructed image.

we present the continuum maps of NGC 1358, and figure 2 illustrates the distribution and kinematics of the gaseous components.

A nuclear spiral of about 1 kpc in diameter is observed in the gas intensity field (Fig 2, middle), but not in the continuum maps (Fig 1). The spiral arms

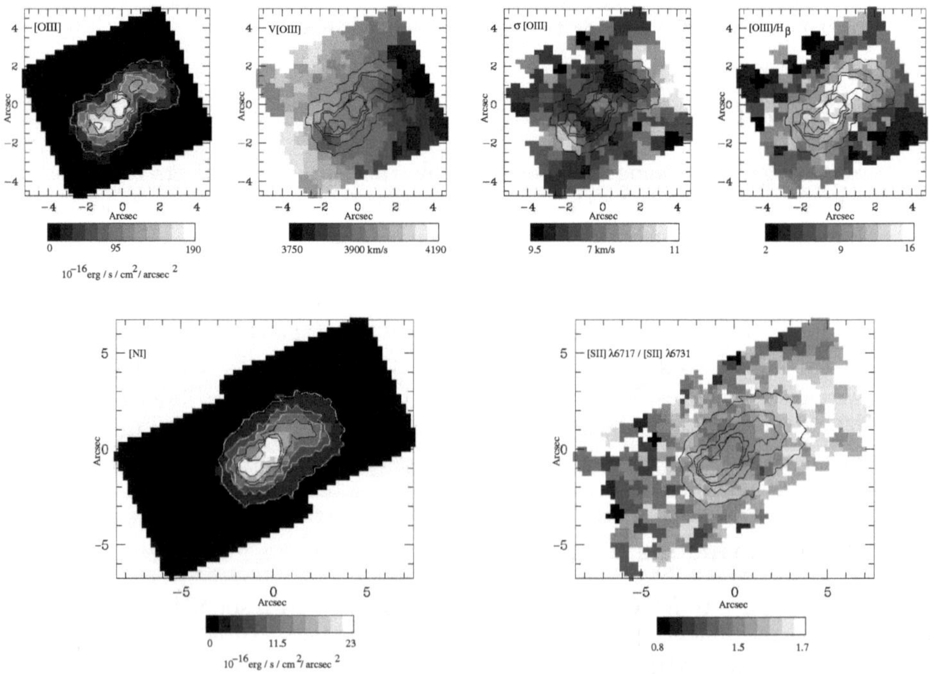


**Fig. 2.** Oasis gas intensity and kinematical maps of NGC 1358. Top panels : MR1 maps. From left to right : [OIII] intensity, velocity and dispersion maps, [OIII]/H$\beta$ map. Contours : [OIII] intensity. Bottom panels : MR2 maps. From left to right : [NI] intensity and [SII]$\lambda$6717/[SII]$\lambda$6731 maps. Contours : [NI] intensity.

correspond to regions of low gas velocity dispersions, and the velocity field is perturbed at the end of the arms, where the dispersion increases (Fig 2, left). Such kinematical features could hint at the presence of an inner bar.

## 3 Modeling

We have therefore constructed a simple static model to constrain the presence and properties of the presumed inner bar (size, pattern speed). This model, based on an orbital description of the gaseous flow within a barred potential, constrains the pattern speed of the inner bar $\Omega_p < 250$ km/s/kpc, the co-rotation radius being then 1.7 kpc $< R_{cor} <$ 4.1 kpc. These results situate the inner bar just inside the ILR of the outer bar of NGC 1358, for which we have $R_{cor}$=5.1 kpc and $\Omega_p$=31 km/s/kpc (2). However, more sophisticated and realistic numerical simulations are required to determine conclusively whether the observed gaseous spiral is being driven by an inner bar or not. It would also be interesting to study the potential coupling between the two bars ((5);(6)) and determine whether the nuclear spiral is due to the inner bar alone, or to a transient phenomenon within the two gravitationally coupled bars.

## Acknowledgments

G.D. wishes to acknowledge C. Mundell (ARI Liverpool UK) for the useful discussions and comments on this work.

## References

[1] P. Erwin, L. Sparke AJ **119**, 536 (2002)
[2] J. Gerssen et al: MNRAS **345**, 261 (2003)
[3] L.C. Ho et al: ApJS **112**, 315 (1997)
[4] R.W. Pogge, P. Martini ApJ **569**, 624 (2002)
[5] P. Rautiainen, H. Salo: A&A **348**, 737 (1999)
[6] P. Rautiainen et al: MNRAS **337**, 1233 (2002)

# VLT/SINFONI Integral Field Spectroscopy of the Superantennae

V. D. Ivanov[1], J. Reunanen[2], L. E. Tacconi-Garman[3] and L. J. Tacconi[4]

[1] European Southern Observatory, Alonso de Cordova 3107, Vitacura, Casilla 19001, Santiago 19, Chile, vivanov@eso.org
[2] Leiden Observatory, P.O. Box 9513, 2300 RA Leiden, The Netherlands
[3] European Southern Observatory, Karl-Schwarzschild-Str. 2, D-85748 Garching bei Muenchen, Germany
[4] MPI fuer Extraterrestrische Physik, Postfach 1312, D-85748 Garching bei Muenchen, Germany

**Summary.** We present $H$ and $K$ SINFONI/VLT spectroscopy of the Superantennae (IRAS 19254-7245) – a late stage merger at ∼250 Mpc, a more powerful analog of the classical Antennae galaxy, harboring two nuclei: a starburst and an AGN. We present rotational and velocity dispersion maps derived from absorption and emission lines. We also show some line index and line ratio maps and discuss their implications for the star formation and the AGN activity in the galaxy.

## 1 Introduction

The Superantennae is an interacting system with nuclear separation of ∼8 arcsec (∼2 Kpc projected separation) and large "antennae" extending up to 5 arcmin (∼350 Kpc), at $z$=0.062 (D∼260 Mpc). The Northern nucleus is in post-starburst stage while the Southern one is a starburst galaxy with an embedded Type 2 AGN.

## 2 Observations and Data Reduction

The observations were obtained in Aug 2004, in a classical ABBA sequences, without AO assistance: 1800 sec in H (R∼2900) under 0.94 arcsec seeing, and 14000 sec in K (R∼4500) with average seeing of 0.77 arcsec. At the time SINFONI was equipped with an engineering grade array. The "scale" was 250 mas px$^{-1}$ giving a field of view of 8×8 arcsec. The individual data reduction steps include: bad pixel masking, distortion correction, sky subtraction, flat fielding, wavelength calibration, telluric correction (Hip099481, B9.5 V) and flux calibration with broad-band $HK$ images.

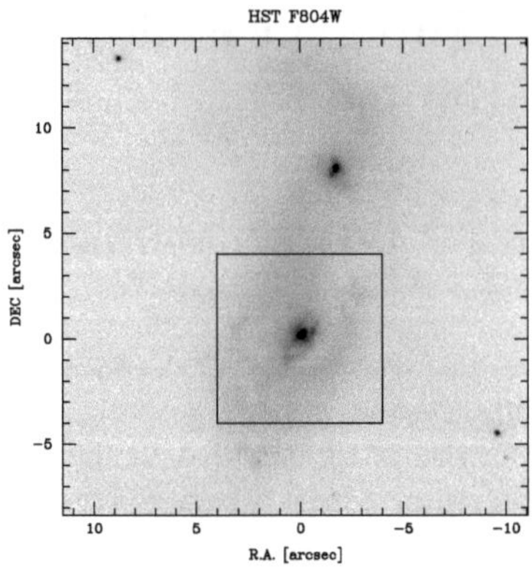


**Fig. 1.** *HST* F804W image of the Superantennae. The SINFONI FoV around the Southern nucleus is indicated with a box. East is to the left and North is up.

# 3 Results

Hydrogen recombination lines in the Southern nucleus suggest SFR$\sim$50 $M_\odot\,yr^{-1}$ within the central 3 Kpc. The coronal [Si VI] originating in the AGN is well detected. Pa$\alpha$ shows complicated velocity structure – it requires at least 3 components, similar to the optical Hydrogen lines. The Pa$\alpha$, [Fe II] and $H_2$ 1–0 S(1) show different line profiles, indicating different velocity structures of the regions where they originate. The $H_2$ line ratios and Br$\gamma$/$H_2$ ratio appear inconsistent with UV fluorescence excitation.

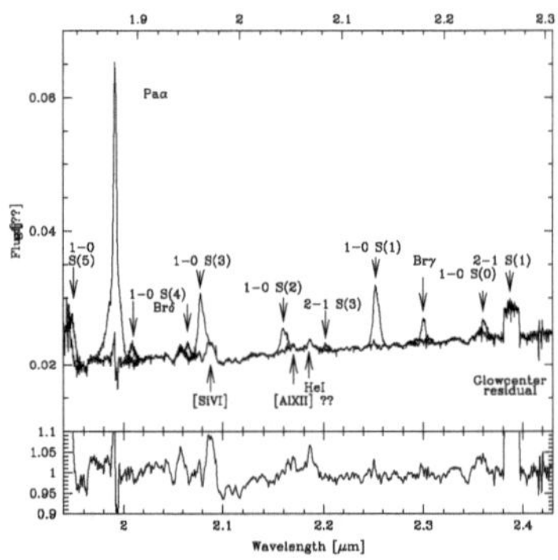


**Fig. 2.** Combined spectrum of the central 5x5 "pixels". The flux is in units of $10^{-13}\,erg\,cm^{-1}\,s^{-1}\,\mu m^{-1}$ (top) and normalized residual spectrum (bottom) after subtracting the strongest lines (Pa$\alpha$, Br$\gamma$, etc.) and smoothing the residual continuum with 5-pixel boxcar.

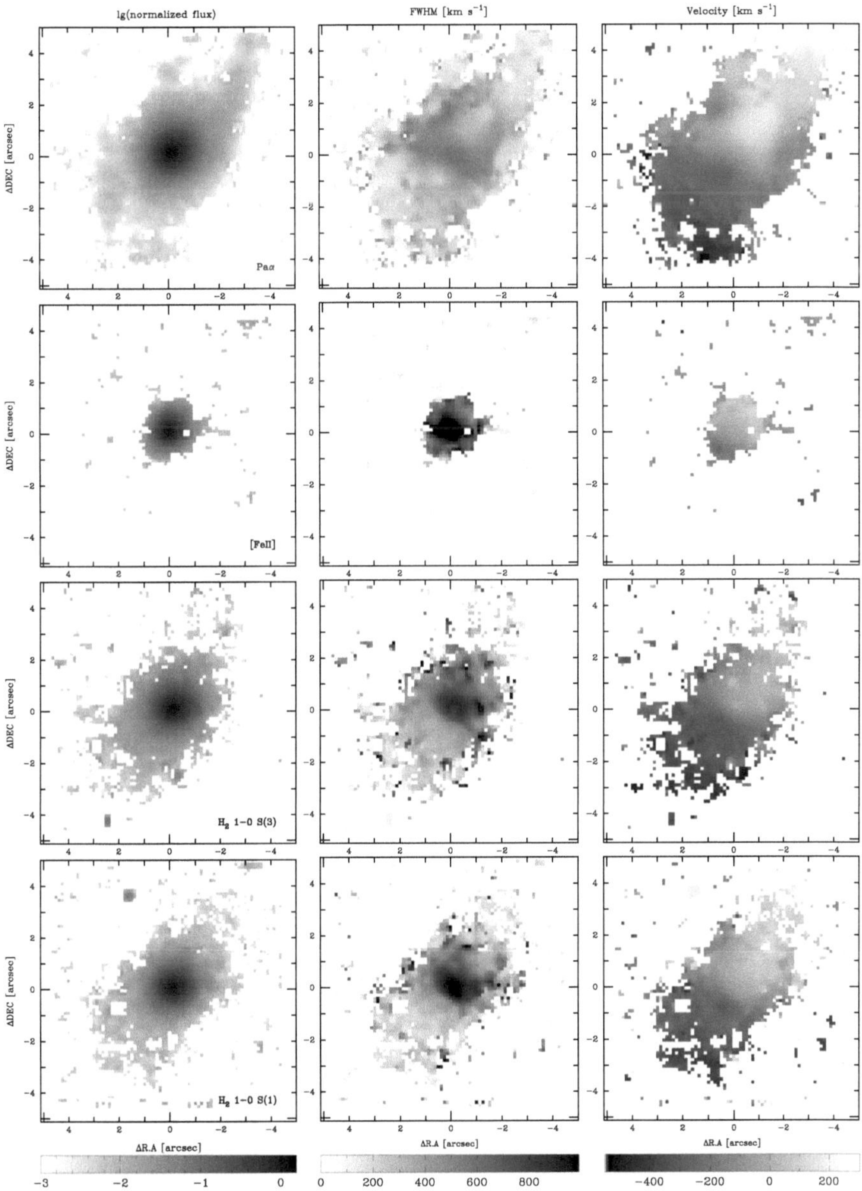


**Fig. 3.** From Left to right: the flux, normalized to the peak flux, Gaussian FWHM (i.e. velocity dispersion) and the velocity field for the emission lines. From top to bottom: Paα (excluding the broad component), [FeII], $H_2$ 1–0 S(3), and $H_2$ 1–0 S(1).

# Integral-Field Observations of Centaurus A Nucleus

D. Krajnović[1], R. Sharp[2] and N. Thatte[1]

[1] Denys Wilkinson Building, University of Oxford, Keble Road, OX1 3RH, UK
dxk@astro.ox.ac.uk; thatte@astro.ox.ac.uk
[2] Anglo-Australian Observatory, Epping NSW 1710, Australia
rgs@aaoepp.aao.gov.au

**Summary.** We observed the nuclear region of Centaurus A in the J band with CIRPASS, mounted on GEMINI South telescope. Here we present the preliminary results focusing on the kinematic structure of the gas disk and present a simple model determining its kinematic orientation and inclination as well as an estimate of the mass of the central black hole.

## 1 Introduction

Due to its proximity, Centaurus A is a truly special galaxy for studies of galaxy formation and evolution. It is the closest giant elliptical with an active galactic nucleus hosting a strong radio source, a radio and an X-ray jet (e.g. (3)). It has a dark, misaligned dust lane, rich in molecular and ionised gas (e.g. (2)) across the whole galaxy, heavily obscuring the nucleus ((10), (6)). Hubble Space Telescope (HST) observations revealed a disk of ionised gas ((11)), while (7) found evidence for a supermassive black hole (SMBH) of mass $M_{BH} = 2 \times 10^8 M_\odot$. Recently, new ground- and space-based studies of the stellar ((12)) and the gaseous component ((8)), as well as ground-based adaptive optics assisted measurements ((4)) refined the mass of the black hole to lie between $8.6 \times 10^7 M_\odot$ and $3.0 \times 10^8 M_\odot$, depending on the assumed inclination of the disk. All these studies used long-slit observations along a few position angles. Two-dimensional observations are, however, needed to robustly constrain the orientation and the inclination of the disk, as well as to resolve possible kinematic substructures.

## 2 Results

Observations in the J band reveal three emission-lines: [FeII], Pa$\beta$ and [PII]. Within the central 0.2 arcsec the relative strength of [FeII]/Pa$\beta$ lines is 1.51 while that of [FeII]/[PII] is 4.5. With increasing aperture size the ratio of [FeII]/Pa$\beta$ decreases while the ratio [FeII]/[PII] increases. We also observe that the [FeII]/[PII] ratio is always higher than the limiting value for non-shock photo-ionisation, but it never reaches the values for a pure shock dominated region (as given by (9)). This suggest a combination of two processes, where the non-shock photo-ionisation dominates in the centre of the compact circum-nuclear disk and the shock towards the edge. This can explain the somewhat disturbed morphology and kinematics of the circum-nuclear gas disk in Centaurs A.

### 2.1 Kinematic Orientation and Inclination of the Disk

Velocity maps of the emission-lines consist of the central region with mostly ordered rotation which becomes less ordered with increasing radius. Different emission-lines show different magnitude of rotation, [PII] being the fastest and Pa$\beta$ the slowest. The zero-velocity curve on the velocity map of [FeII] (Fig. 1), constructed from the highest signal-to-noise ratio data, shows a twist, starting from the north-east towards the south-west corner. The twist is less visible on the Pa$\beta$ velocity map. It is, however, clear that the kinematic orientation of neither of the maps corresponds to the orientation of the Pa$\alpha$ disk ($33^o$, (11)), as well as that it changes with radius. The velocity dispersion is generally low except in the centre (Pa$\beta$) and along a northwest - southeast direction ([FeII]).

Using the kinemetry method ((5)), which utilises the full 2D information from the velocity maps, we measured the global position angle and the inclination of the gas disk. The best fitting values were for i = $23^o$ and PA = -$5^o$ for [FeII], and i = $18^o$ and PA = -$5^o$ for Pa$\beta$ velocity maps. The uncertainties on these parameters are, however, large, excluding inclinations of more than $60^o$ or orientations more than $\pm 15^o$ with only $1\sigma$ significance. Reassuringly, the global parameters of the two emission-lines velocity maps are in good mutual agreement. Similar results were found by (7).

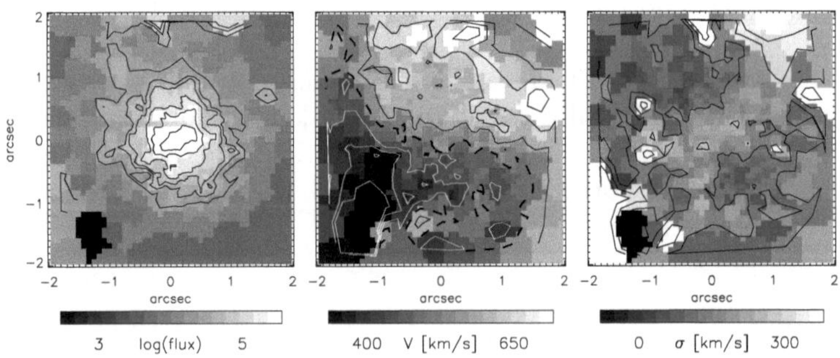


**Fig. 1.** Emission-line [FeII] disk within central 2 arcsec. From left to right: total flux, mean velocity and velocity dispersion of [FeII] emission-line, measured as the area, mean position and dispersion of a Gaussian. The spectra in the data cube were binned using the adaptive scheme of Voronoi tessallation developed by (1).

## 2.2 Simple Dynamical Models

We constructed dynamical models for both [FeII] and Pa$\beta$ distributions using the obtained values for the orientation of the gaseous disk: PA = $-5^o$ and i = $20^o$. We followed a widely used approach assuming that the gas lies in a thin disk and rotates in circular orbits around the central mass concentration, an SMBH. The velocity dispersion is assumed to be solely due to rotation and is ignored in the calculations. The gravitational potential is given by the SMBH. This is a valid approximation since the sphere of influence of the SMBH in Centaurus A is about 1.3 arcsec in radius ((8)), which completely covers our observed velocity maps. The thin disk model is not able to reproduce the kinematic twists visible on the maps. However, our data were obtained by ground-based observations under natural seeing and have low signal-to-noise ratio which does not justify a more complex model. Our 'order of magnitude' estimate of the mass of the SMBH required by the [FeII] velocity map is $M_{BH} = (7 \pm 0.3) \times 10^7 M_{\odot}$. A somewhat larger SMBH, is recovered from the Pa$\beta$ maps: $M_{BH} = (1.3 \pm 0.4) \times 10^8 M_{\odot}$. These estimates are consistent with each other as well as with the values in the literature and firmly rule out a zero mass black hole.

# References


[1] Cappellari, M., & Copin, Y. 2003, MNRAS, 342, 345
[2] Israel, F. P. 1998, A&A Rev., 8, 237
[3] Hardcastle, M. J., Worrall, D. M., Kraft, R. P., Forman, W. R., Jones, C., & Murray, S. S. 2003, ApJ, 593, 169
[4] Haering-Neumayer, N., Cappellari, M., Rix, H.-W., Hartung, M., Prieto M. A., Meisenheimer, & K., Lenzen, R. 2006, ApJ in press, astro-ph/0507094
[5] Krajnović, D., Cappellari, M., de Zeeuw, P. T., & Copin, Y., 2006 MNRAS in press, astro-ph/0512200
[6] Marconi, A., Schreier, E. J., Koekemoer, A., Capetti, A., Axon, D., Macchetto, D., & Caon, N. 2000, ApJ, 528, 276
[7] Marconi, A., Capetti, A., Axon, D. J., Koekemoer, A., Macchetto, D., & Schreier, E. J. 2001, ApJ, 549, 915
[8] Marconi, A., Pastorini, G., Pacini, F., Axon, D. J., Capetti, A., Macchetto, D., Koekemoer, A. M., & Schreier E. J. 2006, A&Ain press, astro-ph/0507435
[9] Oliva, E., et al. 2001, A&A, 369, L5
[10] Schreier, E. J., Capetti, A., Macchetto, F., Sparks, W. B., & Ford, H. J. 1996, ApJ, 459, 535
[11] Schreier, E. J., et al. 1998, ApJ, 499, L143
[12] Silge, J. D., Gebhardt, K., Bergmann, M., & Richstone, D. 2005, AJ, 130, 406

# The Puzzling Case of XBONGs: Will 3D-spectroscopy Explain Their True Nature?

M. Mignoli[1], F. Civano[2], A. Comastri[1] and C. Vignali[2] (on behalf of the *Hellas2XMM* collaboration)

[1] INAF - Osservatorio Astronomico di Bologna marco.mignoli@bo.astro.it
[2] Dipartimento di Astronomia - Università di Bologna

**Summary.** The discovery of luminous hard X-ray sources hosted by "normal" galaxies with optical spectra typical of early-type systems (XBONG; Comastri et al. 2002, (3)) represents one of the most surprising results of *Chandra* and XMM-*Newton* surveys. Why the relatively bright X-ray emission, typical of moderately luminous ($10^{42-43} erg\ s^{-1}$) Active Galactic Nuclei (AGN), does not leave any optical signature of the presence of a nuclear source is still subject of debate. Although several possibilities have been discussed in the recent years, the main concern about the nature of these sources is that the AGN emission lines may be "hidden" by the host galaxy starlight. The already granted **VIMOS-IFU** observations will provide us with a unique opportunity to understand these enigmatic sources. High spatial resolution, integral field spectroscopy in the optical, isolating the nuclei of the XBONGs, would help to settle the AGN line absence vs. starlight dilution controversy. The 3D-spectroscopy will also provide a clear insight into the optical properties of these X-ray emitting galaxies and, therefore, into the nature of the obscured super-massive black holes (SMBH) they host.

## 1 Introduction

Several independent arguments clearly indicate that the bulk of the energy density produced by accretion onto SMBH is absorbed by large amounts of gas and dust. In particular, recent hard X-ray surveys have proven to be very efficient to uncover obscured accreting black holes. A striking result is that these X-ray surveys did not discover a single prevalent population of obscured AGN, but rather a collection of cosmic sources: i) broad emission-line AGN showing significant absorption in the X-ray band (10); ii) type-2 quasars, the long sought-after luminous cousins of the local Seyfert-2 galaxies (4; 9); iii) optically faint ($R > 24 - 25$) sources, with unusually high X-ray-to-optical flux ratio (7): iv) "passive" galaxies with absorption-line optical spectra.

The latter type of sources have been dubbed X-ray Bright Optically Normal Galaxies (XBONG) and do not show any signature of nuclear activity in their optical spectra, which closely resemble those of early-type galaxies (see Figure 1). On the other hand, the relatively high X-ray luminosities ($L_x > 10^{42} erg\ s^{-1}$) and the analysis of hardness ratios – suggestive of (partially) obscured X-ray emission – are evoking the presence of an AGN in the nuclei of these galaxies.

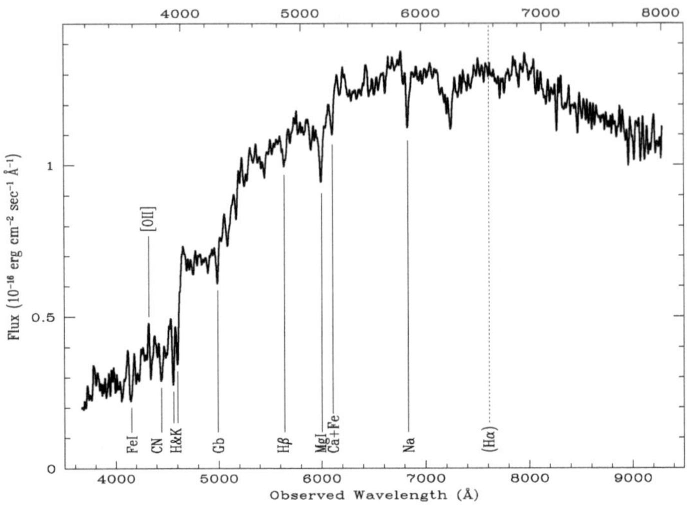


**Fig. 1.** Optical spectrum of the source PKS0312-77 #18 (z=0.159 also known as P3), the archetype of the XBONG class. Note the good S/N level and the absence of any typical AGN spectral features.

## 2 The Enigmatic Nature of the XBONGs

Different physical interpretations of the lack of observational evidence of the AGN activity in the optical have been recently proposed:

• Heavy obscuration by Compton-thick gas covering almost $4\pi$ at the nuclear X-ray source, prevents ionizing optical/UV photons from escaping and hampers the formation of the Narrow Line Regions (3).

• Radiatively inefficient accretion flow (RIAF): a featureless hard X–ray spectrum is expected, with a negligible AGN contribution in the optical and UV bands, which are therefore dominated by the stellar light of the host galaxy (12).

• Extreme BL Lacs objects. Fossati et al. (5) proposed a spectral sequence in which there is an anticorrelation between luminosity and frequency of the synchrotron peak. In this scenario, XBONGs could be the most extreme objects in this sequence. It is worth noting that in one of the selected sources, the rather strong radio emission would favour the BL Lac hypothesis (2).

Whatever the explanation, it is of paramount importance to definitely exclude, or promote, the possibility that the puzzling properties of the XBONGs can be merely attributed to observational effects, such us low S/N spectra and/or insufficient spectral coverage (11) or dilution from the host galaxy light (6; 8). While both these effects could be responsible for some featureless spectra of distant galaxies selected by deep X-ray surveys, the first hypothesis is almost certainly ruled out for the bright nearby XBONGs selected in wide, medium-deep hard X-ray surveys and for which good S/N spectra covering also the H$\alpha$ region are available. Although the upper limits on the optical emission lines expected from the nuclear activity in some of the studied XBONGs are tight enough to place these sources outside the typical AGN properties (Civano et al., in preparation), the possibility that the nuclear emission lines could suffer some dilution by the host galaxy starlight cannot be ruled out on the basis of ground-based, single-slit spectroscopic observations.

## 3 Upcoming Observations and Future Perspective

In order to better understand the nature of the various components of the X-ray background (XRB), we started a program of multiwavelength observations of hard X–ray selected sources discovered in XMM-*Newton* public fields (HELLAS2XMM survey, (1)). Sampling the X–ray sky at relatively bright X–ray fluxes ($> 10^{-14}$ erg cm$^{-2}$ s$^{-1}$), the survey is optimally suited to investigate the bright tail (both in the X–rays and optical bands) of the X-ray source population that produces the unresolved hard XRB. We have performed a complete optical follow-up program (mainly using ESO telescopes) of 122 X-ray sources in five XMM-Newton fields covering about one square degree (the HELLAS2XMM 1dF sample, (4)). From this survey we extracted a "bona-fide" XBONG sample by selecting the five brightest objects, with $z = 0.08 - 0.32$, and all with fairly good S/N optical spectra including the H$\alpha$ region. For this sample we were granted VLT/VIMOS-IFU (Integral Field Unit) observations consisting of 7 hours (in service mode during ESO-period P77) with the high-spatial resolution mode (0.33 arcsec sampling). The overlay of the near-infrared images of the selected XBONGs with the field of view (FoV) of the chosen VIMOS-IFU is shown in Figure 2.

The main objective of these observations is to detect the typical AGN optical emission lines ([O III] 5007Å, H$\alpha$) down to very faint flux limits. The choice of the IFU-VIMOS in high-spatial resolution mode is motivated by the need to reduce as much as possible the contribution from the stellar light of the host galaxy and to finely sample the central regions in order to detect the unresolved active nucleus; furthermore, with the 3D-spectroscopy we will map the extended Narrow Line Region (if present) expected to be ionized by the radiation field from the central AGN. The three faintest XBONGs (R=17.5-18) will be observed with the medium-resolution spectroscopic mode in order to easily deblend the doublet [N II]6548,6583Å from H$\alpha$ and thus apply diagnostic diagrams useful for distinguishing AGN from HII regions. The two galaxies with multiple nuclei, being the brightest selected XBONGs, will be observed in high-resolution spectroscopic mode in order to study in detail the kinematics of the ionized gas and stars in these intriguing systems.

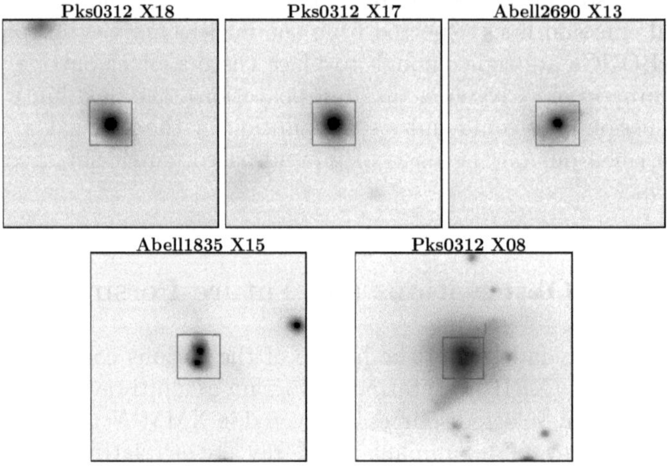


**Fig. 2.** *ISAAC Ks*-band images of our selected XBONG sample, showing the (unexpected) different morphologies. From upper-left to lower right: two ellipticals, a S0/Sa galaxy, a double-nucleus (0.7" apart) object and a system composed of multiple nuclei embedded in a common envelope. In each image the VIMOS-IFU FoV (13"x13") is indicated. The brightest XBONGs with multiple nuclei will be observed in high-resolution spectroscopic mode in order to study in detail their kinematics.

## References

[1] Baldi et al.: ApJ, **564**, 190 (2002)
[2] Brusa et al.: A&A, 409, 65 (2003)
[3] Comastri et al.: ApJ, **571**, 771 (2002)
[4] Fiore et al.: A&A, **409**, 79 (2003)
[5] Fossati et al.: MNRAS, **299**, 433 (1998)
[6] Georgantopoulos & Georgakakis: MNRAS, **358**, 131 (2005)
[7] Mignoli, et al.: A&A, **418**, 827 (2004)
[8] Moran, Filippenko & Chornock: ApJ, **579**, L71 (2002)
[9] Norman, et al.: ApJ, **571**, 218 (2002)
[10] Perola, et al.: A&A, **421**, 491 (2004)
[11] Severgnini et al.: A&A, **406**, 483 (2003)
[12] Yuan & Narayan: ApJ, **612**, 724 (2004)

# Studying of Some Seyfert Galaxies by Means of Panoramic Spectroscopy

A. A. Smirnova, A. V. Moiseev and V. L. Afanasiev

Special Astrophysical Observatory, Russian Academy of Sciences, 369167 Russia
alexiya@sao.ru

**Summary.** We have studied two galaxies with active nuclei using the methods of panoramic spectroscopy. In order to investigate the kinematics of ionized gas and stars, we carried out observations at 6-m telescope of the SAO RAS with the integral field spectrograph MPFS. We have created intensity maps and velocity fields in different emission lines of the ionized gas as well as velocity fields for the stellar component. Diagnostic diagrams have also been made based on the emission lines ratios.

## 1 Observations

We observed Seyfert (Sy) galaxies at the prime focus of 6-m telescope of SAO RAS with the integral field Multipupil Fiber Spectrograph MPFS (1). MPFS takes simultaneous spectra from 256 square lenses, that form an array of 16x16 elements on the sky, with scale 1 arcsec per spaxel. We used a grating provides the spectral resolution about 8Å in the spectral range 4800-7100Å. We reduced the observations using the software developed at the SAO RAS by V. Afanasiev and A. Moiseev and running in the IDL environment. Based on the data cubes we constructed intensity maps and velocity fields in the different emission lines. On the MPFS maps we can see how gas and stars are distributed and move in the central regions of the galaxies. In order to define what is a source of gas ionization, we constructed diagnostic diagrams on the basis of the method by Veilleux and Osterbrock (2). Borderlines separate regions with non thermal (AGN) ionization from regions where thermal ionization dominates (HII) and from regions dominated by shock ionization (LINER).

## 2 Mrk 291

Mrk291 is a Sy1 galaxy at distance of 144 Mpc, the morphological type is SB(s)a (in NED). We have considered velocity fields and intensity maps in the brightest emission lines. In addition to the nucleus on the intensity maps (except map in [O III]$\lambda$5007Å line) we can see that the emission coincides with the bar region. There is only a bright nucleus on the intensity map in the [O III] line and no more details in the MPFS field of view.

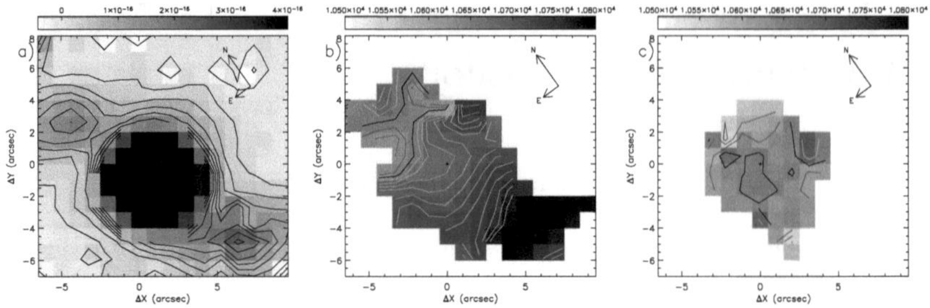


**Fig. 1.** MPFS maps for Mrk291: a) intensity maps in the $H_\alpha$ line; b) velocity field in the [S II] emission line; c) velocity field in the [O III] emission line

Velocity fields in most emission lines show nearly circular rotation, but in case of the [O III] line it seems irregular (see Fig.1). An unresolved second spectral component affects the [O III] emission line.

Using the ratios of the emission lines we chose two regions on our $H\alpha$ image: the nucleus and the region of the bar. The diagnostic diagrams reveal that ionization from hot young stars mainly exceeds an AGN contribution in this region of the galaxy (see Fig.2).

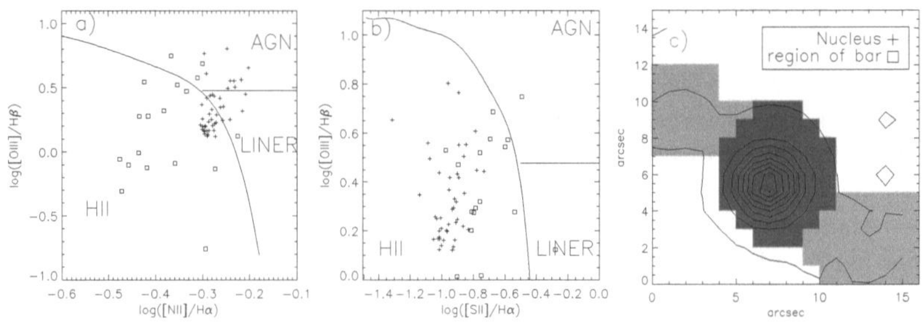


**Fig. 2.** Line ratio diagrams for Mrk291: a) [O III]/$H_\beta$ vs [N II]/$H_\alpha$; b) [O III]/$H_\beta$ vs [S II]/$H_\alpha$; c) the map of the two studied regions and isophotes in the $H\alpha$ line

## 3 Mrk 1469

Mrk1469 is a Sy1.5 galaxy at 124 Mpc that is seen almost edge-on. The morphological classification is SBb in NED. The elongated bar is detected on the continuum map. On the maps in different emission lines the brighter region connects with the active nuclei. Our velocity fields shows almost circular rotation with S-shape distortion of the isovelocities, which is a possible bar signature (see Fig.3).

In order to define a type of source ionization in Mrk1469, we again plot ionizing diagrams. We used the image in the [O III]$\lambda$5007Å for distinguishing

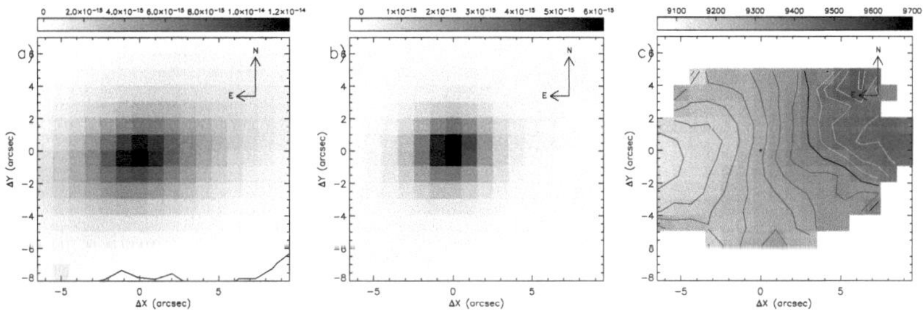


**Fig. 3.** Maps for Mrk1469: a) the red continuum map; b) intensity map in [O III] emission line; c) velocity field in [O III]

between the two regions: nuclear and external. All points fall into the AGN region. Therefore gas ionization comes from a non-thermal source - the Sy nucleus (see Fig.4).

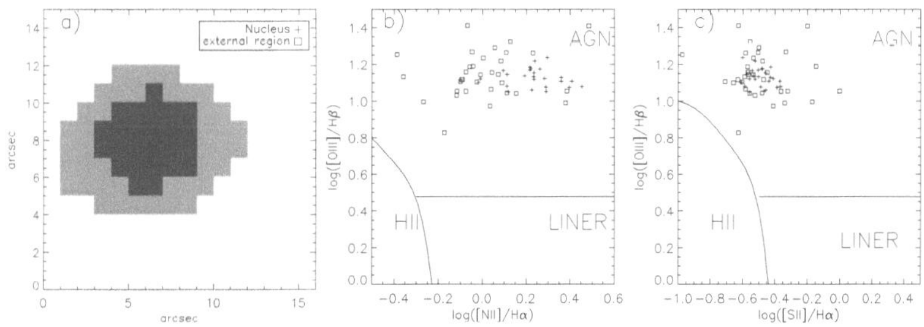


**Fig. 4.** Line ratio diagrams for Mrk1469: a) the mask in the [O III] emission line; b) [O III]/$H_\beta$ vs [N II]/$H_\alpha$; c) [O III]/$H_\beta$ vs [S II]/$H_\alpha$.

We thank the Organizing Committee for financial support and efficient organization of the conference.

## References

[1] V.L. Afanasiev, S.N. Dodonov, A.V. Moiseev: In: *Stellar dynamics: from classic to modern*, ed by L.P. Osipkov, I.I. Nikiforov (Saint Petersburg 2001) pp 103-109

[2] S. Veilleux, D.E. Osterbrock: ApJS, **63**, 295 (1987)

# Part VII

# Emission Nebulae and resolved Populations

# 3D Spectroscopy of Emission Nebulae

M. M. Roth

Astrophysikalisches Institut Potsdam
An der Sternwarte 16
D-14482 Potsdam, Germany
mmroth@aip.de

**ABSTRACT**

Emission nebulae like H II regions, Planetary Nebulae, Novae, Herbig Haro objects etc. are found as extended objects in the Milky Way, but also as point sources in other galaxies, where they are sometimes observable out to very large distances due to the high contrast provided by some prominent emission lines. It is shown how 3D spectroscopy can be used as a powerful tool for observations of both large resolved emission nebulae and distant extragalactic objects, with special emphasis on faint detection limits.

## 1 Introduction

Unlike the areas of extragalactic Astronomy and Cosmology, where 3D spectroscopy has been used quite extensively for a number of years (and, one should add, with great success, see e.g. Emsellem (2007) as well as other extragalactic papers in these proceedings), research on extended gaseous nebulae has, as yet, made suprisingly little use of the new observing technique. This may be linked to the fact that, obviously, instrument builders have designed 3D spectrographs preferentially for their interests in specific science cases, but also to the observation that it often requires considerable effort to become knowledgeable enough to reduce and analyze the data coming from those spectrographs. Newcomers would typically feel that there was a hurdle to overcome before one could hope to make efficient use of 3D spectroscopy data. It was precisely the goal to help alleviate this kind of a problem which formed the basic motivation and objective of the Euro3D Research Training Network (Walsh & Roth 2002).

Nevertheless, in the past, astronomers have indeed applied the technique for several studies on nebulae: e.g. Jacoby et al. (1998), who used the DensePak IFU at WIYN (Barden et al. 1998) to perform intermediate to high resolution integral field spectroscopy over several epochs on the "born again" planetary nebula associated with V4334 Sgr, also known as Sakurai's object. This work allowed to observe both the central star and the nebula, and yielded information on the precise size of the nebula, its expansion velocity, distance, and the cooling history of the star.

As another example, the goal to obtain high spatial resolution spectroscopy in the environment of pre-main-sequence stars in order to study in detail the origin of the jet phenomenon has motived several studies. For example Lavelley et al. (1997) used TIGER (Bacon et al. 1995) to investigate the jet of the T Tauri star DG Tau in the emission line of [O I] 6300Å with an effective spatial resolution of 0.75 arcsec. A subsequent study by Lavalley-Fouquet et al. (2000) with OASIS achieved an even higher resolution of 0.5 arcsec thanks to the PUEO adaptive optics system.

In the following, two extreme and, in a sense, quite opposite applications of 3D spectroscopy for planetary nebulae (PNe) will be reviewed, namely: (1) the use of spatial binning to obtain useful signal-to-noise ratios in extremely low surface brightness haloes of planetary nebulae; and (2) the use of PSF fitting techniques in the case of extragalactic planetary nebulae, which appear as point sources for galaxies more distant than the Magellanic Clouds.

# 2 Haloes of Galactic Planetary Nebulae

## 2.1 Motivation

The physics of planetary nebulae (PNe) is interesting for many reasons, such as: they are excellent laboratories for plasma physics, including the opportunity to measure elemental abundances; they allow insight into stellar evolution and nucleosynthesis; they are used as test particles to study the kinematics in external galaxies, to name just a few. The following is concerned with the radial structure of PNe; in particular the extended faint halo around the bright inner part of a nebula can be used as a record of the mass-loss history, when the predecessor star on the tip of the asymptotic giant branch (AGB) was experiencing pulsations and strong mass loss due to dust-driven winds. As this latter period is short-lived and therefore very difficult to observe directly, the analysis of PN haloes offers the opportunity to study this phase indirectly. Observational data on AGB mass-loss rates is urgently needed, as theory is presently making diverging predictions. Also, quantitative estimates concerning the recycling of matter from stellar populations to the interstellar medium is an important parameter for models of galaxy evolution. However, previous attempts to assess the electron temperature and density in PN haloes through plasma diagnostic analysis have been largely unsuccessful because of the extremely low surface brightness of important diagnostic emission lines, which are typically as faint as $10^{-17}$erg/cm$^2$/s/arcsec$^2$ and below, i.e. comparable to the flux of high redshift Ly$_\alpha$ emitters, whose spectra normally need 8-10m class telescopes for detection of the faint emission line.

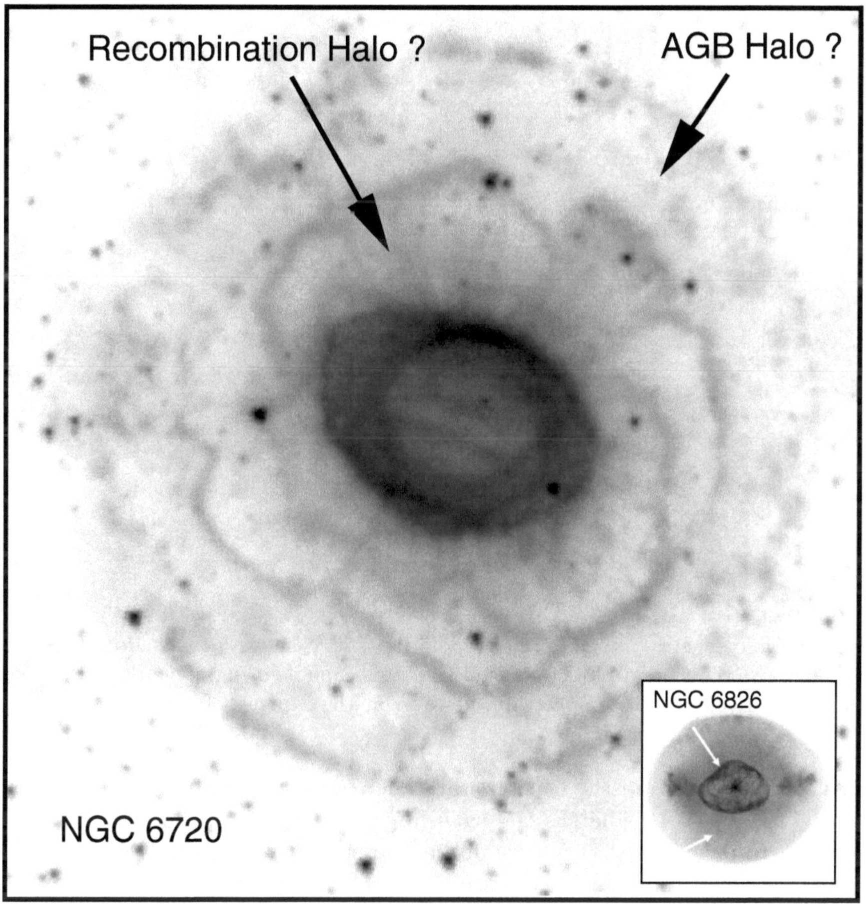


**Fig. 1.** Spitzer image of the "Ring Nebula" NGC 6720, showing the high surface brightness part of the nebula, and two low surface brightness haloes.

## 2.2 NGC 6720, the Ring Nebula

An example of a PN halo is shown in Fig.1 for NGC 6720, an image obtained with the Spitzer Observatory in the NIR. Surrounding the bright inner part of the nebula, there are actually *two* haloes with more or less regular circular symmetry. It is hypothesized that these haloes represent the remnants of the shell and the AGB halo from a previous stage of evolution, when the central star was still luminous, while it is now evolving towards the white dwarf stage ($T_{eff}$=130000 K, L=500 $L_{sun}$). The insert shows another PN at the earlier stage (NGC 6826), where rim and shell are clearly seen as distinct features – the halo is too faint to be visible on this image. In the picture of the hydrodynamical and photoionization evolution of the nebula, the inner bright structure represents the hot bubble, which is shock-heated by the

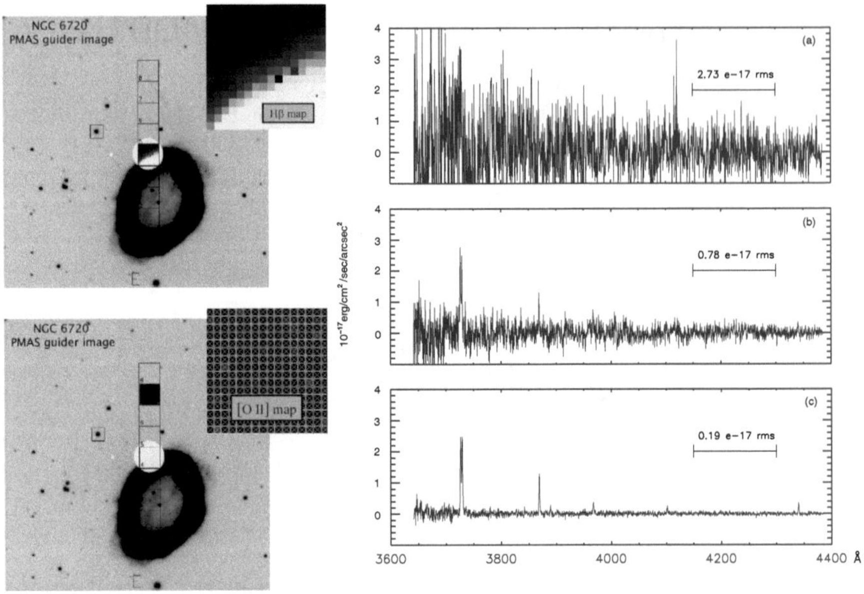


**Fig. 2.** Halo spectra of NGC 6720 for pointing 7. It is shown how coadding 1, 9, or 256 spaxels (right hand plots a-c respectively) is capable of significantly decreasing the noise of the combined spectrum, however at the expense of spatial resolution.

fast stellar wind from the central star. The somewhat fainter shell is heated by photoionization, whereas the even fainter, undisturbed halo structure is thought to contain the signature of mass-loss history on the AGB.

Fig. 2 shows how 3D spectroscopy can help to significantly improve the signal-to-noise ratio in very faint halo regions through the use of spatial binning. This technique was first introduced and is now routinely being used for observations in the faint outer parts of elliptical galaxies (Cappellari & Copin 2003). Since most present-day IFUs span a field-of-view (FOV) of at most a few tens of arcseconds, it was necessary to take a series of several exposures with offset pointings, in order to sample the radial structure of NGC 6720 for a sufficiently large interval from the bright part into the halo. Observations were performed with the PMAS lens array IFU at the Calar Alto 3.5m Telescope (Roth et al. 2005), providing a FOV of 16×16 arcsec and a spatial sampling of 1 arcsec/lens. The upper left panel in Fig. 2 shows the composite of a direct image of NGC 6720 with a reconstructed map of one such pointing in the wavelength of $H_\beta$. The region is bright enough to yield a sufficiently high S/N per spaxel, i.e. S/N>100 in each individual spectrum. The lower panel shows how in the halo this is by far not the case: a map in [O II] 3727/3729 Å reveals nothing but a random noise pattern. A single spectrum corresponding to this halo pointing is shown to the upper right. However, when coadding 9 (middle panel), or all 256 spectra of the IFU (lower panel), the S/N improves dramatically, down to an rms of $1.9\times10^{-18}$erg/cm$^2$/s for the averaged spectrum over an area of 256 arcsec$^2$.

## 2.3 M2-2

Another more recent example is shown with preliminary data for the object M2-2 in Fig. 3. The PN was observed on Sep.13, 2005, as a mosaic of 4 adjacent pointings using the PMAS lensarray in the 0.5 arcsec sampling mode, thus covering a total area of 32"×8". The mosaic is outlined in the PMAS A&G Camera R-band image (left). The rim/shell structure according to the scheme as explained in Corradi et al. (5), and is clearly visible, but no halo can be discerned from the direct image. While Tile 1 is centered on the nebula, Tiles 2. . .4 extend further out in the radial direction, roughly three times the radius of the shell (note that a field star is also contained in Tile 4). The panels to the right show the appearance of the PN in different emission lines, most strikingly the (expected) smaller size of the He II Stromgren sphere, compared to the size of the H sphere. In an attempt to confirm the presence

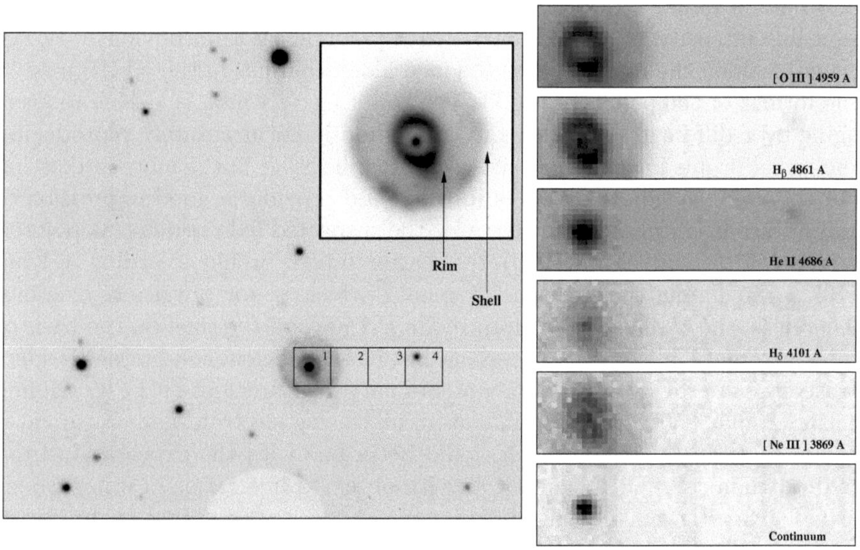


**Fig. 3.** Monochromatic maps of M2-2 in the wavelengths of [O III] 4959Å, $H_\beta$ 4861Å, He II 4686Å, $H_\delta$ 4101Å, [Ne III] 3869Å, and in the continuum (right, top to bottom respectively). The maps are extracted from a mosaic datacube, combined from 4 offset pointings 1-4 with exposure times of 600, 1800, 1800 and 3600s, respectively, as indicated in the direct R-band image (left). Data were obtained with the Calar Alto 3.5m Telescope, using the PMAS IFU and A&G camera. Insert image from Guerrero et al. (1998) for comparison, rim and shell edges are indicated with arrows.

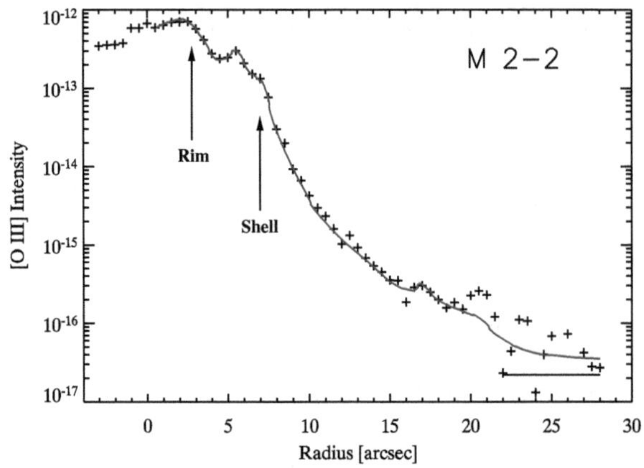


**Fig. 4.** M2-2 surface brightness profile in [O III] 5007 Å, derived from preliminary data reduction and analysis. Note the edges of rim and shell (see also Fig. 3).

of a halo, a radial surface brightness profile was constructed from the emission line intensity in [O III] $\lambda$5007. In a preliminary (and obviously rather crude) fashion the intensity was coadded along columns of the [O III] map in the format of the panels of Fig. 3. The resulting S/N gain is 4. Due to averaging over different radii, this simple method is not accurately reproducing the radial profile in the inner part of the PN; however in the more interesting outer parts, the simplification is justified and provides a good impression of the surface brightness decline towards the suspected halo regime. As pointed out by Middlemass et al. (1989), the scattered light profile according to King (1971) can mimic the presence of a halo, which is not physically existing. Therefore, the extended appearance of a PN beyond the shell on the basis of monochromatic images is a necessary, but not a sufficient condition. 3D spectroscopy can provide a solution: a preliminary measurement of the [O III] line ratio $\lambda\lambda(4959{+}5007)/\lambda4363$ hints at an increasing electron temperature at a radius of 10…20 arcsec, which would be in line with the expectation from hydrodynamical models, as well as with observed line ratios in other haloes (NGC 3587: Fig. 5; NGC 3242: see below). A more careful final analysis is in progress (Sandin et al. 2006, in prep.).

### 2.4 NGC 3242

With yet another gain in light collecting area and field-of-view (FOV), the binning method was employed for the measurement of NGC 3242 and NGC 4361 using the ESO-VLT with the VIMOS-IFU. This instrument is the largest 3D spectrograph to date, providing a total of 6400 spectra with a FOV as large as 54×54 arcsec$^2$ (Le Fevre et al. 2003).

From the results of Monreal-Ibero et al. (2005), Fig. 6 shows maps of NGC 3242 in [O III] and $H_\beta$ at different contrast levels (left), and various halo spectra (right). The most noisy spectrum (second from top) corresponds to a single spaxel at a radial distance from the central star of ≈1 arcmin. The spectrum above is from the same region, however averaged over 4547 spaxels, and shifted by 1 intensity unit for clarity. The third plot is the same spectrum, however sky-subtracted, revealing the emission lines of [O III] $\lambda\lambda$4959,5007, of $H_\beta$, $H_\gamma$, and of [O III] $\lambda$4363 at high S/N. Note that the intensity scale is on units of $10^{-18}$erg/cm$^2$/s/Å/arcsec$^2$. An interesting result from the [O III] line ratio is the detection of a gradient in the electron temperature over the halo from 15,700 K in the inner part to 20,300K near the edge. Another result of this work is the confirmation of the rings as reported by Corradi et al. 2004. No halo was detected for NGC 4361 down to the detection limit of $5 \times 10^{-18}$erg/cm$^2$/s/arcsec$^2$.

## 3 Extragalactic Planetary Nebulae

Extragalactic PNe (XPN) are luminous tracers of intermediate age stellar populations, which are relatively easy to detect owing to the good background contrast of the bright [O III] $\lambda$5007 emission line. Next to supergiants, XPN provide the only way to probe individual stars in galaxies beyond the Local Group. With narrowband imaging techniques they are e.g. abundantly found in galaxies of the Virgo cluster, in the Virgo intracluster space, and they have been even detected as far away as in the Coma cluster at a distance of 100 Mpc (Gerhard et al. 2005). An obvious goal would be the plasma diagnostic analysis of XPN emission line spectra for the purposes of the determination of electron temperature and density, and, ultimately, chemical abundances (Walsh et al. 2002). Using mostly 8m-class telescopes, there is significant progress to this end with XPN in local group galaxies (see ESO

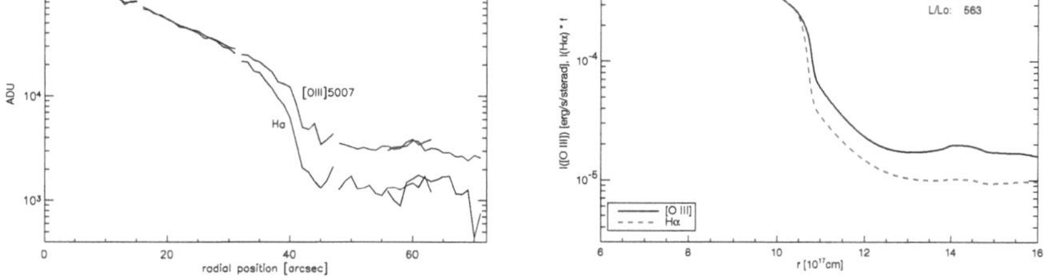


**Fig. 5.** Radial surface brightness profile in Hα and [O III] as determined for NGC 3587 from PMAS 3D spectroscopy (left) is compared to numerical model predictions (right).

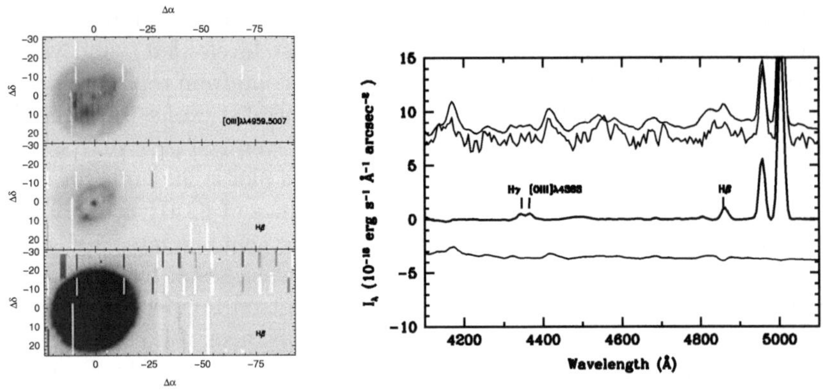


**Fig. 6.** Monochromatic maps of NGC 3242, created from VMOS IFU datacube.

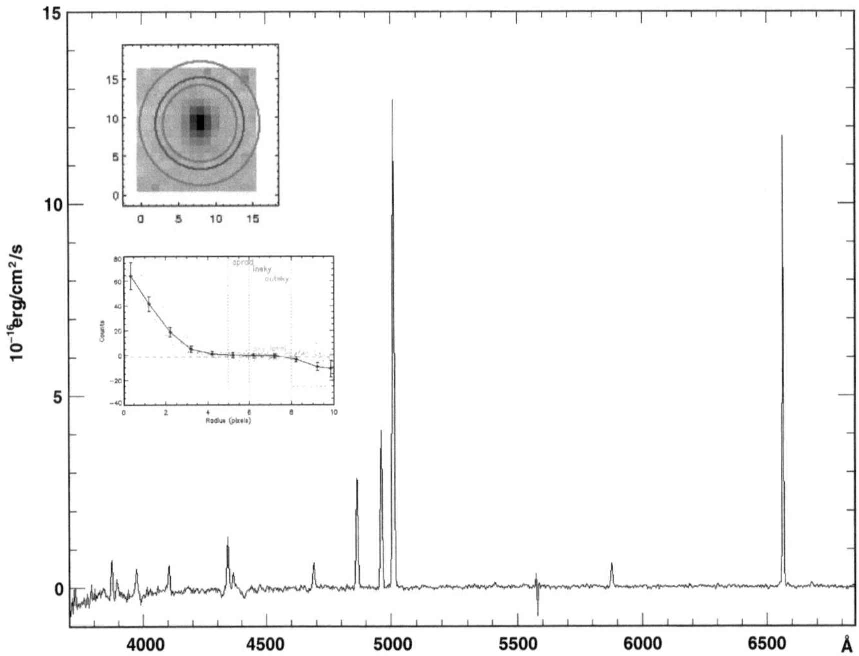


**Fig. 7.** Spectrum of PN in the Leo A dwarf galaxy. The inserts show a map in [O III] 4959 Å as extracted from a PMAS datacube, and a corresponding radial plot for the point source.

Workshop Proceedings "*Planetary Nebulae beyond the Milky Way*", eds. L. Stanghellini, J. R. Walsh, N. Douglas, 2006, for a wealth of detail on the subject).

However, as demonstrated with a pilot study in the nuclear region of M31 (Roth et al. 2004), the accurate spectrophotometry of XPN is significantly hampered by systematic errors of background subtraction in high surface brightness regions of their host galaxy. Such errors can be overcome by using 3D spectroscopy with similar techniques as in use for crowded field CCD photometry packages, i.e. by fitting a model PSF to the monochromatic slices of a datacube simultaneously. As an example, Fig. 7 illustrates how monochromatic maps share essentially the same properties with direct images, albeit over a much smaller FOV than what is state-of-the-art with modern CCD cameras and infrared arrays.

Fig. 8 shows how the PSF-fitting technique is capable of disentangling quite accurately the object spectrum from the bright background, which is both variable in the spatial and in the spectral domain (absorption line spectra from unresesolved stars, emission line spectra from gaseous ISM emission). This is generally not possible with conventional slit spectroscopy; more technical details can be found in Roth et al. 2004 (17)). Courtesy of J. Falcćon-Barroso, an S0 galaxy template spectrum is also shown for comparison, demonstrating that essentially all of the absorption line features of the template are also found in the background spectrum of the M31 bulge.

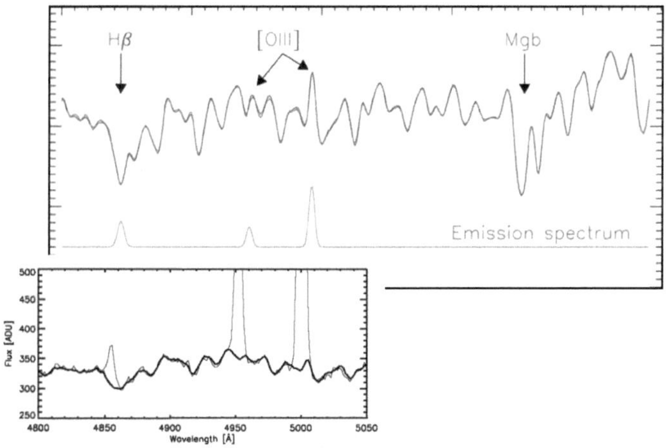


**Fig. 8.** XPN spectrum from the bulge of M31 with background spectrum (PMAS data, upper plot), overlayed on an S0 galaxy template spectrum for stellar population analysis (lower, template from the SAURON project, courtesy Jesús Falcón-Barroso).

Note that the PN emission line spectrum is blue-shifted due to its negative radial velocity.

## 4 Conclusions and Outlook

Although not yet common place for the study of emission nebulae in all of the astronomical community, the potential of 3D spectroscopy should be quite obvious and is now gradually being recognized. The examples of low-surface brightness 3D spectroscopy on the one hand, and high spatial resolution work on the other hand, have highlighted only two possible applications of many more opportunities for planetary nebulae, H II regions, jets of YSO's, and so forth. In particular the study of extragalactic objects warrants further attention. With the impressive suite of powerful 3D spectrographs in the optical and NIR as facility instruments at essentially all major 4-8m class telescopes, it is to be expected that research in the field will significantly grow and benefit from the new capabilities. While the first generation of instruments was generally limited by small FOV, there are now 3D spectrographs available with a coverage of $\approx$1 arcmin in diameter, e.g. VIMOS, PMAS-PPak (Kelz et al. 2005), and other new facilities under development, e.g. MUSE (Bacon et al. 2002, Roth & Bacon these proceedings). A major breakthrough has come through the combination of IFUs with adaptive optics, e.g. SINFONI at the VLT, OSIRIS at Keck. Deployable IFUs are providing multiplex advantage for the study of well separated, but extended individual objects (e.g. FLAMES/VLT). It is now up to the observers to make best use of these facilities.

## References


[1] Bacon, R. et al. 1995, A&A, 113, 347
[2] Bacon, R., et al. 2002, in *Scientific Drivers for ESO Future VLT/VLTI Instrumentation*, eds. J. Bergeron, G. Monnet, Springer, Berlin, p. 108
[3] Barden, S.C., Sawyer, D.G., Honeycutt, R.K. 1998, SPIE 3355, 892
[4] Cappellari, M., & Copin, Y. 2003, MNRAS 342, 345
[5] Corradi, R. L. M. et al. 2003, MNRAS 340, 417
[6] Corradi, R. L. M. et al. 2004, A&A 417, 637
[7] Emsellem, E. 2006, these proceedings
[8] Jacoby, G.H., de Marco, O., Sawyer, D.G. 1998, AJ 116, 1367
[9] Gerhard, O. et al. 2005, ApJ 621, L93
[10] Kelz, A., et al. 2006, PASP 118, 129
[11] King, I.R. 1971, PASP 83, 199
[12] Lavalley, C., Cabrit, S., Dougados, C., Ferruit, P. 1997, A&A 327, 671
[13] Lavalley-Fouquet, C., Cabrit, S., Dougados, C. 2000, A&A 356, L41

[14] LeFevre, O., et al. 2003, SPIE 4841, 1670
[15] Middlemass, D., Clegg, R. E. S., & Walsh, J. R. 1989, MNRAS 239, 5
[16] Monreal-Ibero, A. et al. 2005, ApJ 628, L139
[17] Roth, M. M., Becker, T., Kelz, A., & Schmoll, J. 2004, ApJ 603, 531
[18] Roth, M. M., et al. 2005, PASP 117, 620
[19] Walsh, J.R., Jacoby, G.H., Peletier, R.F., Walton, N.A. 2000, SPIE 4005, p.131
[20] Walsh, J. R. & Roth, M. M. 2002, The Messenger, 109, 54

# GMOS-IFU Observations of LV2

M. J. Vasconcelos[1,2], A. H. Cerqueira[1,2], H. Plana[1], A. Raga[2] and C. Morisset[3]

[1] Laboratório de Astrofísica Teórica e Observacional - LATO/UESC - Brazil
mjvasc@uesc.br,hoth@uesc.br,plana@uesc.br
[2] Instituto de Ciencias Nucleares - ICN/UNAM - México
raga@nucleares.unam.mx
[3] Instituto de Astronomía - IA/UNAM - México morisset@astroscu.unam.mx

**Summary.** We present GMOS-IFU spectroscopic observations of the proplyd 167-317 (LV2) near the Trapezium cluster in the Orion nebula, obtained at the Gemini South Observatory. The proplyd was observed in optical wavelengths ranging from $\simeq$ 5515 Å to $\simeq$ 7630Å. The ionized photo-evaporated flow was detected together with a microjet and possibly the blue-shifted jet that appear in most of the detected emission lines as multiple peaks at different velocities. Typically, there is a low velocity peak at $\sim$ 30 km s$^{-1}$, a high velocity red-shifted peak at $\sim$ 100 km s$^{-1}$ and a very faint blue-shifted peak at $\sim -60$ km s$^{-1}$. The observations also allowed us to determine spatially the emitting regions. The red-shifted jet is located to the SE of the central emission. We find evidences that the red-shifted jet has a variable velocity, which slowly drops with increasing distance from the proplyd

## 1 Introduction

Proplyds are extended objects found in regions of star formation near OB associations. They were initially detected as unresolved emission line objects by (1) and then as compact radio sources by (2). However, it was not until recently that the real nature of proplyds could be understood, when (3) (4), using HST, detected spatially resolved emission in H$\alpha$, [O III]$\lambda$5007 Å , [O I]$\lambda$6300 Å and some other emission lines. With this information, Johnstone et al. (5) formulated a model in which the proplyds are explained as YSOs exposed to the dissociating and ionizing radiation field from an O star placed nearby. The radiation field initially photo-dissociates (6) mainly $H_2$ from the surface of the accretion disk generating a neutral photo-evaporated flow that, after passing the ionization front, becomes ionized. The size of the photo-dissociating region, the distance between the ionization front and the surface of the disk and the velocity distribution of the photo-evaporated flow changes with distance from the ionizing star. Some other models (7) were also able to reproduce the shock between this photo-evaporated flow and the stellar wind that appears as a bow shock visible in in H$\alpha$ and in [O III]$\lambda$5007 Å (8) and the cometary tail commonly seen in proplyds (9).

In particular, the proplyd 167-317 or LV2 is located in the Orion Nebula at a projected distance of $\sim 5 \times 10^{16}$ from $\theta^1$ Ori C, the main ionizing source for this $H_2$ region. LV2 presents the main features seen in other proplyds: a cometary shape, a bright ionization front and the bow shock generated by the wind-wind collision. It presents also a microjet with a red-shifted component which is readily detected (10; 11) and a fainter blue-shifted component (12; 11).

In this work, we present IFU - 3D spectroscopic observations of LV2 taken at the Gemini South observatory. The 3D spectroscopy is very useful for proplyds because it permits one to have a detailed spatial and spectral information. In particular, the resolution provided by the GMOS-IFU spectrograph has permitted us to isolated the emission from the photo-evaporated flow and the microjet.

## 2 Observations and Data Reduction

The data were taken during the System Verification run of the GMOS-IFU at the Gemini South Telescope (GST) on 2004 February $26^{th}$ and $27^{th}$. The science field of view (FOV) is $3''\!.5 \times 5''$ with an array of 1000 lenslets of $0''\!.2$. Our sampling is 0.34Å per pixel ($\approx$ 15 km s$^{-1}$ per pixel at H$\alpha$) and the spectral coverage is between $\simeq$ 5515 Å to $\simeq$ 7630Å. The instrumental profile is 47 km s$^{-1}$ $<$ FWHM $<$ 63 km s$^{-1}$, decreasing with increasing wavelength. The data were reduced using the standard Gemini IRAF v1.6 [4] routines. We maintain the IFU original resolution of $0''\!.2$ px$^{-1}$ although an interpolation was performed in order to turn the IFU hexagonal lens shape into squared pixels. In order to extract high energy events we constructed an IDL routine described in more details in (11). The background subtraction is done through the fitting of an intensity versus position plane in a semi-rectangular field located $\approx 1'' - 1''\!.5$ away from the peak proplyd emission in each of the individual velocity channel maps which is subsequently subtracted from each pixel. In Figure 1 we show the observed field, the object orientation in the plane of the sky, and four regions relevant to this work: region R1 that includes the the proplyd peak emission (region R2) and the jet peak emission (region R3). For this region R1 we have defined an arbitrary $xy$-coordinate system limited by $1 \le x \le 11$ (in the East direction) and $1 \le y \le 17$ (in the North direction).

## 3 Results

The integrated spectra obtained for each one of the above defined regions show 38 emission lines. Many of them are not detected in region R3 but this

[4]IRAF is distributed by the National Optical Astronomy Observatories, which are operated by the Association of Universities for Research in Astronomy, Inc., under cooperative agreement with the National Science Foundation.

could be caused by its low signal-to-noise. Other lines, like those of Si III, Si II λ5979, [Ni II] and [Fe II] are stronger in the background indicating that they are at least partially absorbed by the dust in the proplyd.

Near the central region, the line profile is broad and can be fitted by 2 or 3 Gaussians, as can be seen in Figure (2). In this Figure, we plot the Hα line profile for pixel (x,y) = (6,7), located inside region R2 (see Figure 1). In the top right panel, we show the observed line profile as a solid line superimposed by the fit, in this case, a combination of 3 Gaussians, represented by crosses. In the top left panel, we show the only the data and the respective fit corresponding to the main, low velocity component of the line profile. It peaks at $\sim$ 30 km s$^{-1}$ and it is related to the photo-evaporated flow. The plot in the left bottom panel shows the data subtracted from the fit for the main component. We can see that there is a clearly visible high velocity ($\sim$ 100 km s$^{-1}$) component that can be related to the red-shifted microjet. This high velocity component appears in other regions, mainly in the SE, between regions R2 and R3 (see Figure 8 of (11)). Also, a fainter component can be seen, with a peak velocity of $\sim -75$ km s$^{-1}$. This component appears in other pixels at the NW of region R2 and can be related to the blue-shifted microjet of LV2 (see also (12)). In the right bottom panel we show the data subtracted from the composed 3 Gaussian fit. It gives us a measure of the goodness of the

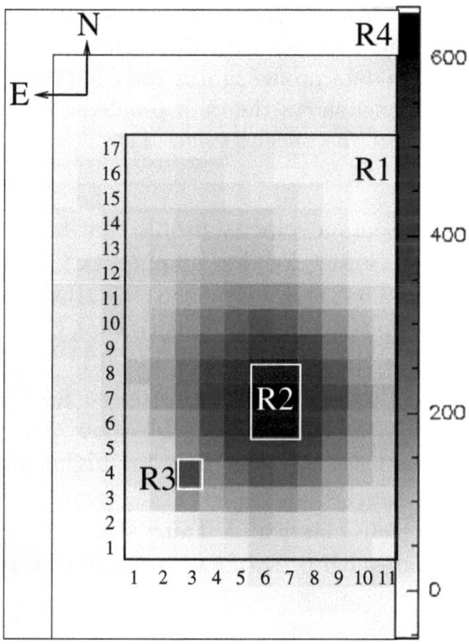


**Fig. 1.** Hα map of the proplyd LV2 obtained with the GMOS-IFU spectrograph. Superposed is a sketch showing the relevant regions for this work. Figure extracted from Vasconcelos et al. (11).

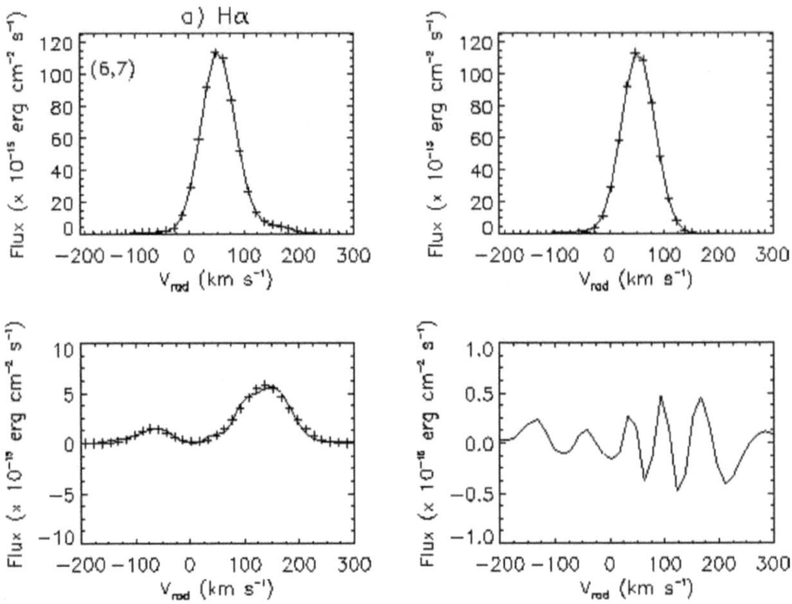


**Fig. 2.** Hα line profile at pixel x,y = 6,7 (see Figure 1). Observed data is represented by the solid line while Gaussian fits by crosses. The left upper panel shows the line profile superimposed by the three Gaussian fits. The right upper panel shows the main, low velocity component together with the corresponding Gaussian fit. The left bottom panel depicts the data profile subtracted from the main component fit. Finally, the right bottom panel shows the data profile subtracted from the three components fit. Adapted from Vasconcelos et al. (11).

fit. The red-shifted jet presents a velocity profile that falls off with increasing distance from the proplyd center, located at pixel (x,y) = (6,7) (11).

Using the ratio ([N II]λ6548 + [N II]λ6583)/[N II]λ5754, we were able to obtain the spatially distributed electron densities for LV2. We find $n_e > 10^5$ $cm^{-3}$ with an increasing of the density at region R3.

Also, we derived mass loss rates for the proplyd itself and the microjet. The values are, respectively, $(6.2\pm0.6)\times10^{-7}$ $M_\odot$ $year^{-1}$ and $(2.0\pm0.7)\times10^{-8}$ $M_\odot$ $year^{-1}$. The jet mass loss rate is a little bit higher than typical values found for jets around T Tauri stars ($\sim 10^{-9}$ $M_\odot$ $year^{-1}$) an indication that this YSO is more active than classical T Tauri stars. The proplyd mass loss rate, however, is consistent with previous values found in the literature (10).

## 4 Conclusions

We have used the GMOS-IFU capabilities to observe the proplyd 167-317 or LV2. This instrument has proved to be very useful for this object because it has permitted us to spatially resolve the emission and in this way to isolate contributions from the different components of this proplyd.

In particular, we were able to study the emission and kinematics of the photo-evaporated flow and the red-shifted microjet associated with LV2. Also, we could detect, although marginally, the blue-shifted component of the jet. The red-shifted jet presents a behavior consistent with optical jets related to non irradiated sources, with a mass loss rate value indicating an active accretion object.

The derived proplyd mass loss rate is consistent with previous values in the literature and it imposes an age of at most $10^5$ years for $\theta^1$ Ori C with this rate is constant.

## 5 Acknowledgement

This work was supported by FAPESB, UESC, the Milenium project (CNPq Project n. 62.0053/01-1-PADCT III/Milênio), by the CONACYT grants 41320-F and 43103-F and DGAPA (UNAM) grants IN112602 and IN111803.

## References

[1] P. Laques, J. L. Vidal: A&A **73**, 97 (1979)
[2] E. Churchwell, M. Felli, D. O. S. Wood, M. Massi: ApJ, **321**, 516 (1987)
[3] C. R. O'Dell, Z. Wen: ApJ, **436**, 194 (1994)
[4] C. R. O'Dell, K. Wong: AJ, **111**, 846 (1996)
[5] D. Johnstone, D. Hollenbach, J. Bally: ApJ, **499**, 758 (1998)
[6] H. Störzer, D. Hollenbach: ApJ, **502**, L71(1999)
[7] F. García-Arredondo, W. J. Henney, S. J. Arthur: ApJ. **561**, 830 (2001)
[8] J. Bally, R. S. Sutherland, D. Devine, D. Johnstone: AJ, **116**, 293 (1998)
[9] S. Richling, H. W. Yorke: ApJ, **539**, 258 (2000)
[10] W. J. Henney, C. R. O'Dell, J. Meaburn, S. T. Garrington, J. A. Lopez: ApJ, **566**, 315 (2002)
[11] M. J. Vasconcelos, A. H. Cerqueira, H. Plana, A. C. Raga, C. Morrisset: AJ, **130**, 1707 (2005)
[12] T. Doi, C. R. O'Dell, P. Hartigan: AJ, **127**, 3456

# The Enigma of Ultraluminous X-ray Sources may be Resolved by 3D-Spectroscopy (MPFS Data)

S. Fabrika and P. Abolmasov

Special Astrophysical Observatory, Russia
fabrika@sao.ru

**Summary.** The ultraluminous X-ray sources (ULXs) were isolated in external galaxies for the last 5 years. Their X-ray luminosities exceed 100-10000 times those of brightest Milky Way black hole binaries and they are extremely variable. There are two models for the ULXs, the best black hole candidates. 1. They are supercritical accretion disks around a stellar mass black hole like that in SS433, observed close to the disk axes. 2. They are Intermediate Mass Black Holes (of 100-10000 solar masses). Critical observations which may throw light upon the ULXs nature come from observations of nebulae around the ULXs. We present results of 3D-spectroscopy of nebulae around several ULXs located in galaxies at 3-6 Mpc distances. We found that the nebulae to be powered by their central black holes. The nebulae are shocked and dynamically perturbed probably by jets. The nebulae are compared with the SS433 nebula (W50).

## 1 Introduction

The main properties of the ultraluminous X-ray sources (ULXs) – huge luminosities ($10^{39-41}\ erg/s$), diversity of X-ray spectra, strong variability, connection with star-forming regions, their surrounding nebulae. ULXs may be supercritical accretion disks observed close to the disk axis in close binaries with a stellar mass black hole or microquasars ((1), (2), (3)). Another idea is that ULXs may be intermediate-mass black holes (IMBHs) with "normal" accretion disks ((4), (5)). It is also possible that ULXs are not a homogeneous class of objects.

It was suggested originally by Katz (6) that SS433 being observed close to the jet axis, will be extremely bright X-ray source. Fabrika & Mescheryakov (1) discussed observational properties of face-on SS433-like objects and concluded that they may appear as a new type of extragalactic X-ray sources. In (7) we discussed possible properties of the funnel in the supercritical accretion disk of SS433. We predicted X-ray spectra and temporal behaviour of the funnel in "face-on SS433" star in application for ULXs. Here we continue to develop this idea and consider nebulae surrounding the ULXs sources.

The main difference between SS433 and other known X-ray binaries is highly supercritical and persistent mass accretion rate ($\sim 10^{-4}\,M_{\odot}/y$) onto the relativistic star (a probable black hole, $\sim 10M_{\odot}$), which has led to the formation of a supercritical accretion disk and the relativistic jets. SS433 properties were reviewed recently by (8).

Similar to SS 433 the ULXs are connected with nebulae. They are frequently located in bubble-like nebulae. New data (9) show that the nebulae are expanding with a velocity $\sim 80\,km/s$ (up to $\sim 250\,km/s$). The nebula sizes are from 20 to a few hundred parsecs, such nebulae are easy for observations even from "megaparsec distances". Here we compare the gas nebula around SS 433 with nebulae of ULXs in Holmberg II, NGC 6946 and IC 342 galaxies observed recently by the Integral-field spectroscopy methods.

## 2 Observations and data reduction

We discuss results of observations on the Russian 6-m telescope with the integral field spectrograph MPFS (10). The integral field unit of 16×16 square spatial elements covers a region of 16"×16" on the sky. Integral field spectra were taken in the spectral range 4000 – 6800 Å with a seeing 1.0-1.3″ (FWHM). Data reduction was made using procedures developed in IDL environment (version 6.0) by V. Afanasiev, A. Moiseev and P. Abolmasov and include all the standard steps.

## 3 Nebulae surrounding ULXs and SS433

The object SS 433 is surrounded by the elongated radio nebula W 50, which was produced (or distorted) by SS433 jets due to the jet interaction with interstellar medium (11). The radio emission is synchrotron, relativistic electrons appear at the jet deceleration. Bright optical filaments are observed (12) in places of the jets termination, they radiate in HI, [OI], [NII], [SII] lines. Optical filaments in the bipolar nebula are located at $\pm 0.5°$ or $\pm 50\,pc$ from SS433. A total energy of the nebula is $E_k \sim 2 \cdot 10^{51}\,erg$ (12), which corresponds to the jet kinetic luminosity $L_k \sim 3 \cdot 10^{39}\,erg/s$ for 20000 years. The observed velocity dispersion in the filaments is $\sim 50\,km/s$, however [NII]/H$\alpha$ line ratio corresponds to dispersion $\sim 300\,km/s$ (12). SS433 is an edge-on system, $i = 79°$. If one takes into account this factor, the velocity dispersion may reach $250 - 300\,km/s$. The ULXs nebulae studied by us, have about the same sizes, the same line luminosities and about the same total energy budget $10^{51-52}$ erg (13).

In Fig 1 we present emission line maps of the nebula MF 16 ("a peculiar SN remnant"), surrounding ULXs in the galaxy NGC 6946. In [OIII] $\lambda 4959, 5007$ lines we found a radial velocity gradient across the nebula in East-West direction. In later observations using spectrograph SCORPIO (14) in a long-slit mode we confirmed this gradient. The radial velocity gradient reaches 100 km/s across the whole nebula (20 pc).

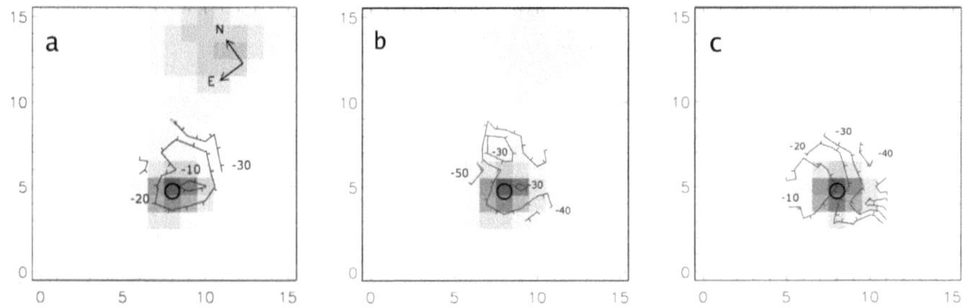

**Fig. 1.** Hα (a), [SII] λ6717, 6734 (b) and [OIII] λ4959, 5007 (c) 15 × 15" maps of MF 16 nebula surrounding the NGC 6946 ULX-1. Circles show location of the X-ray source from GHANDRA's data. Equal radial velocity lines are shown. Marks show a direction of increasing of the velocity absolute value.

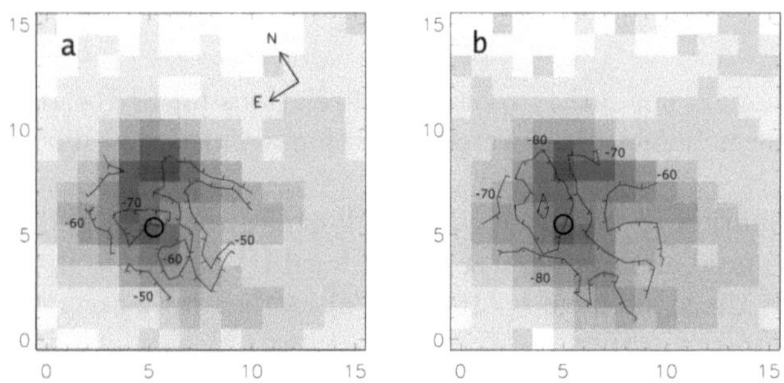

**Fig. 2.** Hα (a) and [SII] λ6717, 6734 (b) maps of the nebula surrounding the IC 342 ULX-1. Designation are the same as in Fig 1.

In Fig 2 we present results of observations of the nebula surrounding the ULX-1 in the galaxy IC 342 using the same device. In this nebulae we have also detected a radial velocity gradient as a whole expansion of the nebula ±20 km/s. This nebula is not bright because of strong light absorption in direction of IC 342.

Fig 3 presents results of observations of Holmberg II ULX-1 nebula with the MPFS and SCORPIO spectrographs. Bright He II λ4686 emission line is observed in this nebula. We have also detected a radial velocity gradient ±50 km/s on a spatial scale ±50 pc in the nebula (15). The radial velocity across the nebula has been measured in more details in observations with SCORPIO.

In all three studied ULXs nebulae diagnostic line ratios indicate collisional excitation of the gas (see (14) for more details). We obtain two more conclusions which may be principal for understanding of the ULXs:
1. The radial velocity gradient of 50 – 100 km/s on spatial scales 20 – 100 pc testify that the nebulae are dynamically perturbed. The IMBHs can not perturb the interstellar gas on such big scales, the capture Bondi radii are not greater than 0.1 pc. These nebulae can not be SNRs, they are too big and

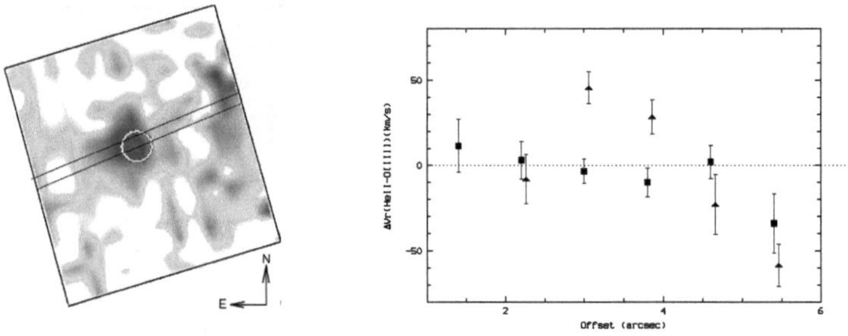


**Fig. 3.** HeII $\lambda\,4686$ map of the nebula surrounding the Holmberg II ULX-1 (left). Slit positions in LS-spectroscopy are also shown. Radial velocities (right) of this line measured along the slits relative to the [OIII] $\lambda\,5007$ line are shown by squares for the upper slit and by triangles for the bottom slit.

energetic, they do not satisfy to standard relations for SNRs. It is very probable that the nebulae are powered by stellar wind or jets like that it is in SS 433.
2. For explanation of luminosities in high excitation lines and of the whole spectrum we need an additional source of hard UV radiation (14). The source luminosity is the same huge ($\sim 10^{40}$ erg/s) as that in X-rays.

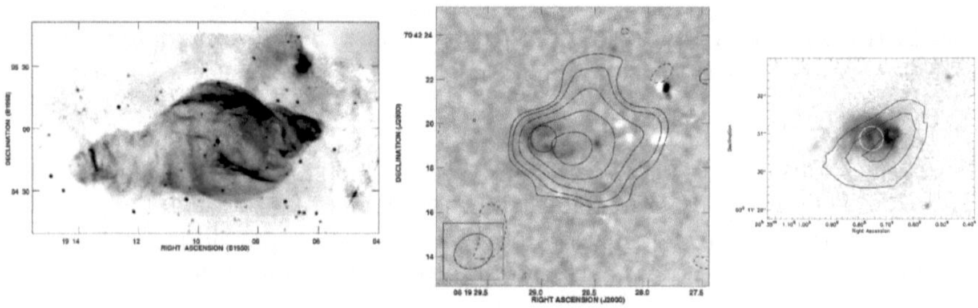


**Fig. 4.** Three nebulae in the same scale in parsecs. VLA image of W 50 (11) with SS 433 in center, left; Holmberg II ULX-1 in HST HeII image with VLA isophotes, middle (16) and NGC 6946 ULX-1 in HST H$\alpha$+[SII] image with VLA isophotes (17), right. Circles show X-ray Chandra positions.

In Fig. 4 we present in the same linear scale the nebula W 50 together with nebulae surrounding ULXs in Holmberg II and NGC6946 galaxies (15), (13). The nebulae in Holmberg II and NGC 6946 have circle-like features in the line-images. In both cases the radio sources are shifted to a brighter circle-like feature. The radio sources are not resolved. In the both cases the part of the nebulae coinciding with radio source is approaching, the opposite part is receding (15), (13). At some imagination one may conclude that the nebulae around these two ULXs are face-on versions ($i = 10° - 30°$) of the SS 433 nebula. We need to continue observations to take more representative sample of nebulae connected with ULXs.

## Acknowledgements

This work has been supported by the RFBR grant 04–02–16349 and RFBR/JSPC grant N 05–02–19710. The authors are grateful to the SOC and LOC of the Workshop "Science Perspectives for 3D Spectroscopy" for support.

## References

[1] S. Fabrika, Mescheryakov: In: *Galaxies and their Constituents at the Highest Angular Resolution*, IAU Symp. N205, ed by R.T. Schilizzi, (Manchester, United Kingdom 2001) p. 268
[2] A.R. King, M.B. Davies, M.J. Ward, G. Fabbiano, M. Elvis: ApJ (Letters) **552**, L109 (2001)
[3] E. Koerding, H. Falcke, S. Markoff, R. Fender: Astron. Gesells. Meet. Abstr. **18**, 176 (2001)
[4] E.J.M. Colbert, R.F. Mushotzky: ApJ **519**, 89 (1999)
[5] J.M. Miller, A.C. Fabian, M.C. Miller: ApJ (Letters) **614**, L117 (2004)
[6] J.J. Katz: ApJ **317**, 264 (1987)
[7] S. Fabrika, S. Karpov, P. Abolmasov, O. Sholukhova: In *Populations of High Energy Sources in Galaxies*, IAU Symposium N230, ed by E.J.A. Meurs & G. Fabbiano, 2006 in press
[8] S. Fabrika: *Astrophysics and Space Physics Reviews*, **12**, 1 (2004)
[9] M. Pakull: In *Populations of High Energy Sources in Galaxies*, IAU Symposium N230, ed by E.J.A. Meurs & G. Fabbiano, 2006 in press
[10] V.L. Afanasiev, S.N. Dodonov, A.V. Moiseev: In *Stellar dynamics: from classic to modern*, ed by Osipkov L.P., Nikiforov I.I. (Saint Petersburg 2001) p. 103
[11] G.M. Dubner, M. Holdaway, W.M. Goss, I.F. Mirabel: AJ **116**, 1842 (1998)
[12] W.J. Zealey, M.A. Dopita, D.F. Malin: MNRAS **192**, 731 (1980)
[13] S. Fabrika, P. Abolmasov, O. Sholukhova: 2006, in preparation
[14] P. Abolmasov, S. Fabrika, O. Sholukhova: 2006, in preparation
[15] I. Lehmann, T. Becker, S. Fabrika et al: A&A **431**, 847 (2005)
[16] N.A. Miller, R.F. Mushotzky, S.G. Neff: ApJ (Letters) **623**, L109 (2005)
[17] S.D. van Dyk, R.A. Sramek, K.W. Weiler et al: ApJ (Letters) **425**, L77 (1994)

# Spatially Resolved Spectroscopy of Abell 30

K. M. Exter[1] and L. Christensen[2]

[1] IAC, La Laguna, Tenerife, Spain kme@star.ucl.ac.uk
[2] ESO, Santiago, Chile lichrist@eso.org

**Summary.** Spatially resolved spectroscopy on a sample of planetary nebulae (PNe) has been carried out using INTEGRAL (WHT) and PMAS (Calar Alto). We present here our preliminary results for Abell 30, a nebula with a system of outflowing, H-deficient knots surrounding the central star. We have investigated the spectra of these knots, and find distinct differences to the surface brightness distribution in different emission lines (different ions), particularly interestingly so for the C and O recombination lines.

## 1 Introduction

PNe are the penultimate evolutionary stage for low-intermediate mass (LIM) stars. The outer layers of the precursor asymptotic giant branch star is expelled through the action of a wind (and other activities), to form a visible nebula when the stellar core (how a hot subdwarf) becomes hot enough to ionise it. PNe are bright emission line objects, especially in the forbidden lines of [O III] (e.g. 5007 Å), H$\alpha$ and [N II]. As such, they are often used in galactic studies: as tracers of the underlying (and much less visible) LIM stellar population; in measuring a galaxy's kinematics; and in galactic metallicity studies (especially for the oxygen abundances).

A planetary nebula, one would assume, should essentially be spherically symmetric (as it was expelled from a spherically symmetric star), but this, in fact, is not so (see, eg. catalogues of (1) and (6) or the *HST* pictures web pages). Many PNe are spherically symmetric, of course, but they also come in a wide variety of other shapes: highly elongated bipolar, elliptical, multi-polar, butterfly, point-symmetric, with jets... Deep imaging has shown that they can be surrounded by enormous faint halos and rings. There is also much substructure within the nebulae themselves (clumps, knots, filaments, arcs). In addition, while the expansion velocities of the nebula are generally not high (20 km/s is typical), there are those with outflows ('jets') travelling in the 100's km/s.

With such interesting (and still puzzling) overall shapes, and the significant level of substructure, PNe are worthy of study with integral field spectroscopy (IFS) in their own right. Not only could IFS allow for detailed studies of their reasonably complex velocity fields (especially when multiple lobes are superimposed on the line-of-sight, or when various outflowing structures have different velocities and directions), but also of studies of the changing physical and chemical conditions in the substructures. It is this point that forms the basis of our first studies of PNe with IFS.

## 2 Goal of our present study

It has been known for some time that for most PNe there is a difference in the chemical abundances derived from optical recombination lines (ORLs) and collisionally excited lines (CELs; forbidden lines), with the ORLs giving the lower electron temperatures ($T_e$) and higher chemical abundances. For most PNe this is only a factor of 2, but for 5–10% it is over a factor 10 (9),(5). Various explanations have been offered (see (8), for a good overview), but we are concentrating on the idea (5) that for at least those with with large differences, this is caused by the presence of dense, cool, metal-rich inclusions (knots). The ORLs arise from the core of these knots, and the CELs from the surrounding envelope (and/or the rest of the nebula). This has been researched with slit spectroscopy (e.g. for Abell 30 by (10),(3)), and we begin our IFS investigation with Abell 30, a PN for which metal-rich knots are easily visible. Our long-term goal is to extend this project to other PNe, particularly those for which the ORL-CEL discrepancy is high but the inclusions are not so obviously visible (and thus more difficult to find with slit spectroscopy).

## 3 Abell 30

Abell 30 is an old planetary nebula, with a large and faint spherically symmetric nebula. Surrounding the H-deficient, hot central star, however, is a system of 'polar' and 'equatorial' outflowing knots (7),(11), which are also H-poor. The ORL-CEL discrepancy factor for the polar knots (named J3 and J1) is known to be very large (factors 100's; (10)), but the equatorial knots (J4 and J2) have not been studied in much detail. As well as using photoionisation analysis (e.g. (3)) to show that the spectra of all the knots can be fit with the inclusions model, we will use the spatial resolution to investigate differences within the knot systems, and for differences between the polar and equatorial systems (e.g. were they ejected in different events?).

## 4 Spectral-maps

The data reduction included a correction for the differential atmospheric refraction (using the central star as a guide), which spatially shifts the blue and red ends of the spectra (3700–6600 Å) by just under one INTEGRAL spaxel (0.9 arcsec non-contiguous circles, covering 12x16 arcsec). The data

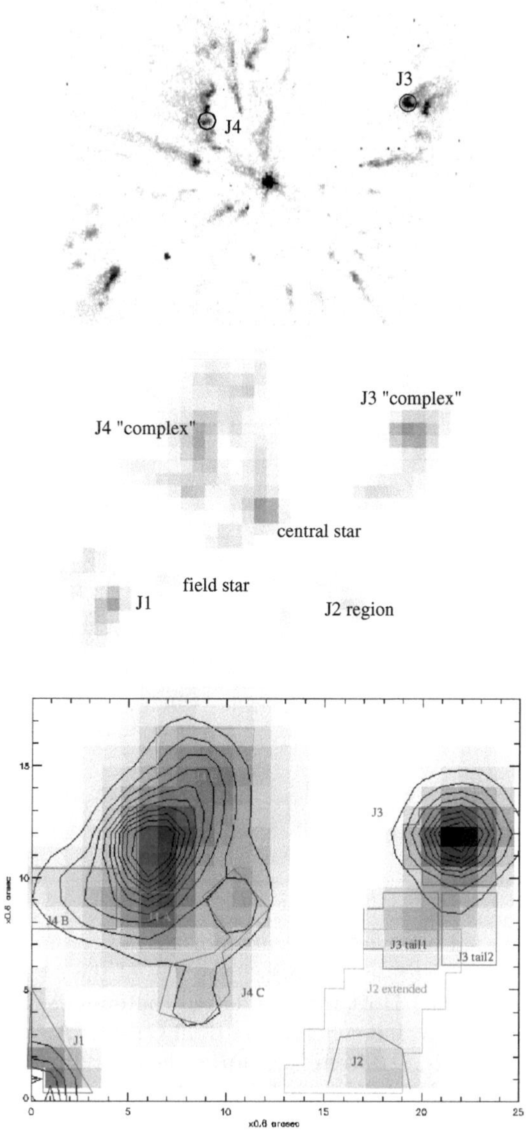


**Fig. 1. Top** An image made in [O III] 5007 Å from our INTEGRAL data, compared to the *HST* image, reduced to the same seeing and pixelation. The various knot complexes are identified. *HST* image is based on that of Borkowski et al. **Bottom** In greyscale is [O III] and over-plotted are contours (from 10% of the peak flux in J4) of [N II].

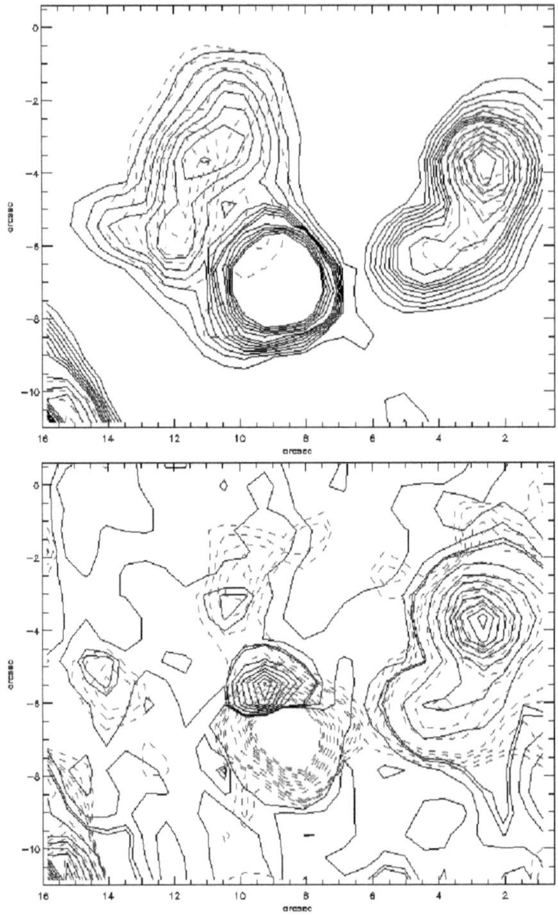

**Fig. 2. Top** In dotted contours is [O III] from 30% of the peak flux in J4, and in full, the contours for He II 4686 Å, also from 30% of the J4 peak. (The central feature is the central star.) **Bottom** The dotted contours are for the wavelength slice centred on C II 4267 Å, full contours are from O II 4076 Å. The C II map scales from 20% of the peak flux, O II from 30%, up to the peak flux, with a jump in the contour levels to reduce the crowding in knot J3. (The central feature is the central star.)

were spatially re-binned to 0.6 arcsec continuous square spaxels. That this reproduces the true appearance well is shown in Fig. 1, where the re-binned *HST* image (taken from (2)) is compared to the [O III] 5007 Å spectral image.

In Figs. 1–2 are the overlaid contours of the emission line fluxes of [N II] 6584 Å and [O III], He II 4686 Å and [O III], and C II and O II. Our detailed analysis of these (and other) images, as well as the measured lines fluxes extracted from the various distinct parts of the knot system, will be presented in Exter et al. (A&A), however it is already obvious from inspection of these maps that the surface brightness distribution between the compared ions differs. The map of the C and O ORLs is particularly interesting; comparison to Fig. 1 shows that the peak for C II in J4 (faint, but real) lies not in the brightest clumps of that knot complex, but rather in the fainter (and less dense) tails. If the ORLs arise from cool, dense knots, this is not what one would expect.

The different surface brightness distributions mapped probably simply mean that the physical conditions for the knots vary; which does not seem, perhaps, such a revelation. *However*, it is something that is unlikely to have ever been be picked up with slit spectroscopy, and to our knowledge, rarely has. This highlights the advantage of IFS for the study of such nebulae. Indeed, from our data we have even found it possible to extract the spectra of the knots very close to the central star (which can be seen in as 'J4C' in Fig. 1). These data, therefore, will allow us to study variations in the physical and chemical conditions in the various knots, *within* a single complex and *between* different complexes, at a level of spatial detail not possible before. It is our hope this this will shed light on how and where/when the knots complexes were formed.

## References

[1] A. Acker, F. Ochsenbein, B. Stenholm, R. Tylenda, J. Marcout, C. Schohn, Strasbourgh-ESO Catalogue of Galactic Planetary Nebulae. European Southern Observatory, 1992
[2] K.J. Borkowski, J.P. Harrington, Z. Tsvetanov, R.E.S. Clegg, ApJ, **449**, 143 (1995)
[3] B. Ercolano, M.J. Barlow, P.J. Storey, X.-W Liu, T. Rauch, K. Werner, MNRAS, **344**, 1145 (2003)
[4] X.-W Liu, P.J. Storey, M.J. Barlow, I.J. Danziger, M. Cohen, M. Bryce, MNRAS, **312**, 585 (2000)
[5] Y Liu, X.-W Liu, M.J. Barlow, S.-G Luo, MNRAS, **353**,1251 (2004)
[6] A. Manchado, M. Guerrero, L. Stanghellini, M. Serra-Ricart. The IAC Morphological Catalogue of Northern Galactic Planetary Nebulae. (La Laguna: IAC), 1996.
[7] J. Meaburn, J.A. Lopez, ApJ, **472**, 45 (1996)
[8] R.H. Rubin, P.G. Martin, R.J. Dufour, G.J. Ferland, K.P.M. Blagrave, X.-W Liu, J.K. Nguyen, J.A Baldwin, MNRAS, **340**, 362 (2003)
[9] Y.G. Tsamis, M.J. Barlow, X.-W Liu, I.J. Danziger, P.J. Storey, MNRAS, **353**, 953 (2004)
[10] R. Wesson, X,-W Liu, M.J. Barlow, MNRAS, **340**,253 (2003)
[11] Y. Yadoumari, S. Tamura, PASP, **106**, 165 (1994)

# Constraining the Galactic Structure with Stellar Clusters in Obscured H II Regions

M. Messineo[1], M. Petr-Gotzens[1], K. M. Menten[2], F. Schuller[2] and H. J. Habing[3]

[1] ESO, Garching, Germany mmessine@eso.org
[2] MPIfR, Bonn, Germany
[3] Leiden Observatory, The Netherlands

**Summary.** We present first results of a program with the integral field spectrometer SINFONI on the ESO-VLT (Very Large Telescope) Yepun to observe obscured stellar clusters in H II regions in the infrared H and K bands with the goals of obtaining a more complete and less biased picture of the Milky Way's spiral structure, and of improving our understanding of the spatial distribution of Galactic H II regions. The classification in spectral subclasses of early type stars together with the extinction determination and the stellar apparent magnitudes enable us to obtain spectrophotometric distances, which are independent and complementary to kinematic distances and free from the ambiguity inherent to the latters' determination. Most importantly, our method is much less limited by interstellar extinction than previous optical programmes, allowing studies along lines of sight where optical measurements are impossible.

## 1 Introduction

There is multiple evidence that our Milky Way is a barred spiral galaxy, but the properties of, both, spiral and bar remain poorly constrained.

Regions of star formation are useful probe of Galactic structure, in particular of the spiral arms. But their distance determinations are extremely problematic and often rely on "kinematic" estimates, from radio spectral line measurements, which have an inherent ambiguity ("near" or "far") and are based on the assumption of circular orbits. (10) in their historical work on the spiral structure of the Milky Way used, both, kinematic and spectrophotometric distances, the latter from visual observations of OB stars. A re-analysis of the spatial distribution of H II regions by (20) shows that despite the availability of new radio and optical data no further improvements can be achieved, mainly because H II regions at a large distance from the Sun are typically not optically observable. Thus, the actual number (estimates range from 2 to 4) and location of spiral arms remain still uncertain. Also, there is a puzzling apparent dearth of H II regions within 3 kpc of the Galactic centre (e.g. (6)). A real lack of H II regions would suggest a "hole" in the gaseous distribution of the inner Galaxy. However, their census could be extremely incomplete because of high interstellar extinction. Note that the distribution of late-type stars shows strong signature of a Galactic bar extending to 3.5 kpc and of a nuclear stellar disk with a radius of about 800 pc ((18), (12)).

The recent availability of large mid- and near-infrared surveys, such as 2MASS, ISOGAL and MSX, has led to the discovery of several hundreds of new Galactic HII regions, e.g. (11), and candidate infrared stellar clusters. The study of their spatial distribution by combining kinematic and infrared spectrophotometric information will greatly improve our knowledge on the structure and therefore formation and evolution of the Milky Way.

### 1.1 Spectrophotometric distances of stellar clusters in HII regions

We have compiled a list of 647 Galactic young clusters which were mostly selected by inspecting 2MASS images at the position of known HII regions by (2) and (1), (7), (8), (15), (5), (14). Their young nature is confirmed by associated continuum emission at 1.4 GHz (NRAO VLA Sky Survey) – for 85% of them. Only for some of these clusters infrared follow-up studies exist.

H-band and K-band spectra allow a spectral classification of early type stars, by using photospheric hydrogen, H, and helium, He, lines and other infrared lines such as those of atomic carbon and nitrogen – e.g. the He I(4-3) line at 1.700 $\mu$m, the Brackett H I(11-4) transition (1.681$\mu$m), the Brackett $\gamma$, $Br_\gamma$ line at 2.16$\mu$m, the NIII(2.1155$\mu$m) and CIV(2.078$\mu$m), and the HeI atomic lines at 2.058$\mu$m and 2.1126$\mu$m. A medium resolution (R$\approx$2000-5000) spectrum in H and K-band with a signal-to-noise ratio above 100 yields a classification of early type stars consistent with that derived in the visual within 1 or 2 subclasses (13; 19). Luminosity classes are more uncertain; they can be estimated by the equivalent widths of the Brackett lines, where dwarfs show EW(H I(11-4)) $\approx$ 60 Åand supergiants $\approx$ 40 Å((4)), and by the presence of certain emission lines.

The classification of the spectral type together with the extinction estimate and the apparent luminosity yields the stellar *distance* (e.g. (16)). Although the uncertainty of the spectrophotometric distance to each star can be up to 20 – 30%, the accuracy of the distance to the cluster can be improved by averaging the distances to several cluster stars.

### 1.2 Observations with SINFONI and first results

We have used SINFONI (9) to observe eight stellar clusters (W31a, W31b, W42, W43, W51a, W51b, W51A IRS2, and Dutra-8) in order to determine their spectrophotometric distance. The program includes observations of candidate OB stars in both the H (1.45 – 1.85$\mu$m) and K (1.95 – 2.45) bands at the highest SINFONI resolution (R $\sim$ 3000 and 4000, respectively). For each cluster, when resolved in the 2MASS K-band image, candidate stars were selected having 2MASS K $<$ 12 mag and (H–K) $>$ 0.6 mag – the latter constraint to exclude foreground objects. Furthermore, in almost all cases the targets are near the cluster's centre. Since high-mass stars are known to form at such locations, the identification of the stars thus selected as cluster members is unambiguous. Typically, 4 – 6 stars per cluster were chosen.

A first look at the data taken during last August confirms that we have successfully reached the initial aim of detecting photospheric lines of OB stars which yield a stellar spectral-classification and therefore distance estimates. The four brightest stars detected in K-band in the W31 cluster show NIII and CIV in emission, $Br_\gamma$ and HeI and HeII in absorption (Fig. 1). Such lines are typical of dwarfs of spectral type 05 ±1. Therefore we confirm the distance to W31 of 3.4 kpc ((3)). By performing integral field spectroscopy with SINFONI, we simultaneously image and resolve the core of the stellar clusters and obtain spectra. Since often in one field more than one star can be measured, it is typically possible to obtain a complete census of the cluster's brightest stars by employing a few telescope offsets in a single 1 hr observation sequence (Fig. 1). Furthermore, SINFONI enables us to detect simultaneously nebular emission (Fig. 2) (such as molecular hydrogen lines, HI recombination lines and ionized iron) and the spectrum of the candidate ionizing star and allows distinguishing between these contributions. Finally, the detection of $Br_\gamma$ emission from the HII region provides an estimate of the extinction to the HII region when compared with radio continuum flux measurements.

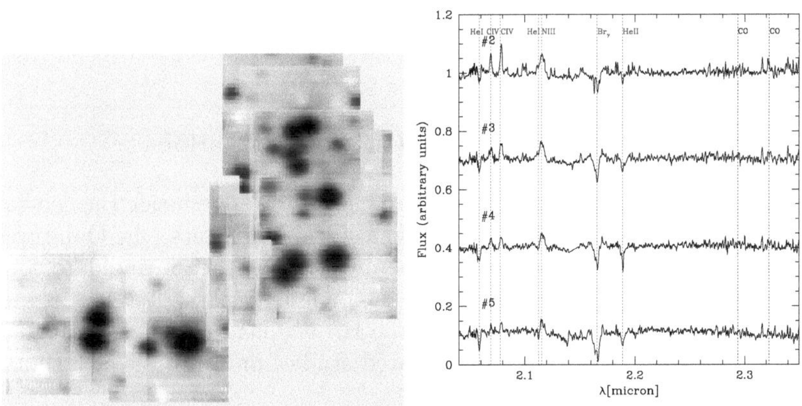


**Fig. 1.** SINFONI K-band data of the stellar cluster W31. The mosaic (left) was obtained with one single observing block (about 1 hour). On the right K-band spectra of the 4 brightest stars are shown.

Figure 2 shows the Galactic stellar cluster Dutra-8, successfully resolved with SINFONI. A late O dwarf (09 – B1) star and a nebula surrounding it were detected. The spectral type and the determination of apparent magnitudes yield an estimate of interstellar extinction ($A_K = 3.2$ mag) and a spectrophotometric distance of 2kpc ± 0.8 kpc (Messineo et al. in preparation).

The derived distance agrees well with one of the two gas complex distances detected along the line of sight ((21)).

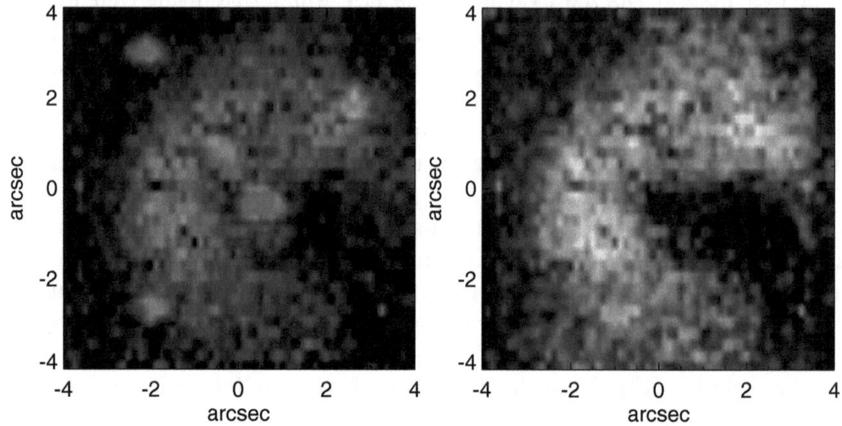


**Fig. 2.** *Left:* Three colour image of the Galactic stellar cluster, Dutra-8. Blue represents the H-band image, green the Br-gamma image, and red the K-band image. The cluster center is resolved and four bright stars are detected in both H and K band. The central star appears surrounded by a nebula. *Right:* Three colour image of the nebula. Here, blue is $H_2$ emission at 2.12 $\mu$m, green $Br_\gamma$ emission, and red [FeII] emission at 1.644 $\mu$m.

### 1.3 Outlook and complementary (radio) observations

Only few young and massive clusters are known to reside inside the central 3 kpc, all of them concentrated within the central 50 parsecs (the Quintuplet, Arches, and Sgr A clusters), though the number of 2MASS candidate young stellar clusters located in the direction of the inner Galaxy $|l| < 60°$ is quite large (241). Kinematical properties of the HII regions can help selecting candidate inner Galactic clusters for infrared studies on the basis of forbidden and peculiar velocities.

We have collected line-of-sight velocities of the HII regions from literature (e.g. (17)) and started a program with the Effelsberg 100 meter radio telescope to detect radio recombination lines. The $l - v$ diagram of those 155 clusters (see Fig. 3) does not show forbidden velocities for pure circular rotation, like those seen in the CO map, e.g. at negative velocities between $20° > l > 0°$. There is a lack of high velocities at zero longitudes which are typical of the nuclear disk. From the $l - v$ diagram we conclude that almost all of these clusters are outside of the co-rotation radius (3.5 kpc) on quasi-circular orbits. So far, using infrared and kinematic information we have identified two new candidate inner Galaxy clusters. We expect to find many

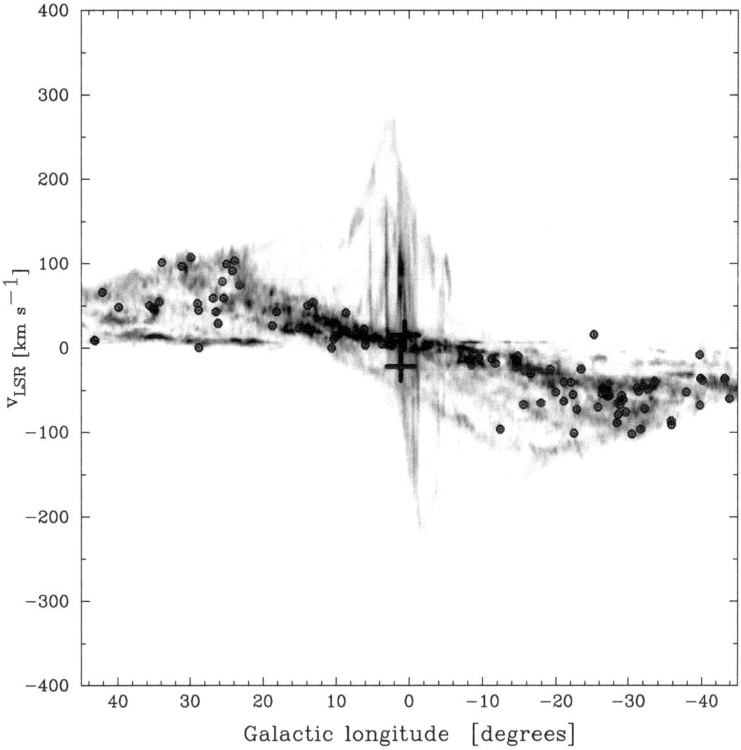


**Fig. 3.** Longitude-velocities of 155 young stellar clusters in direction of the inner Galaxy, overlaid on the CO map (grayscale) of Dame et al. (2001). The distribution of line-of-sight velocities of the stellar clusters suggests that the majority of the known clusters are located outside of co-rotation (3.5 kpc) on quasi-circular orbits. Candidate inner Galactic clusters are marked with crosses.

more such clusters in the near future, when UKIDSS, a deeper near-infrared survey of the Galactic plane, becomes available.

## References

[1] Bica, E., Dutra, C. M., Barbuy, B., 2003, A&A, 397, 177
[2] Bica, E., Dutra, C. M., Soares, J., Barbuy, B., 2003, A&A, 404, 223
[3] Blum, R. D., Damineli, A., & Conti, P. S. 2001, AJ, 121, 3149
[4] Blum, R. D., Ramond, T. M., Conti, P. S., et al., 1997, AJ, 113, 1855
[5] Borissova, J., Pessev, P., Ivanov, V. D., et al., 2003, A&A, 411, 83
[6] Comeron, F., Torra, J., 1996, A&A, 314, 776
[7] Dutra, C. M., Bica, E., 2000, A&A, 359, L9
[8] Dutra, C. M., Bica, E., Soares, J., Barbuy, B., 2003, A&A, 400, 533
[9] Eisenhauer, F., Abuter, R., Bickert, K., et al., 2003, SPIE, 4841, 1548

[10] Georgelin, Y. M., Georgelin, Y. P., 1976, A&A, 49, 57
[11] Giveon, U., Becker, R. H., Helfand, D. J., White, R. L., 2005, AJ, 129, 348
[12] Habing, H. J., Sevenster, M., Messineo, M., et al., 2005, submitted to A&A
[13] Hanson, M. M., Kudritzki, R.-P., Kenworthy, M., 2005, ApJS, 161, 154
[14] Ivanov, V. D., Borissova, J., Bresolin, F., Pessev, P., 2005, A&A, 435, 107
[15] Ivanov, V. D., Borissova, J., Pessev, P., et al., 2002, A&A, 394, L1
[16] Koornneef, J., 1983, A&A, 128, 84
[17] Kuchar, T. A. & Clark, F. O., 1997, ApJ, 488, 224
[18] Messineo, M., Habing, H. J., Sjouwerman, et al., 2002, A&A, 393, 115
[19] Repolust, T., Puls, J., Hanson, et al., 2005, A&A, 440, 261
[20] Russeil, D., 2003, A&A, 397, 133
[21] Russeil, D., Georgelin, Y. M., Georgelin, Y. P., et al., 1995, A&AS, 114, 557

# Integral Field Spectroscopy of a Peculiar Supernova Remnant MF16 in NGC6946

P. Abolmasov[1,2], S. Fabrika[2], O. Sholukhova[2] and V. Afanasiev[2]

[1] Moscow State University
[2] Special Astrophysical Observatory RAS

**Summary.** We present a study of a peculiar Supernova Remnant MF16, associated with the Ultraluminous X-ray Source (ULX) NGC6946 ULX-1. Observations were taken with the Multi-Pupil Fiber Spectrograph (MPFS) on 6-m telescope in January 2005. The nebula is found to be highly asymmetric, one of the parts being much denser and colder. Two-component structure of the emission lines and radial velocity gradient in some of them argue for a non-spherical shell, expanding with a velocity of about $100\,km\,s^{-1}$. Neither shock models nor X-ray emission can adequately explain the actual emission line spectrum of MF16, so we suggest an additional ultraviolet source with a luminosity of about $10^{40} erg\,s^{-1}$. We confirm coincidence of the ULX with the central star and identify radio emission observed by VLA with the densest part of the nebula.

## 1 Introduction

The attention to MF16 was first drawn by Blair & Fesen (1), who identified the object as a Supernova Remnant (SNR), according to the emission-line spectrum with bright collisionally-excited lines. It was long considered an unusually luminous SNR, because of its huge optical emission-line ($L_{H\alpha} = 1.9 \times 10^{39} erg\,s^{-1}$, according to (1), for the tangential size $20 \times 34pc$) and X-ray ($L_X = 2.5 \times 10^{39} erg\,s^{-1}$ in the $0.5 - 8$keV range (2)) luminosities.

However, it was shown by Roberts & Colbert (2), that the spectral, spatial and timing properties of the X-ray source do not agree with the suggestion of a bright SNR, but rather suppose a point source with a typical "ULX-like" X-ray spectrum: cool Multicolor Disk (MCD) and a Power Law (PL) component. So, apart from the physical nature of the object, MF16 should be considered a *ULX nebula*, one of a new class of objects, described by Pakull & Mirioni, (3). For introduction on ULX-counterparts see (4).

## 2 Observations and Data Reduction

We analyse spectra obtained on the SAO 6m telescope using Multi-pupil Fiber Spectrograph (MPFS, see (6)), with a spectral resolution $\Delta\lambda \simeq 8$ Åin $4000 - 7000$ Åspectral range. MPFS field consists of $16 \times 16$ $1'' \times 1''$ spaxels. For the nebula size $1'' \times 1.''5$ it means rather poor spatial resolution, yet one can resolve the structure of the object, considering offset pixels, were the PSF wings of the closest parts of the nebula contribute.

Data reduction system was written in IDL environment, using procedures, written by V. Afanasiev, A. Moiseev and P. Abolmasov. We also added atmospheric dispersion correction, in order to calculate correctly the nebula barycenters.

## 3 Results

More complete version of the results will be published in (7). Integral fluxes, velocities, FWHMs and barycenters for selected lines are listed in Table 1. The integral unreddened flux in $H_\beta$ line is $F_{H\beta} = (2.00 \pm 0.06)10^{-14} erg\,cm^{-2}s^{-1}$. Assuming $H_\alpha/H_\beta = 3$, we find interstellar absorption $A_V = 1.^m24$. Unreddening was produced using Cardelli et al. (5) algorithm. In the Table 1 we present the most important parameters of selected emission lines, including barycenter shifts versus *Chandra* source coordinates (2). An interesting result can be seen that the barycenter is shifted to the West for low-excitation lines (Balmer lines, [SII], [OI]), and to the East for the most high-excitation line HeIIλ4686. That can be understood as a global physical conditions gradient: the western part of MF16 is denser and colder than the eastern.

**Table 1.** MF16 selected emission line properties.

| *line* | $F(\lambda)/F(H_\beta)$ | $F(\lambda)/F(H_\beta)$ (unreddened) | $V, km\,s^{-1}$ | $FWHM$,Å | $\Delta\alpha_c, arcsec$ | $\Delta\delta_c, arcsec$ |
|---|---|---|---|---|---|---|
| *HeII* λ4686 | 0.20±0.03 | 0.22±0.03 | -5±18 | 6.8±0.7 | 0.302±0.05 | 0.146±0.03 |
| $H_\beta$ | 1.00±0.03 | 1.00±0.03 | -28±4 | 7.0±0.2 | -0.11±0.02 | -0.10±0.02 |
| *[OIII]* λ5007 | 7.42±0.15 | 6.94±0.14 | -17±2 | 6.7±0.1 | 0.058±0.015 | 0.011±0.015 |
| *[OI]* λ6300 | 1.42±0.08 | 0.93±0.05 | 25±9 | 10.1±0.4 | -0.196±0.015 | -0.237±0.015 |
| $H_\alpha$ | 4.73±0.09 | 2.96±0.06 | -16±2 | 7.8±0.1 | -0.111±0.013 | -0.08±0.013 |
| *[NII]* λ6583 | 4.16±0.08 | 2.59±0.05 | -19±2 | 7.6±0.1 | -0.061±0.013 | -0.064±0.013 |
| *[SII]* λ6717 | 2.46±0.02 | 1.496±0.015 | -38±2 | 7.7±0.1 | -0.116±0.014 | -0.081±0.014 |

For some of the line profiles, that seem to have distinct two-component structure, we also produce two-gaussian fit (see Table 2). The velocity shift between the line components is about $120 km\,s^{-1}$ for all the lines resolved into two components. This value is consistent with the velocity gradient seen for [OIII]λ5007 line, exceeding $50 km\,s^{-1}$ (7) and directed along the major axis of the nebula. Recent observations with higher spectral resolution (8) confirm the result, that the nebula expands with either $\sim 120$ (if we suggest just one shock wave with precursor), or $\sim 60 km\, s^{-1}$ (if we see the both receding and approaching parts of the shell) velocity. In all the profiles the blue components are significantly broader, corresponding to velocity dispersion $300 - 400 km\, s^{-1}$.

**Table 2.** MF16 two-component emission lines properties

| line | $F_1(\lambda)/F(H_\beta)$ | $V_1, km\, s^{-1}$ | $FWHM_1$,Å | $F_2(\lambda)/F(H_\beta)$ | $V_2, km\, s^{-1}$ | $FWHM_2$,Å |
|---|---|---|---|---|---|---|
| $HeII$ $\lambda 4686$ | 0.107±0.007 | -142±20 | 11.6±0.6 | 0.093±0.004 | 10.±5. | 4.26±0.15 |
| $H_\beta$ | 0.424±0.037 | -122±17 | 10.24±0.50 | 0.58±0.02 | -2±5 | 5.67±0.16 |
| $[OIII]$ $\lambda 4959$ | 0.875±0.014 | -122±3 | 8.5±0.1 | 1.486±0.012 | 8±1 | 5.78±0.12 |
| $[OIII]$ $\lambda 5007$ | 2.696±0.015 | -115±1 | 8.77±0.04 | 4.455±0.012 | 11±2 | 5.63±0.05 |

Having the spectral lines' fluxes and kinematics, one can judge about the nature of the nebula. It is well-known (10; 2) that the optical emission-line luminosity of MF16 exceeds very much that of a SNR with comparable size, so one should suggest either a more powerful shock or an additional photoionizing source. If we consider pure shock-wave excitation, the $H_\beta$ flux density will be connected to the shock parameters as was stated by Dopita & Sutherland (11):

$$F_{H\beta} = F_{H\beta,shock} + F_{H\beta,precursor} = 7.44 \times 10^{-6} \left(\frac{V_s}{100 km\, s^{-1}}\right)^{2.41} \times \left(\frac{n_2}{cm^{-3}}\right) + 9.86 \times 10^{-6} \left(\frac{V_s}{100 km\, s^{-1}}\right)^{2.28} \times \left(\frac{n_1}{cm^{-3}}\right) erg\, cm^{-2} s^{-1}$$

Here $n_1$ and $n_2$ are pre- and postshock number densities, correspondingly. Basing on this formula and assuming the nebula a sphere with 13pc radius, and $n_2 \simeq 500 cm^{-3}$ (10) one can see the potential shock velocity vary from $150 km\, s^{-1}$ (for $n_1 = 100 cm^{-3}$, adiabatic shock) up to $350 km\, s^{-1}$ (for $n_1 \rightarrow 0 cm^{-3}$, fully radiative shock).

## 4 High-excitation lines

[OIII] and HeII lines in MF16 appear to be unusually bright for a shock-ionized nebula. Observed ratio of [OIII]$\lambda 5007$ / $H_\beta \sim 7$ requires a shock velocity about $400 km\, s^{-1}$ (12). As for HeII / $H_\beta \sim 0.2$ value, it cannot be explained by existing shock+precursor models (12; 11)

HeII ionization suggests a quite specific photo-ionizing source – extreme ultraviolet (EUV) with $\lambda < 228$ Å. An estimate of the EUV luminosity can be made using ionizing quanta number estimate (13):

$$L_{\lambda<228\mathring{A}} \geq \frac{4\,\mathrm{Ry}}{E_{\lambda 4686}} \frac{\alpha_B}{\alpha^{eff}_{HeII\,\lambda 4686}} \times L_{HeII\lambda 4686} \simeq 10^{39} erg\, s^{-1}$$

# 5 Photoionization Modeling

We have computed a grid of CLOUDY96.01 (14) photoionization models in order to fit MF16 spectrum avoiding shock waves. We have fixed X-ray spectrum known from Chandra observations (2), assuming all the plasma is situated at 10pc from the central isotropic point source, and introduced a blackbody source with the temperature changing from $10^3$ to $10^6$K and integral flux densities from 0.01 to 100 $erg\, cm^{-2}\, s^{-1}$.

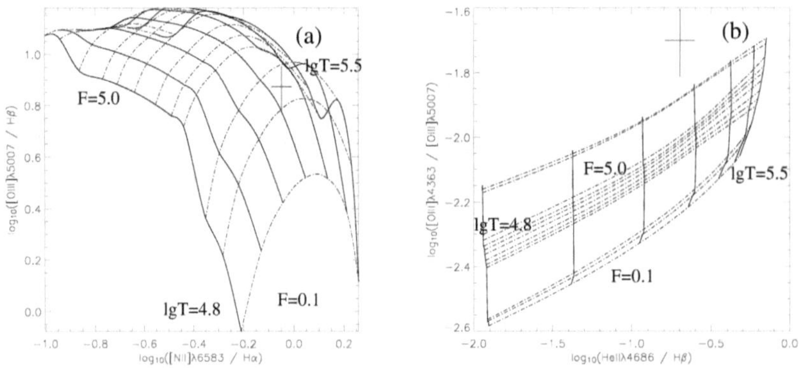


**Fig. 1.** $[OIII]\lambda 5007/H_\beta$ versus $[NII]\lambda 6583/H_\alpha$ (a) and $HeII\lambda 4686\ /\ H_\beta$ versus $[OIII]\lambda 4363\ /\ [OIII]\lambda 5007$ (b) ionization diagrams with the integral spectrum of MF16 shown by a cross of error bars. Photoionization grid described in the text is also shown. Solid lines trace the constant temperature levels, dot-dashed lines correspond to constant black body flux densities. Only models with $\lg T(K) = 4.80 - 5.50$ and $F = 0.1 - 7.0 erg\, cm^{-2}\, s^{-1}$ are shown.

The best fit parameters are $\lg T(K) = 5.15{\pm}0.05$, $F = 0.6{\pm}0.1 erg\, cm^{-2}\, s^{-1}$, that suggests quite a luminous ultraviolet source: $L_{UV} = (7.5 \pm 0.5) \times 10^{39} erg\, s^{-1}$, more than 100 times brighter then what can be predicted by extrapolating the best-fit mode for X-ray data (2).. More completely the modeling results and methods will be presented in (8).

## 6 Conclusions

MF16 has a nontrivial emission line spectrum, similar to those observed in NLRs of Seyfert galaxies. The nebula is highly asymmetric, with a dense cold western part, possibly connected with the radio-bright region (15). Kinematical properties suggest an expanding non-spherical shell. The sources of ionization acting in different parts of the nebula are probably different: strong radiative shock at the western end and in the outer regions, and hard UV/EUV radiation source in the inner/eastern regions.

Basing on the HeII4686 luminosity, we should suggest a $\sim 10^{39} erg\, s^{-1}$ EUV ($\lambda \leq 228\,\text{Å}$) source, responsible for the second ionization of He. CLOUDY simulations suggest a $L_{UV} \sim 10^{40} erg\, s^{-1}$ FUV luminosity needed to produce the actual [OIII]$\lambda$5007 / $H_\beta$ ratio.

The research has been supported by RFBR grants 04-02-16349 and 03-02-16341, and also by joint RFBR/JSPS grant 05-02-19710.

## References

[1] Blair, W. P., Fesen, R. A. ApJ, **424**, 103 (1994)
[2] Roberts, T. P., Colbert, E. J. M.: MNRAS, **341**, 49 (2003)
[3] Pakull, M. W., Mirioni, L.: RevMexAA (Serie de Conferencias), **15**, 197 (2003)
[4] Fabrika, S., Abolmasov, P., Sholukhova, O.: *this issue* (2005)
[5] Cardelli, J. A., Clayton, G. C., Mathis, J. S.: ApJ, **345**, 245 (1989)
[6] Afanasiev V.L., Dodonov S.N., Moiseev A.V.: In *Stellar dynamics: from classic to modern*, ed by Osipkov L.P., Nikiforov I.I., Saint Petersburg, p 103 (2001)
[7] Fabrika, S., Abolmasov, P., Sholukhova, O., Afanasiev, V.: *in preparation* (2006)
[8] Abolmasov, P., Fabrika, S., Sholukhova, O., Afanasiev, V.: *in preparation* (2006)
[9] Walsh, W., Beck, R., Thuma, G., Weiss, A., Wielebinski, R., Dumke, M.: A&A, **388**, 7 (2002)
[10] Blair, W. P., Fesen, R. A., Schlegel, E. M.: The Astronomical Journal, **121**, 1497 (2001)
[11] Dopita, M. A., Sutherland, R. S.: ApJSS, **102**, 161 (1996)
[12] Evans, I., Koratkar, A., Allen, M., Dopita, M., Tsvetanov, Z.: ApJ, **521**, 53 (1999)
[13] Osterbrock, D. E.: *Astrophysics of Gaseous Nebulae* (ed by W. H. Freeman and Company, San Francisco 1974)
[14] Ferland, G. J. Korista, K.T. Verner, D.A. Ferguson, J.W. Kingdon, J.B. Verner, E.M.: PASP, **110**, 761 (1998)
[15] Van Dyk, S. D., Sramek, R. A., Weiler, K. W., Hyman, S. D., Rachel, E. V.: ApJ, **425**, 77 (1994)

# SN 1987A: A Complex Physical Laboratory

K. Kjær and B. Leibundgut

European Southern Observatory
Karl-Schwarzschild-Str. 2
D-85748 Garching bei München, Germany
kekjaer@eso.org

**Summary.** SN 1987A is located in the Large Magellanic Cloud and thus close enough to study the very late time evolution of a supernova and its transition to a supernova remnant. After almost two decades the supernova ejecta have now reached the pre-explosion circumstellar ring and started to interact with it. The system presents a complex structure involving forward and reflected shocks with highly different physical conditions in close spatial proximity.

The supernova ring system is currently observed in X-rays (Park et al. 2004, 2005), optical (Graves et al. 1996, Michael et al. 2003), mid-infrared (Bouchet et al. 2004) and radio (Manchester et al. 2005). The near-infrared data presented here complement all this.

Based on science verification observations with SINFONI we present the supernova-ring interaction as it appears in the near-infrared. SINFONI's adaptive optics supported integral field spectrograph spatially resolves the ring and the data thus provide new information and details for this unique object.

## 1 Introduction

### 1.1 Background

SN 1987A is surrounded by an inner and two outer circumstellar rings (Panagia et al. 1996), which combined are thought to outline an hourglass shape, with the inner ring as the waist of the hourglass. We only observe the inner circumstellar ring, which has an apparent size on the sky of 1.7"x1.2". SN 1987A was a type II supernova (SNe II) with a brightness of only 10% of normal SNe II. It took the supernova 80 days to reach its maximum, which is a relatively long time for a SN II. These atypical features arise from the abnormal progenitor, since it was a blue supergiant that exploded and not a red supergiant, which is the most common progenitor of type II supernovae (Arnett et al. 1989). Since the blue supergiants are more compact than the red supergiants, some of the energy that would have powered the lightcurve was needed to overcome the deeper gravitational potential.

### 1.2 Motivation

The inner ring around SN 1987A was ionized by the UV flash from the supernova explosion and has been observable ever since as the ionised gas is recombining (Fransson et al. 1987). The ring is thought to be a product of the progenitor evolution. There are several models for the formation of the triple rings. Some predict that the three rings have a common origin, other models involve separate events. The most probable scenarios include either a stellar wind from the supergiant phases of the progenitor or a binary merger where the common envelope had to be expelled because of too much angular momentum. For more details on the models see the reviews by Arnett et al. (1989) and McCray (1993).

In the last 6 years the inner ring around SN 1987A has undergone a rebrightening as the ejecta hit protrusions of the circumstellar ring. This collision was first indicated in 1995 by a 'bright spot' appearing on the inner ring (HST: Michael et al. 1998, 2000). Now the bright spots are numerous covering the whole ring (Sugerman et al. 2002). Their light curves are continuously monitored in X-rays (using the Chandra Observatory: Park et al. 2004), optical (with HST and from the ground: Graves et al. 2005, Michael et al. 2003), mid-infrared (Bouchet et al. 2004) and radio (Manchester et al. 2005). Near-infrared observations provide a window into the complex shock physics of forward, reverse and oblique shocks, dust formation and heating of pre-explosion dust (McCray et al. 1993) and complement observations in the other wavelengths. With this collision affecting the emission, the system can be seen in new light and thus we can constrain physical parameters for the explosion and the inner ring.

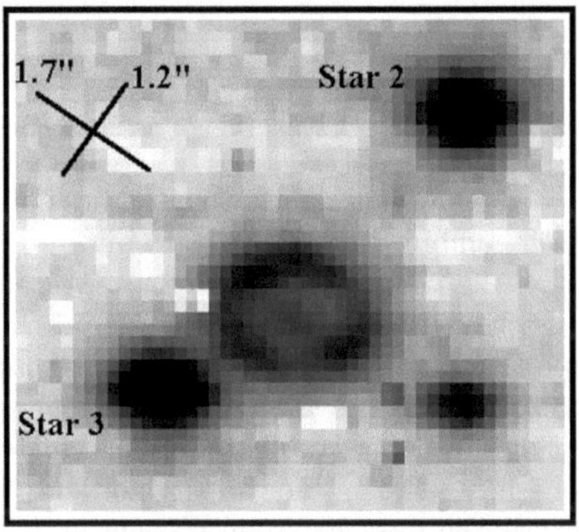


**Fig. 1.** The SINFONI FOV of SN 1987A 8"x8", seen here in the K-band. Stars 2 and 3 are labeled. The object at the lower right is another star.

## 2 The Data

The data presented here are SINFONI Science Verification observations of SN 1987A carried out in late November 2004 (preliminary results were published in Gillesen et al. 2005). Integral-field spectroscopy was obtained in J, H and K. The observations were AO supported using star 3 as a reference source.

The spectral resolution corresponds to a velocity resolution of J: 150 km/s, H: 100 km/s and K: 75 km/s.

Figure 1 shows the SINFONI FOV of the SV data. The data have been reduced and recombined into data cubes using the SINFONI pipeline version 1.0.0 (available at: http://www.eso.org/observing/dfo/quality/SINFONI/pipeline/pipe_gen.html). The pipeline constructs a datacube, which is a rebinning of the dithered observations (each with a sampling of 0.25" per spaxel). Figure 1 shows an image of such a rebinned collapsed cube. Using star 3 in this image we find the spatial resolution to be 0.8".

## 3 Analysis

### 3.1 Line Strengths

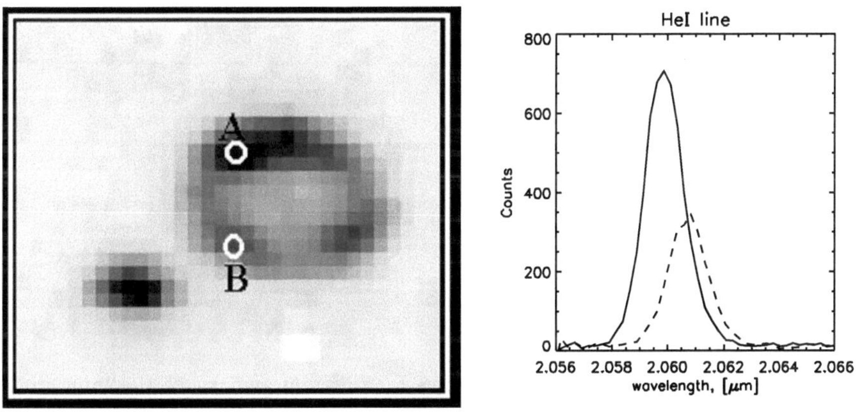


**Fig. 2.** The HeI line ($\lambda_0 = 2.058\mu$m). Left the collapsed cube over the wavelengths 2.057-2.063 $\mu$m. Right the spectrum at position A (full curve) and position B (dashed curve) from the left figure.

The SINFONI data provide a unique view of SN 1987A. The adaptive optics supported 3D data allow us to explore the supernova and its ring at improved spatial resolution, while the wavelength information yields spatially resolved spectroscopy of the ring. In our case the resolution of the inner ring is not good enough to resolve the individual bright spots (see McCray 2005 for a recent HST image). The ring system emits exclusively through emission lines. The material excited by the shock moving into the enhanced density of the inner ring cools through emission lines as well. Since we are not fully resolving the ring and shock regions, we observe emission lines averaged over different regions.

The HeI image of the supernova ring (left panel of Fig. 2) is a combination of all the frames in the wavelength range 2.057-2.063 microns. This wavelength range covers the HeI emission line ($\lambda_0 = 2.058\mu$m). The image reveals that the flux is very asymmetric across the ring, which is also seen at other wavelengths, most prominently in X-rays, (Park et al. 2005). This is because the bright 'spot 1' accounts for the majority of the emission in the upper NW part of the ring at all wavelengths. Why this spot is so much brighter than the other spots is still not resolved. It is likely that this asymmetric distribution arises from an asymmetry in the supernova explosion and/or asymmetry in the ring itself.

### 3.2 Line profiles

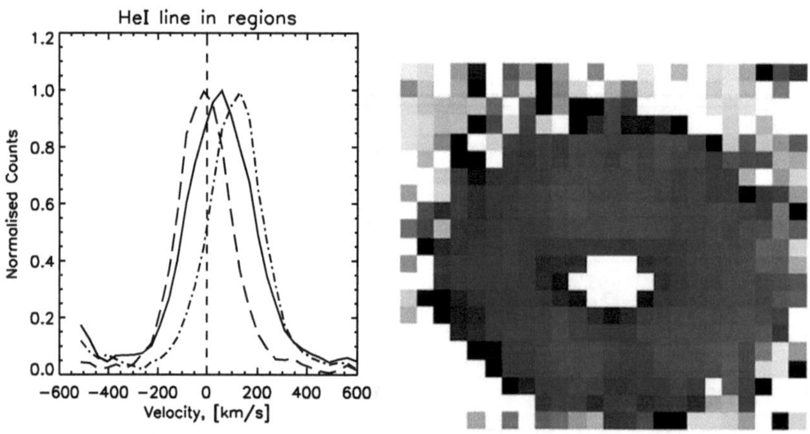


**Fig. 3.** Left panel: The HeI line profiles for different regions of the inner ring. The distributions are very similar. The lines correspond to a slit going through the center (full line), the spectrum at position A (dashed line) and at position B (dash-dotted line). (Positions A & B are shown in Figure 2). Right panel: An image of the FWHM of the HeI line for the whole ring.

Stepping through the data cube, we can see where the emission line is most blue-shifted (position A in the left panel of Fig. 2) and where it is most red-shifted (position B). In position A the shift of the line is $\Delta\lambda = 0.0019$ $\mu m$ and in B $\Delta\lambda = 0.0027$ $\mu m$. This corresponds to a line of sight velocity of the matter of $v_A$ = -6 km/s and $v_B$ = 110 km/s, when using a redshift for the LMC of $z = 0.000944$. The difference in central wavelength for this line shows a velocity dispersion across the ring of 100 km/s. With a spectral resolution in the K band of 75 km/s, this velocity dispersion is little more than the resolution. This indicates that the gas emitting the HeI line has a similar velocity all over the ring.

The HST observations have identified three different velocity components (Pun et al. 2002). A narrow component, nearly Gaussian (FWHM $\sim$ 10 km/s) coming from the unshocked ring. An intermediate velocity component arising from shocked cooling gas, with a clearly non-Gaussian form extending to $\sim$ 300 km/s. A very broad component from the freely expanding ejecta, extending to $\sim$ 15 000 km/s.

The strength of the SINFONI observations is that we can extract the FWHM of the lines at spatially different positions. The right panel of Figure 3 shows a spatial distribution of the FWHM for the HeI line. The FWHM is of the order of $\sim$ 200-250 km/s all over the ring. We can conclude that the line emission of HeI belongs to the intermediate velocity component and therefore comes from the shocked cooling gas.

## References


[1] Arnett, D., Bahcall, J., Kirshner, R. P., Woosley, S. E., ARA&A, **27**, 629 (1989)
[2] Bouchet, P., et al. ApJ, **611**, 394 (2004)
[3] Fransson, C., et al. A&A, **177**, L15 (1987)
[4] Manchester, R., et al. ApJ, **628**, L131 (2005)
[5] McCray, R., ARA&A, **31**, 175 (1993)
[6] McCray, R., in Cosmic Explosions, IAU Colloquium 192, eds. J. Marcaide, K. Weiler, New York: Springer Verlag, 77 (2005)
[7] Michael, E., et al. ApJ, **509**, L117 (1998)
[8] Michael, E., et al. ApJ, **542**, L53 (2000)
[9] Sugerman, B., et al. ApJ, **572**, 209 (2002)
[10] Gillesen, S., et al. Messenger, **120**, 26 (2005)
[11] Graves, G., et al. ApJ, **629**, 944 (2005)
[12] Panagia, N., et al. ApJ, **459**, L17 (1996)
[13] Park, S., et al. ApJ, **610**, 275 (2004)
[14] Park, S., et al. ApJ, **634**, L73 (2005)
[15] Pun, J., et al. ApJ, **572**, 906 (2002)

# 3D-Spectroscopy of Extragalactic Planetary Nebulae as Diagnostic Probes for Galaxy Evolution

A. Kelz, A. Monreal-Ibero, M. M. Roth, C. Sandin,
D. Schönberner and M. Steffen

Astrophysikalisches Institut Potsdam, An der Sternwarte 16, D-14482 Potsdam, Germany akelz@aip.de

## 1 Introduction

In addition to the study of extragalactic stellar populations in their integrated light, the detailed analysis of individual resolved objects has become feasible, mainly for luminous giant stars and for extragalactic planetary nebulae (XPNe) in nearby galaxies. A recently started project at the Astrophysical Institute Potsdam (AIP), called "XPN–Physics", aims to verify if XPNe are useful probes to measure the chemical abundances of their parent stellar population. The project involves theoretical and observational work packages. Amongst other techniques, the use of 3D-spectroscopy is applied, despite the fact that XPNe are point-like sources, as it allows an accurate recording and hence subtraction of the complex underlying background.

## 2 XPN Physics

Due to their bright emission in [O III]$\lambda$5007 and H$\alpha$, extragalactic Planetary Nebulae have been found in large numbers in the Local Group and out to galaxies in the Virgo and Coma clusters. Also, XPN were detected in the intra-cluster space of Virgo (1) and Coma (2). As XPNe can be individually resolved, they may serve as tracers for the low to intermediate mass stellar populations (9), (11) that otherwise can be studied only in the integrated light of its unresolved stars.

**Observational aspects:** For diagnostic purposes, spectroscopy of much weaker lines, such as [O III]$\lambda$4363 or [O II]$\lambda$3727, is needed. Because of the underlying, highly variable background which contaminates the spectra, 3D-spectroscopy has proven to be the preferred observational technique (6). It avoids slit-losses, and the resulting data cubes offer the possibility to correct for effects caused by atmospheric dispersion (3). Furthermore, the 2-dimensional field-of-view of the integral-field-unit avoids the instrumental effects that are otherwise caused by slit orientation. Figure 1 presents various re-constructed, monochromatic maps of a planetary nebula (PN29) in the Andromeda galaxy. Clearly, the crowded, unresolved background is not only complex, but also varies with wavelength. By extracting various 'slit'-spectra from the 3D-data cube, it could be shown that the resulting intensities for (weak) lines are dominated by the background subtraction (7). The background coverage is a function of the slit orientation, even if the PN is always centered on the virtual slit. However, using a deconvolution algorithm, the 3D data can be separated into a background and an object channel, thus yielding improved results in the determination of the line intensities.

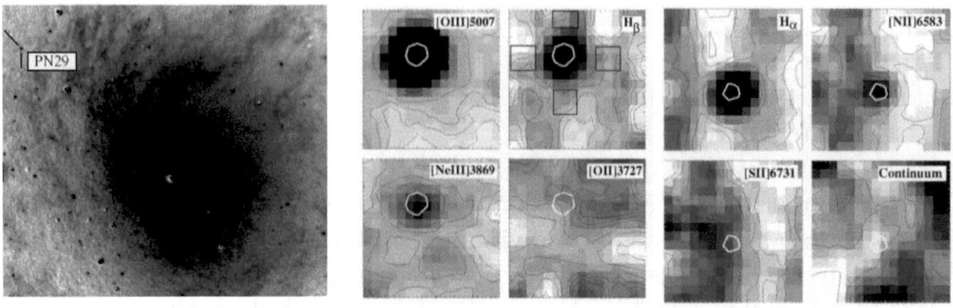


**Fig. 1.** Left: Hα narrow-band image of the central 3 arcminutes of M31, obtained with Calar Alto 3.5 m prime focus camera+Fabry-Pérot (7). The continuum has been subtracted using an off-band image to reveal the complex and fuzzy emission structure of the interstellar medium. The black dots mark XPN candidates, while spots with white substructures are indicative of residuals from imperfectly subtracted continuum sources (i.e. stars). Right: Re-constructed monochromatic maps from PMAS (8) observations of PN29 (7), illustrating the problem of a highly variable background surface brightness distribution that changes with wavelength. 2D-sampling is crucial as otherwise the sky subtraction depends on slit orientation (as indicated in the Hβ map). North is to the top, and east to the left.

**Theoretical aspects:** Many aspects, such as the photoionization of the circumstellar gas from the hot central star, the hot stellar wind and the occurrence of shock waves, play an important role in shaping the ionized nebula (5). Conventional diagnostic analyses of PNe often neglect the time-dependence of these important mechanisms, treating the nebula as a homogeneous and static object in energy and ionization equilibrium. Within the "XPN–Physics" project, time-dependent hydrodynamical simulations (10) are used in connection with selected observations, also of resolved galactic PNe (see Fig.2) as to assess systematically, how far common plasma diagnostic techniques are effected by density structure, dynamical evolution, and metalicities.

Using the combined approach of model simulations tested by specific observations, the XPN-Physics programme aims to address topics such as:
- How are the structure and expansion of a PN determined by its metalicity?
- What is the influence of the strong winds from the central stars?
- Does a minimum metalicity exist, necessary for the formation of a PN?

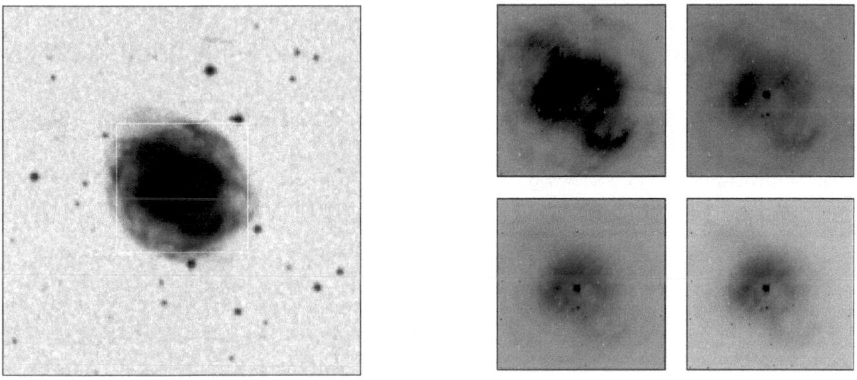

**Fig. 2.** NGC 4361 observed with the VIMOS-IFU at the VLT on April 18, 2004. Left: A Palomar image for comparison of the observed region (white square). Right: The panels show a 2x2 mosaic of four snapshot pointings as monochromatic maps obtained from the data cube in the wavelengths of [O III] 5007, [O III] 4959, H$\beta$, He II 4686 (4). This object is a metal-poor PN from the halo-population of the Milky Way and will be used as a reference to assess the theoretical model calculations.

– What are the quantitative differences if hydrodynamic simulations are used for the plasma diagnostic instead of static photoionization codes?
– Which ionization factors follow from hydrodynamic models?
– Are XPNe useful probes of resolved stellar populations in galaxies?

Beyond the research related to planetary nebulae in particular, it is expected that the results from this project will make a contribution to the overall topic of spectral analysis of resolved stellar populations which is one of the science cases for future ELTs.

**Acknowledgements:** The XPN-Physics project is funded by the German Research Society DFG with grant number SCHO 394/26.

## References

[1] Feldmeier, J. J., Ciardullo, R., Jacoby et al. In: IAU Symposium **217**, 64 (2004)
[2] Gerhard, O., Arnaboldi, M., Freeman, K. C., et al: ApJ **621**, L93 (2005)
[3] Kelz, A.: AN, **325**, 673 (2004)
[4] Monreal-Ibero, A., Roth, M. M., Schönberner, D., Steffen, M., Böhm, P.: ApJ **628**, L139 (2005)
[5] Perinotto, M., Schönberner, D., Steffen, M., Calonaci, C.: A&A **414**, 993 (2004)
[6] Roth, M. M.: Review of Nebular Integral Field Spectroscopy. In: this issue

[7] Roth, M. M., Becker, T., Kelz, A., Schmoll, J.: ApJ **603**, 531 (2004)
[8] Roth, M. M., Kelz, A., Fechner, T., Hahn, T. et al.: PASP **117**, 620 (2004)
[9] Roth, M. M., Schönberner, D., Steffen, M., Becker, T.: ANS **325**, 46 (2004)
[10] Schönberner, D., Steffen, M.: In: IAU Symposium **209**, 147 (2003)
[11] Walsh, J.R., Jacoby, G.H., et al.: Proc. SPIE Vol. **4005**, 131 (2000)

# 3D Spectroscopy of Herbig-Haro Objects

R. López[1], S. F. Sánchez[2], B. García-Lorenzo[3], R. Estalella[1], G. Gómez[3], A. Riera[4] and K. Exter[3]

[1] Dept. d'Astronomía i Meteorología, U. Barcelona. rosario,robert@am.ub.es
[2] Centro Astronómico Hispano-Alemán de Calar Alto. sanchez@caha.es
[3] Instituto de Astrofísica de Canarias bgarcia,ggv,katrina@ll.iac.es
[4] Dept. Física i Enginyería Nuclear (EU Vilanova i la Geltrú , UPC) angels.riera@upc.edu

## 1 Introduction and Observations

HH 110 (in L1267, $d$ = 460 pc (5)) and HH 262 (in L1551, $d$ = 140 pc (1)) are two Herbig-Haro (HH) objects that share a peculiar, rather chaotic morphology. In addition, no stellar source powering these jets has been detected at optical or radio wavelengths. Both, previous observations (2) (3) and models (4), suggest that the jet emission reveals an early stage of the interaction between a supersonic outflow and the dense outflow environment. These HHs are thus suitable to search for observational signatures of supersonic outflow/dense environment interaction.

We mapped these HHs with the Integral Field Instrument PMAS (Postdam Multi-Aperture Spectrophotometer) at the 3.5m CAHA telescope, under the PPAK configuration (331 science fibers, of 2.″7 each, in an hexagonal grid of ~ 72" of diameter). We used the J1200 grating (spectral resolution ~ 15 km s$^{-1}$ for H$\alpha$; wavelength range ~ 6500-7000 Å that includes the emissions from the characteristic red HHs lines: H$\alpha$, and the [N II] and [S II] doublets). Mosaics from several overlapping pointings (4 for HH 110 and 8 for HH 262) were made in order to cover the whole area of the emission of the HHs.

## 2 Results

From the PMAS data, we obtained maps of the morphology (monochromatic flux) and kinematics (radial velocity field and velocity dispersion) for the H$\alpha$, [N II] and [S II] line emissions of these two HHs. In addition, we computed line-ratio maps in order to explore the 2D structure of the electron density and excitation conditions, looking for the behaviour of the excitation and density spatial structures as a function of the velocity field. We show in the next section some of the maps obtained as an example of the PMAS science output.

## 2.1 HH 110

The morphology of the emission is very similar for all the lines mapped, although some differences among the three lines can be appreciated. In particular, the H$\alpha$ emission appears slightly more extended across the jet beam than the [N II] and [S II] emissions (a low-emission component surrounding the knots is only detected in the H$\alpha$ line).

The radial velocities derived from all the lines mapped appear blue-shifted (in the rest frame of the cloud) and show a similar behaviour in all the lines along the jet axis: first, $V_R$ becomes more red-shifted as we move from the northern edge of the jet (knot A, with $V_R \simeq -30$ km s$^{-1}$ in H$\alpha$) to the south, up to a distance of $\sim$ 40" from knot A, where the most red-shifted $V_R$ values are found ($V_R \simeq -15$ km s$^{-1}$ in H$\alpha$); it changes further out towards more blue-shifted values; the highest blue-shifted values of $V_R$ ($\simeq -35$ km s$^{-1}$ in H$\alpha$) are found at $\sim$ 100" from knot A.

In general, electron densities, as derived from the sulphur line ratio, decrease along the jet axis with distance from knot A. However, the highest values ($n_e \simeq 700$ cm$^{-3}$) are found at $\sim$ 40" from knot A, around the positions where $V_R$ reaches the reddest values for all the mapped lines.

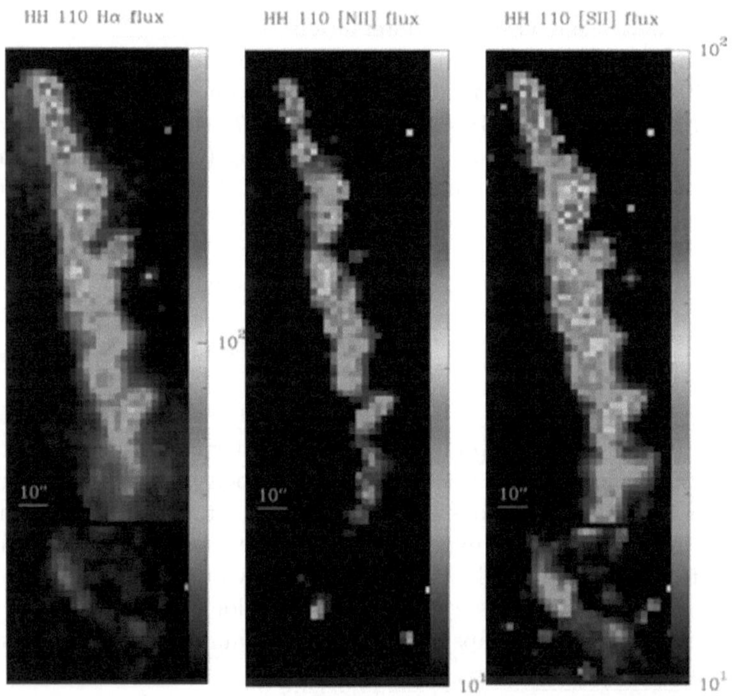


**Fig. 1.** HH 110: Monochromatic maps of the integrated flux in the H$\alpha$, [N II] and [S II] emission lines. North is up and East is to the left.

### 2.2 HH 262

The morphology of the Hα and [S II] emissions are similar. In contrast, emission from the [N II] line is only found in the northwestern region of HH 262.

The radial velocities derived from the Hα and [S II] lines appear red-shifted, in the rest frame of the cloud, for all the HH 262 knots, and present a rather complex pattern. As a general trend, $V_R$ increases from the north to the center of HH 262 and decreases from the center to the south. For the HH 262 knots, significant differences between the Hα and [S II] $V_R$ values are found, and the velocity difference changes for each knot. FWHMs of these two lines appear significantly wider (by 20–30 km $s^{-1}$) towards the center of HH 262 relative to the rest of the knots.

Electron densities derived from the sulphur line ratios are close to the low-density limit for most of the HH 262 knots. The highest values ($n_e \simeq 240$ $cm^{-3}$) are found towards the northwestern HH 262 knots, coinciding with the loci where emission from the [N II] lines is detected. For the rest of the object, electron density is less than 100 $cm^{-3}$, decreasing slightly from north to south.

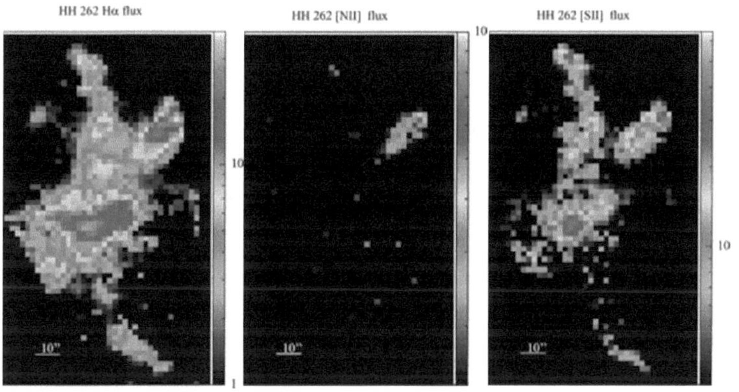


**Fig. 2.** HH 262: Same as Fig. 1 but for HH 262

# References

[1] J.A. Graham, M.H. Heyer: PASP **102**, 972 (1990).
[2] R. López, M. Rosado, A. Riera et al: AJ **116**, 845 (1998).
[3] R. López, R. Estalella, A.C. Raga et al: A&A **432**, 567 (2005).
[4] A.C. Raga, E.M.de Gouveia Dal Pino, A. Noriega-Crespo et al: A&A **392**, 267 (2002).
[5] B. Reipurth, M. Olberg: A&A **246**, 535 (1991).

# High Mass Star Formation in the Light of SINFONI

D. E. A. Nürnberger

European Southern Observatory, Casilla 19001, Santiago 19, Chile
dnuernbe@eso.org

**Summary.** As part of the SINFONI science verification, we have performed AO supported integral field spectroscopy of the M 17 silhouette disk. The results demonstrate that integral field spectroscopy at infrared wavelengths can add important pieces to the puzzle of high mass star formation.

## 1 Introduction

High angular resolution infrared (IR) imaging of the Galactic H II regions NGC 3603 and M 17 led to the discovery of high mass protostellar candidates at the interface between the H II region and the adjacent molecular cloud (Nürnberger 2003, Chini et al. 2004). This observing technique takes advantage of the presence of a cluster of early-type main sequence stars, which provide a huge amount of energetic photons and powerful stellar winds, capable of evaporating and dispersing the surrounding interstellar medium. This way, nearby young stars — which otherwise would be deeply embedded in their natal molecular cloud cores — are set free at a relatively early evolutionary stage and become detectable at near and mid IR wavelengths. Here we focus on the case of a tremendous opaque silhouette seen in absorption against the background emission of the M 17 H II region (Chini et al. 2004).

This silhouette, unambiguously seen in deep near IR data taken with ISAAC and NAOS-CONICA at ESO's Very Large Telescope (VLT), is associated with an hourglass shaped optical nebula and surrounded by a large disrupted envelope. From $^{13}$CO data obtained with the IRAM Plateau de Bure interferometer, the gas and dust mass of the entire disk + envelope system was estimated to be up to 110 $\mathcal{M}_\odot$. Although the kinematics of the innermost part of the disk remains unresolved by the $^{13}$CO data, there is indication that the outer part of the disk + envelope system slowly rotates at a velocity of about 0.85 km s$^{-1}$. Assuming Keplerian rotation, one can deduce a mass of up to 15 $\mathcal{M}_\odot$ for the central protostar. This value is corroborated by the absolute IR brightness ($M_K \sim -2.5$) of the central, slightly elongated source which would correspond to a main sequence star of $\mathcal{M}_* \sim 20\,\mathcal{M}_\odot$ and $T_{\rm eff} \sim 35\,000$ K. Optical spectroscopy of the bipolar nebula suggests that accretion processes are still active. Therefore, one might argue that a massive protostar is on its way to become an O-type star.

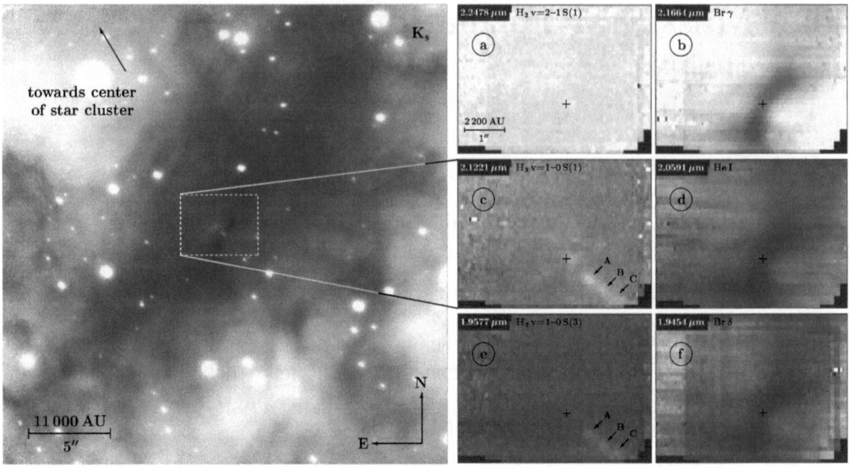


**Fig. 1. Left:** Finding chart based on high angular resolution $K_s$ band data previously obtained with VLT + NAOS-CONICA (Chini et al. 2004). It is centered on the suspected protostellar source located at the very center of the silhouette disk, and shows a field-of-view of roughly $27'' \times 27''$, which corresponds to about 60 000 AU × 60 000 AU at the adopted M 17 distance of 2.2 kpc. The dashed box outlines the area of our VLT + SINFONI study. **Right (panels a–f):** Selected image planes extracted from the SINFONI data cube, each comprising a field-of-view of $4\farcs8 \times 3\farcs6$, i.e. 10 560 AU × 7 920 AU. The spatial pixel scaling is $0\farcs05$ in right ascencion and $0\farcs10$ in declination. In the upper left corner of each panel the corresponding wavelength and the most likely atomic / molecular line designation is denoted. In addition, the position of the suspected protostellar source at the disk center is indicated by a cross. In panels c and e the knots of the newly discovered $H_2$ jet are labeled A, B and C.

## 2 Observations

To further characterize and understand the on-going physical processes associated with this large circumstellar disk, we have performed Adaptive Optics (AO) supported near IR integral field spectroscopy with SINFONI at the VLT as part of the science verification (Gillessen et al. 2005) of this instrument. We used the 100 mas pixel scaling together with the K band grating, which comprises the wavelength range from 1.933 $\mu$m to 2.476 $\mu$m at the spectral resolution of 0.5 nm. For more details we refer to Nürnberger et al. (2006).

## 3 Results: the Silhouette Disk

As derived from the absorption signatures in the Br $\gamma$, Br $\delta$ and He I lines, both the diameter (about 4 000 AU) and sub-structures of the innermost, densest part of the flared disk closely resembles those seen in high angular resolution $K_s$ band imaging data obtained previously with NAOS-CONICA. The disk is seen almost edge-on and appears shaped wing-like. The latter is most likely caused either by the slight inclination (about 10°) of the disk plane against the line-of-sight, or by the pressure exerted on the disk surface by both ionizing photons and strong winds originating from the nearby main sequence O stars of the M 17 cluster. In the same sense, the outflow cavity towards the north-east (i.e., on the O star facing side of the disk) apparently has a much wider opening angle than the outflow cavity towards the south-west (i.e., facing away from the O stars). This points towards a stronger interaction of outflowing material with the incoming wind / ionization front on the disk side which is directly exposed to the nearby massive main sequence stars. The counter outflow might be less affected by these interactions as it is largely shielded by the disk.

## 4 Results: the $H_2$ Jet

Most importantly, based on $H_2$ v=1–0 S(1) and $H_2$ v=1–0 S(3) emission maps we report the discovery of a $H_2$ jet which apparently arises from the suspected protostellar source located at the very center of the disk. Three knots — marked A, B and C in panels c and e of Fig. 1 — of $H_2$ emission are detected at distances of about 0.″7, 1.″2 and 1.″5 from the disk center. The corresponding S(1)/S(3) intensity ratios are roughly 2.6, 1.8 and 1.2. These ratios suggest, together with the non-detection of the three knots in the $H_2$ v=2–1 S(1) line (hence, $H_2$ v=1–0 / $H_2$ v=2–1 values much larger than unity), that shocks are the most likely excitation mechanism. No counter jet has been found on the other side of the circumstellar disk. This is not surprising as any potential counter jet is ejected straight towards the neighbouring main sequence O stars and, hence, is expected to be affected significantly (deflected and squeezed back to the disk surface) by the incoming shock and ionization front. Since ejection of material through a jet and / or outflow is always linked to accretion of gas and dust, either onto a circumstellar disk or onto the central (protostellar) source, then in the case of the M17 disk the presence of the $H_2$ jet provides indirect but unquestionable evidence for ongoing accretion processes.

## References

[1] R. Chini, V. Hoffmeister, S. Kimeswenger et al: Nature **429**, 155 (2004)
[2] S. Gillessen, R. Davies, M. Kissler-Patig et al: ESO Messenger **120**, 26 (2005)
[3] D.E.A. Nürnberger: A&A **404**, 255 (2003)
[4] D.E.A. Nürnberger, R. Chini, F. Eisenhauer et al: A&A submitted (2006)

# Internal Kinematics of IIZw40

H. Plana[1], V. Bordalo[2] and E. Telles[2]

[1] Universidade Estadual de Santa Cruz - Ilheus - Brazil plana@uesc.br
[2] Observatorio Nacional - Rio de Janeiro - Brazil

We are presenting here velocity and dispersion velocity maps of blue dwarf galaxie IIZw40. We used GMOS IFU at Gemini North.

## 1 Introduction

Among the diferent dwarf galaxies, blue dwarf galaxies represent a population with a high star formation rate. IIZw 40 is the prototype of a dwarf blue irregular galaxy (5). It presents a young starburst in its central region (3). Optical spectrum is dominated by very strong emission lines. HI imaging confirms the merging scenario of the formation. It is believed that the merging triggered the high star formation in the center (1). Color gradients also are consistent with the scenario of the ongoing merger of two gas rich dwarf galaxies. The purpose of this study is to understand the inner kinnematics of the large star formation regions in order to understand the history of these regions.

## 2 Observation

Observations have been carried out using the Integral Field Unit (IFU) of the GMOS instrument at Gemini North telescope. IFU will give a spectrum for each lens of the lens array covering the area. Each lens is covering 0.2" on the sky. With the one slit configuration, the total field of the IFU is 5" x 3.5". In order to cover the central part of the galaxy we did 6 fields, with a common area one to the other of 0.6" (See Figure 1a). We have used the R815 grating giving a velocity sampling of 15 km.s$^{-1}$ between 4000 Å and 7000 Å . The exposure time for each field was 5min. The data reduction have been done using standard IRAF package for the extraction of the spectra and special IDL macros for velocity and dispersion maps. The original data cube has been re-sampled to 0.1" per lens. We derived the maps using the H$\alpha$ emission line.

## 3 Results

We are presenting in Figure 1b, the velocity dispersion map of the inner part of IIZw40. The dispersion velocity have been corrected of the thermal and instrumental broadening. The velocity dispersion is varying between 20 km.s$^{-1}$ and 62 km.s$^{-1}$. This confirms the supersonic character of the gas motion in the high dispersion regions. The velocity field (not shown here) does not show ordered motion like rotation. Iso-intensities over-plotted on Figure 1b map represent monochromatic H$\alpha$ emission. We can see that the emission has an extension on the south side and shows a peak intensity in the center. We can clearly see three clouds with high velocity dispersion on Figure 1b. These regions do not correspond to regions with a high emission, they are surrounding the central emitting regions. Figure 2 shows the diagram of the velocity dispersion vs the intensity. This diagram has been introduced by (4) with the Cometary Stirring Model in order to explained the formation and evolution of stellar clusters. This scenario takes into account the empirical correlations between the velocity dispersion and scale and between dispersion and intensity. This scenario shows several steps from the formation of massive stars to the expanding shell around them into the initial molecular cloud. Each step corresponds to a feature in the velocity dispersion vs intensity diagram. On Figure 2, we can clearly see three peaks. According to (2) who applied this method to Giant HII Regions in nearby spiral galaxies, the inclined bands correspond to the signature of massive stars interacting with

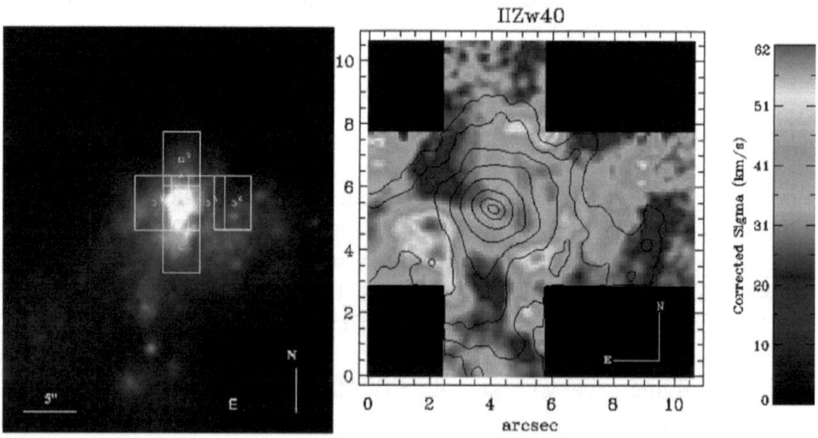


**Fig. 1.** Fig. 1a (left) shows the pre-imaging of GMOS with the 6 fields of the IFU. Fig. 1a (right) shows the velocity dispersion map with the H$\alpha$ over-plotted.

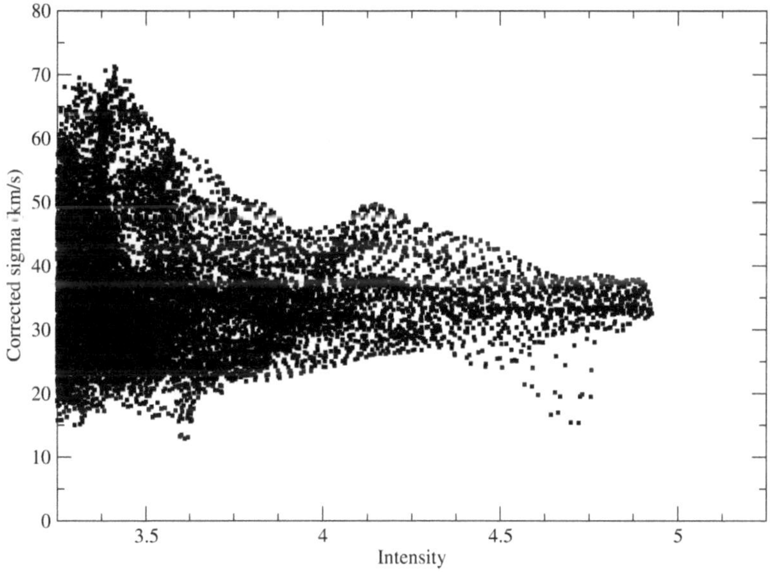


**Fig. 2.** represents the sigma - intensity plot of the inner part of IIZw40, showing two inclined bands.

the ambient medium. On the graphics only two inclined bands appear, the third peak does not show a inclined band.

# 4 Conclusion

Using GMOS IFU, we are presenting velocity, emission and velocity dispersion maps of dwarf HII galaxy, IIZw40. The velocity dispersion gave us informations on history of the star forming clusters using the sigma - intensity plot, despite the crude approximation of a single Gaussian fit. As a preliminary conclusion, we can see say that the star clusters of an HII galaxies, as IIZw40 have the same behaviour as the Giant HII Regions in nearby galaxies.

# References


[1] Brinks & klein 1988 MNRAS 198, 535
[2] Muñoz-Tuñòn et al. 1996 AJ 112, 1636
[3] Searle & Sargent 1972 ApJ 173, 25
[4] Tenorio-Tagle et al. 1993 ApJ 418, 767
[5] Zwicky & Herzog 1963 catalog of galaxies and cluster of galaxies vol.2

# Integral Field Spectroscopy Survey of Classical LBV Stars in M33

O. Sholukhova[1], P. Abolmasov[1], S. Fabrika[1], V. Afanasiev[1] and M. Roth[2]

[1] Special Astrophysical Observatory, Russia olga@sao.ru
[2] Astrophysikalisches Institut Potsdam, An der Sternwarte 16, D-14482, Potsdam, Germany mmroth@aip.de

**Summary.** Five well-known LBV stars in M33 were observed with the Multi-Pupil Fiber Spectrograph (MPFS) on the 6-m Russian telescope. We observed LBVs var A, var B, var C, var 2 and var 83. In three of them, var 2, var 83, var B, large-scale nebulae were found with sizes from 15 pc and larger. The nebula shapes are complex, like one-side tails or conical nebulae. They all are related to their LBV stars. In var 2 and var 83 stars we found radial velocity gradients 15–30 km/s across their nebulae. The stars var A and var C do not show extended nebulae, but nebular lines are certainty present in their spectra.

## 1 Observations and Data Reduction

This observation were carried as continuation of our program of studying of LBV-candidate stars in M33. Here we present results of observations on the Russian 6-m telescope with the integral field spectrograph MPFS (1) in November 2004. The integral field unit of 16×16 square spatial elements covers a region of 16"×16" on the sky. Integral field spectra were taken in the spectral range 4000 – 6800Åwith a seeing from 1.0 to 1.5" (FWHM). Data reduction was made using procedures developed in the IDL environment (version 6.0) by V. Afanasiev, A. Moiseev and P. Abolmasov and include all the standard steps.

## 2 Results and Discussion

Practically all known LBV stars have circumstellar nebulae (4). Typical Galactic LBV nebulae have sizes in the range of 0.1–4 pc, expansion velocities 15–100 km/s, and their dynamical times are in the range 100–$5 \times 10^4$ years (6). Numerical models of the nebular expansion around massive stars (3) have shown that those nebulae may reach diameters up to 20–40 pc in the main-sequence and pre-LBV stages.

Fabrika et al. (2) have studied two LBV-like stars and their nebulae in M 33. The first one, B416 is a B[e]-supergiant with an expanding ring-like nebula 20×30 pc. In the second, v532 (LBV/Ofpe star) they found an elongated nebula. Both stars' nebulae show radial velocity gradients of about a

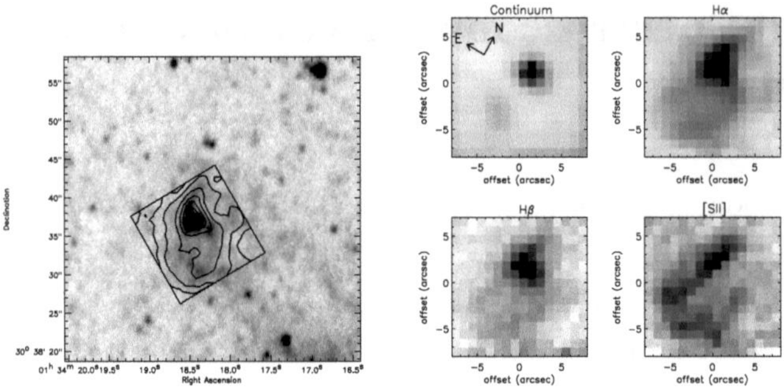


**Fig. 1.** Left: Hα image of the var 2 region taken by (5) and the MPFS field with the line isophotes superimposed. Right: Monochromatic and continuum MPFS maps of var 2. A bipolar nebula is clearly seen in the Hα and Hβ emission line maps. It is non-symmetrical and presents different morphology in lines of different excitation. The star itself is a source of emission in permitted lines. The bipolar nebula 20x40 pc is shock excited. An Hα radial velocity gradient of ±30 km/s was detected along the bipolar structure.

few tens km/s. We started a special study of gas environments around classical LBV stars in M 33 to confirm the presence and to study the large-scale nebulae in these objects.

We found the large-scale nebulae around LBV stars var B, var 2 (see Fig.1) and var 83 (Fig.2). The structure of the nebulae indicates that they were formed by the LBV (or pre-LBV) winds (see Fig.3). The nebulae are kine-

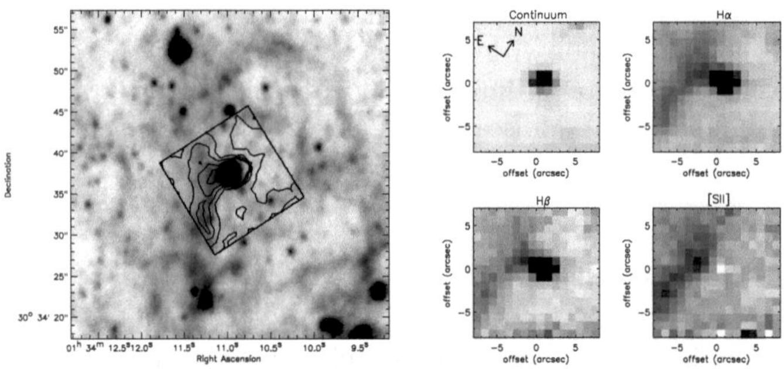


**Fig. 2.** Same as in Fig. 1 for var 83. In Hα, Hβ, [S II] lines we see a double-tail-like nebula, which is a part of a bigger nebulosity. Velocity maps in Hα show the common nature of the star and the nebula. The double tail approaches us with a velocity ≈ 15 km/s.

matically connected with the host stars. Their physical extension is about 15 – 30 pc, and their dynamical times are in the range of $10^5 - 10^6$ years.

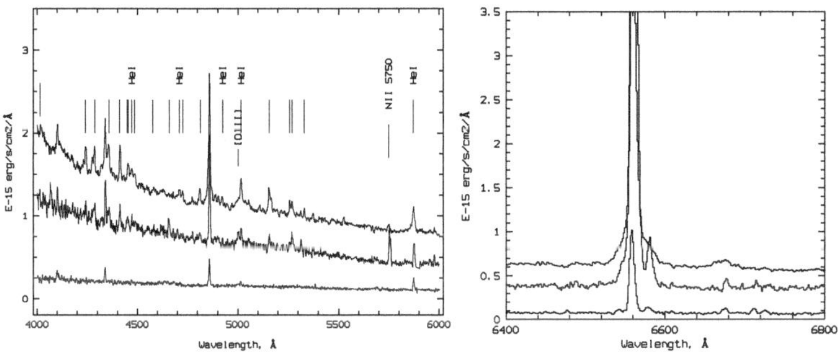


**Fig. 3.** Fragments of the spectra of var 2, var B and var 83 are shown from bottom to top (left panel). The brightest spectral lines are hydrogen lines and He I. Fe II, [FeII] lines shown by unlabeled vertical lines. In H$\alpha$ lines we see broad components (right panel), that indicate the stellar wind.

The stars var A and var C do not show extended nebulae, but nebular lines are certainty present in their spectra.

Nebulae of such dimensions around LBV-stars in our Galaxy cannot be easily studied, because their diameters would exceed 1 degree. Detection of large-scale nebulae around LBV-stars is important, as a study of these nebulae can give information about the earliest phases of evolution of massive stars.

## 3 Acknowledgements

This work has been supported by the RFBR grants N 03–02–16341, 04–02–16349 and RFBR/JSPC grant N 05–02–19710. O. Sholukhova, P. Abolmasov, S. Fabrika are grateful to the SOC and LOC of the Workshop “Science Perspectives for 3D Spectroscopy” for support.

## References

[1] Afanasiev, V.L., Dodonov, S.N., Moiseev, A.V., 2001, in “Stellar dynamics: from classic to modern”, eds. Osipkov L.P., Nikiforov I.I., Saint Petersburg, 103
[2] Fabrika, S., Sholukhova, O., Becker T., Roth, M., & Sanchez, S.F., 2005, A&A, 437, 217
[3] Garcia-Segura, G., Mac Low, M. & Langer, N., 1996, A&A, 305, 229
[4] Humphreys, R.M. & Davidson, K. 1994, PASP, 106, 1025
[5] Massey, P., Hodge, P. W., Holmes, S., Jacoby, G., King, N. L., Olsen, K., Saha, A. & Smith, C. 2001, AAS 199th meeting, BAAS, 33, 1496
[6] Weis, K. 2003, A&A, 408, 205

# Unravelling the Chemical Inhomogeneity of Planetary Nebulae with VLT FLAMES

Y. G. Tsamis[1] and J. R. Walsh[2]

[1] Dept. of Physics & Astronomy, University College London ygt@star.ucl.ac.uk
[2] ST-ECF, ESO jwalsh@eso.org

In this contribution we present the rationale of a collaborative study which also involves D. Péquignot (Obs. de Meudon), M. J. Barlow (UCL), and X.-W. Liu (Beijing Univ.).

## 1 Introduction

Planetary nebulae (PNe) are the evolved remnants of stars whose progenitors were quite similar to the Sun and, as such, they are emblematic objects to study. Most stars in late-type galaxies (such as the Milky Way) will live through the PN phase in their final evolutionary stages. The study of PNe is therefore very important in an (extra-)galactic context. It is well known that before and during the PN phase the interstellar medium (ISM) is enriched with heavy elements from the evolving star, some of which (e.g. He, C, N) have been modified by nucleosynthetic processes so that they do not reflect the make-up of the stellar progenitor. These elements, along with freshly formed dust species and molecules, mix with the ambient ISM and alter its composition. The ionized PN envelopes are ideal laboratories for the study of the exact chemical composition of the mass ejected from the dying star, and for the determination of the abundance of species relative to hydrogen. The most powerful technique for this purpose involves the observation of emission lines from He, C, N, O, Ne, Mg, S, Cl, Ar etc. ions in the UV, optical and IR spectral domains. The emission lines involved belong in two distinct categories: collisionally excited lines (CELs; e.g. [O III] $\lambda\lambda$4959, 5007) and optical recombination lines of heavy ions (ORLs; C II, N II, O II). Abundances derived from ORLs should in principle be more accurate than those derived from CELs from the same ions, due to the exponential dependence upon the adopted nebular temperature of the latter lines.

## 2 The Problem

Extensive surveys of a large number of PNe that were recently published (7),(8),(4),(10) show that for most nebulae the abundances of C, N, O, and Ne derived from analyses of their faint ORLs (e.g. C II $\lambda$4267, O II $\lambda$4650) are systematically higher than those obtained from the classical method which utilizes the bright CELs, with abundance discrepancy factors (ADFs) of 1.6–3.0 obtained for most nebulae. For about 10% of nebulae however the ADFs have values of 4–80. A closely related problem (3),(8) involves the fact that the hydrogen Balmer jump electron temperatures of PNe are lower, on average, than those derived from the [O III] λ4363/λ5007 line ratio (5); the latter is the most widely used nebular thermometer. A similar situation prevails in HII regions where the ADFs are relatively low (2–5; (6)). This 'thermal and abundance discrepancy' problem is important since, if we were to casually adopt the high ORL abundances for CNONe, this would have profound implications for studies which rely on accurate PN and HII region abundances in the Galaxy, and beyond, in order to set constraints on cosmic chemical evolution models.

## 3 A Neat Solution: the Dual-Abundance Paradigm

A proposition that aspires to explain both the temperature and abundance discrepancies is one according to which PNe are chemically inhomogeneous objects, incorporating two (or more) gas phases of very different elemental composition. This description assumes the existence of hydrogen-poor clumps or filaments within a PN, embedded in material of solar-like abundances. It predicts that the overwhelming fraction of heavy-element ORL emission arises from H-poor regions, which due to their enhanced metallicity have efficiently cooled down to ~1000 K by emitting IR fine-structure lines (1),(2). Observational evidence in support of the dual-abundance model comes from direct measurements of the average temperatures in the inferred H-poor regions, by means of He I and O II ORL ratios, which yield very low electron temperatures indeed (8),(4),(10). At the moment the exact nature and origins of the inferred H-poor regions in PNe remain unclear. Hypotheses include: (i) the evaporation of primitive material left over from the formation of the PN precursor star (e.g. planetesimals); (ii) incompletely mixed dredged-up material ejected along with the rest of the star's outer envelope during the PN formation phase. For HII regions one could appeal to incompletely mixed supernova ejecta which occur frequently in OB-type stellar associations (9).

## 4 Testing the Solution with VLT FLAMES Observations

The proposed solution lacks further direct observational support. According to our analyses from self-consistent photoionization models, a few jovian masses of H-poor material immersed in gas of normal metallicity are enough to explain the CNONe ORL emission and resolve the discrepancy paradox. The small mass and low filling factor of the inferred inclusions means that their sizes will be similarly small and large telescopes providing high spatial resolution and excellent sensitivity are required to detect them. The VLT 8.2-m/Kueyen is a leader in this class. We have thus acquired during Jun–Aug '05 FLAMES integral-field unit (IFU) spectroscopy in Argus mode of 3 Galactic PNe (NGCs 5882, 6153, 7009) which show a wide range of ADF behaviour (~2–10).

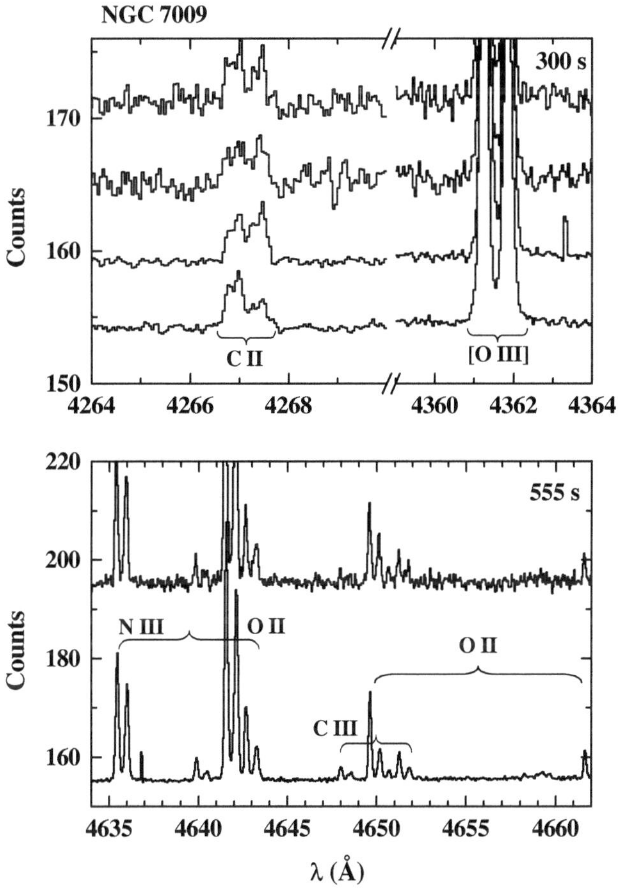


**Fig. 1.** The emission line spectrum of the planetary nebula NGC 7009, taken with VLT FLAMES/Argus at 0.3″ per fibre, showing spectral regions including the C II (upper plot) and O II (lower plot) optical recombination lines at a spectral resolution of 0.15Å. The seeing was 0.5″.

A first glimpse of the secured raw spectra of NGC 7009 is provided in Fig. 1 taken with the 6.6″ × 4.2″ IFU at a spectral resolution, $\lambda/\delta\lambda$, of 32 500 (i.e. 9.2 km $s^{-1}$).

The plotted spectra have been extracted over 5 and ∼80 spatial pixels (top and bottom tracings in each of the two panels respectively) in order to highlight the change in the S/N ratio of the C II and O II ORLs. Note the split line profiles due to nebular expansion, and how the components of the C II λ4267 doublet are clearly resolved in each of the blue- and red-shifted nebular portions (Fig. 1, top panel). Several exposures will be combined so that the increased S/N per spatial pixel will allow us to map the fibre-to-fibre (spatial) variations of the velocity profiles of ORLs versus CELs (e.g. C II λ4267 vs. [O III] λ4363) and determine whether the two types of line arise from kinematically distinct regions. This might be one of the signatures of a discrete H-poor gas phase embedded within the nebulae.

In order to elucidate further the nature of the inferred H-poor regions we aim to: (i) construct temperature and density maps across the IFUs and determine the variations of chemical composition (and of the ADF) relative to the position of the PN nuclei; (ii) resolve or set strict upper limits to the sizes of the inferred H-poor clumps by comparing C II λ4267/Hγ vs. [O III] λ4363/Hγ intensity ratio maps at the highest (0.3″) spatial resolution; (iii) create images of the nebulae in the light of heavy-element C II λ4267 and O II λ4649 ORLs. All this will help us to pinpoint the emitting regions of the elusive H-poor nebular component, constrain its physical properties and discriminate amongst the various hypotheses regarding its origin.

## References


[1] Harrington J. P. & Feibelman W. A., 1984, ApJ, 277, 716
[2] Liu X.-W., Storey P. J., Barlow M. J., et al., 2000, MNRAS, 312, 585
[3] Liu X.-W., Luo S.-G., Barlow M. J., et al., 2001, MNRAS, 327, 141
[4] Liu Y., Liu X.-W., Barlow M. J., Luo S.-G., 2004, MNRAS, 353, 1251
[5] Peimbert M. 1971, Bol. Inst. Tonantzintla, 6, 29
[6] Tsamis Y. G., Barlow M. J., Liu X.-W. et al, 2003a, MNRAS, 338, 687
[7] Tsamis Y. G., Barlow M. J., Liu X.-W., et al, 2003b, MNRAS, 345, 186
[8] Tsamis Y. G., Barlow, M. J., Liu, X.-W., et al, 2004, MNRAS, 353, 953
[9] Tsamis Y. G. & Péquignot D., 2005, MNRAS, 364, 687
[10] Wesson R., Liu X.-W., Barlow M. J., 2005, MNRAS, 362, 424

# Part VIII

# Extended Emission Line Regions

# Integral Field Spectroscopy of Galaxy Emission Line Haloes

R. J. Wilman

Dept. of Physics, University of Durham, South Road, Durham, DH1 3LE, UK
r.j.wilman@durham.ac.uk

**Summary.** We summarise recent results from two programmes of integral field spectroscopy (IFS) on galaxy emission line haloes. Firstly, we present SAURON IFS of the 100 kpc-scale diffuse Ly$\alpha$-emitting blobs in the SSA22 protocluster at $z = 3.09$ and show how IFS helped uncover evidence for a galaxy-wide superwind outflow. Secondly, we present VIMOS observations of H$\alpha$ emission in the cores of $z \sim 0.1$ cooling flow clusters and illustrate how IFS is an essential tool for understanding these complex regions.

## 1 SAURON IFS of $z = 3$ Ly$\alpha$ Haloes: Probing Massive Galaxy Formation

According to the current theory, galaxies form from gas which has managed to cool in gravitationally collapsed haloes of dark matter. But if cooling alone were the whole story, there would be too many luminous massive galaxies around us in the local Universe today (3). Some mechanism is needed to halt star formation in young massive galaxies and it is widely believed that such 'feedback' is provided by galaxy-wide outflows ('superwinds') driven by supernovae and stellar winds and/or active galactic nuclei. Such winds have the added benefit of transporting metals out into the intergalactic medium where they are seen to be widely distributed just a few billion years after the Big Bang.

Although superwinds have been well-studied in the local Universe, e.g. in the dwarf starburst galaxy M82, far less is known about their counterparts in more extreme massive starburst galaxies at high redshift. Long-slit spectroscopy of $z \simeq 3$ Lyman-break galaxies (LBGs) shows that the ISM absorption lines are blue-shifted by several hundred km s$^{-1}$ (9); but, owing to the limited spatial information, it is unclear whether such outflows are localised to regions of current star formation just a few kiloparsec in extent or whether they are galaxy-wide. Further indirect evidence for LBG superwinds has come from surveying QSO sightlines situated beyond $z \sim 3$ LBG fields (1); these reveal an HI deficit within 1–2 Mpc (projected) of an LBG, possibly due to evacuation by powerful superwinds. However, the relevant sightlines were few in number and the result was not confirmed by a similar study at $z = 2 - 3$ (2).

To test the superwind hypothesis further, we have begun to study the extended Ly$\alpha$ emission line haloes of some of the most massive star-forming galaxies at $z \sim 3$, searching for the signatures of energy and metal injection. Since such haloes can extend beyond 100 kpc (i.e. 12 arcsec or more), IFS is the ideal tool for this. Here we describe our SAURON observations of the two diffuse Ly$\alpha$ blobs (LABs) discovered by (10) in the SSA22 protocluster region at $z = 3.09$. With sizes exceeding 100 kpc and Ly$\alpha$ luminosities of $10^{44}$ erg s$^{-1}$, the LABs resemble the haloes of high-redshift radio galaxies but are in fact radio quiet. Both are, however, sub-mm sources with LAB-1 the stronger and consistent with 1000 M$_\odot$ yr$^{-1}$ of star formation. LAB-2 has an X-ray obscured AGN. SAURON is a high-efficency $33 \times 41$ arcsec IFU on the William Herschel Telescope which can be used for observing Ly$\alpha$ at $z = 2.95 - 3.4$.

### 1.1 LAB-1

LAB-1 was observed by us for 9 h and found to exhibit complex structure with many halo components (5). Variations in the Ly$\alpha$ line profile are inconsistent with a simple outflowing shell and the overall distribution is better modelled as distinct components moving relative to each other at several hundred km s$^{-1}$. If the kinematics are converted into a virial mass estimate, their 500 km s$^{-1}$ line-of-sight dispersion implies a mass of $1.3 \times 10^{13}$ M$_\odot$ within a radius of 75 kpc, as expected for a small cluster. The central sub-mm source of LAB-1 appears to coincide with a small cavity in the Ly$\alpha$ emission, possibly due to a weak or unevolved outflow, although there is no unambiguous evidence for HI absorption in the Ly$\alpha$ profiles. The two LBGs embedded in the LAB-1 halo show distinct haloes: C15 to the north shows a clear bipolar velocity field suggestive of an M82-style outflow (or rotation) at a de-projected speed of $\sim 200$ km s$^{-1}$.

### 1.2 LAB-2

In contrast to the chaotic nature of LAB-1, our 15 h SAURON observation of LAB-2 admits a simpler empirical description (11). As shown in Fig. 1, the Ly$\alpha$ profiles at most positions across the source are double-peaked. The position of the central minimum varies across the galaxy by less than 60 km s$^{-1}$ in its rest-frame.

A plausible interpretation is that part of the Ly$\alpha$ is scattered out of our sightline by HI in a coherent foreground screen. Fits to the Ly$\alpha$ with Voigt profile absorbers atop Gaussian emission imply $N_{\rm HI} \sim 10^{19}$ cm$^{-2}$. The absorber is typically blueshifted by less than 250 km s$^{-1}$ from the peak of the underlying emission, whilst the latter displays a range in absolute velocity of $\sim 290$ km s$^{-1}$ and line-widths of 1000 km s$^{-1}$ FWHM.

A shell of absorbing HI is predicted to form from a superwind bubble of hot gas which has broken out of its host galaxy and swept up gas from

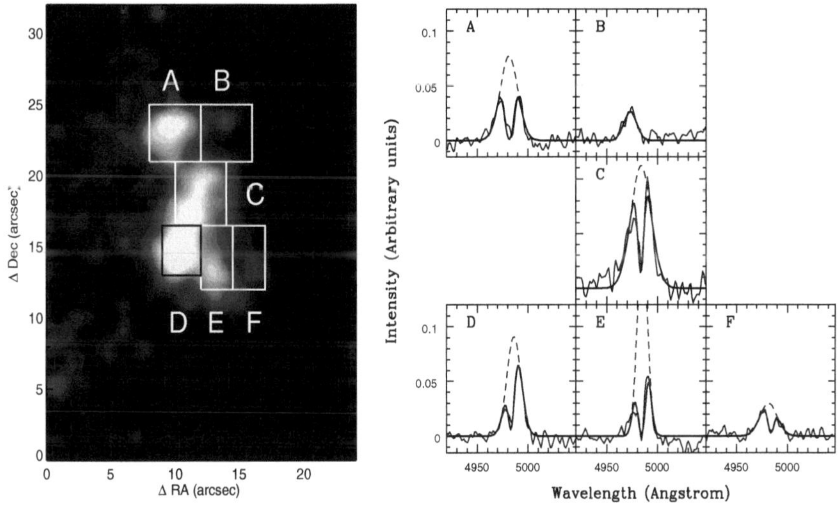


**Fig. 1.** Left: reconstructed Ly$\alpha$ image of LAB-2 in the SSA22 protocluster at $z = 3.09$ from SAURON observations. Right: Ly$\alpha$ profiles for the labelled regions. Thick solid line are fits to model in which the intrinsic emission (dashed lines) is partially absorbed by foreground HI. Measured HI column densities are $\sim 10^{19}$ cm$^{-2}$, but the 4Å resolution implies $1\sigma$ uncertainties of at least a factor of 10. The absorption minimum varies by $< 60$ km s$^{-1}$ across the galaxy. From (11).

its surroundings. Since the projected linear size of the absorber is at least 100 kpc, it must be a comparable distance from the central source. Its kinematic coherence suggests that it may be part of a very much larger structure or, more likely, that the shell has almost stalled. A shell will spend most of its time close to the stalling radius and hence is most likely be observed there. At an initial ejection speed of 1000 km s$^{-1}$, it takes $10^8$ yr to travel 100 kpc. The slowing of the shell due to mass loading from the IGM and work against gravity implies a likely age of several $10^8$ yr. Wilman et al. (11) demonstrated that the shocked IGM is able to cool on this timescale and that the absorber is likely to have a small HI fraction in the range 0.001–0.1. The shell's energy requirements can be met by supernovae in its host galaxy.

The apparent absence of a large-scale outflow in LAB-1 may be due to evolutionary and environmental effects. LAB-1 has a much higher rate of obscured star formation, whereas LAB-2 has an X-ray obscured AGN. LAB-2 may thus be more evolved from obscured starburst to luminous quasar, allowing time for a superwind to propagate to 100 kpc. Environmentally, LAB-1 resides in one of the densest parts of SSA22, and the hydrodynamic and gravitational interaction with its neighbouring LBGs (a few 10s of kpc away) may erase an HI halo like that in LAB-2. Alternative models have since been proposed for the HI absorption, e.g. in an accretion shock from infalling

IGM (6), but regardless of the true physical explanation it is clear that IFS was essential to the discovery of coherent, galaxy-wide absorption.

## 2 VIMOS IFS of H$\alpha$ Emission in Cluster Cores at $z \sim 0.1$

The fate of cooling X-ray gas in the cores of low-redshift galaxy clusters has been a matter of controversy for nearly 30 years ago (Fabian 1994). For much of this time, these 'cooling flow' clusters have been known to harbour large scale H$\alpha$ nebulosity, often in the form of filaments many kiloparsecs long, as in NGC 1275 in the Perseus cluster, but their relation to the cooling flow is unclear. They are too luminous to be simply due to gas cooling from X-rays, so some form of re-excitation is needed, possibly from massive stars in the most luminous systems e.g. (4). Within the last 5 years new X-ray observations have shown that cooling appears to be quenched at lower temperatures, possibly due to some form of heating. Simultaneously, large-scale reservoirs of cool molecular gas have been identified in the cores of cooling flows (7), together with smaller masses of hot $H_2$(8). The challenge now is to understand the relationship between the various phases of ionized and molecular gas to test competing models for X-ray cooling suppression. Central to this effort is spatially resolved spectroscopy of the H$\alpha$ and $H_2$ emission to map their kinematics and ionisation state, and we are conducting several such IFS projects. Here we show results from our VIMOS IFU observations of the cluster Abell 1664 ($z = 0.1276$) (Wilman, Edge & Swinbank, in prep).

The VIMOS IFU observations of A1664 in Spring 2003 were amongst the first obtained in open time. The cluster was observed for $\simeq 1$ hour with the HR-RED grism and the data were reduced with the VIPGI pipeline, but flexure between the arc and science frames necessitated wavelength calibration with sky lines. The complexity of reducing the data was matched by the richness of its scientific content, showing emission in H$\alpha$+[NII], [SII] and [OI] with multiple velocity components spread over several hundred km s$^{-1}$.

In Fig. 2 we show continuum and H$\alpha$ images and the gas kinematics. The H$\alpha$ emission shows a 30 kpc-long filament, with two velocity components split by upto 700 km s$^{-1}$ in the nuclear regions. Our interpretation is that the H$\alpha$ stream is due to the disturbance of the CO-rich cluster core (containing several $10^{10}$ $M_\odot$ of molecular gas; (7)) by the small galaxy located just beyond the end of the filament (obj #1 in the continuum image), whose redshift places it at the same velocity as the gas at position B2. The two H$\alpha$ velocity components are plausibly due to this galaxy having orbited the central galaxy. Despite the complex kinematics and morphology of the H$\alpha$, its ionization state is uniform, with no variations in [NII]/H$\alpha$, [SII]/H$\alpha$ or [OI]/H$\alpha$ as a function of position or H$\alpha$ surface brightness. This implies that the gas is excited by small-scale processes unrelated to the disturbance of the

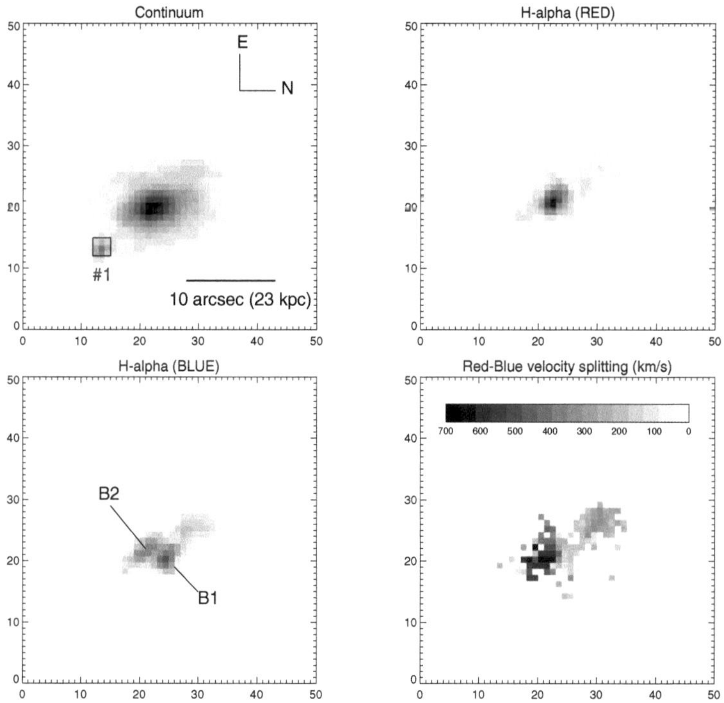


**Fig. 2.** VIMOS IFU observations of the core of the cluster Abell 1664 at $z = 0.1276$. The continuum image was derived over 6700–7000Å (observed frame) and the H$\alpha$ emission comprises two velocity components split by up to 700 km s$^{-1}$. The filamentary H$\alpha$ structure is probably due to the passage of a small galaxy (obj #1 in the continuum image) through the gas-rich cluster core. The redshift of this galaxy ($z = 0.1255$) places it close in velocity to the H$\alpha$ emission at position B2.

gas on kiloparsec scales. The H$\alpha$ kinematics are a striking match to the cluster's (unusually broad) CO(1-0) line profile, suggesting that the CO and H$\alpha$ emission originate in the same clouds. CO interferometry (e.g. with ALMA) will yield a direct 3D IFS comparison with the H$\alpha$ emission.

*Acknowledgements* RJW thanks Euro3D for funds to attend this meeting. The SAURON LAB observations were in collaboration with J. Gerssen, R.G. Bower, S.L. Morris, R. Bacon, P.T. de Zeeuw and R.L. Davies. The VIMOS IFU cluster core observations were in collaboration with A.C. Edge.

## References

[1] Adelberger K.L. et al.: ApJ, **584**, 45 (2003)
[2] Adelberger K.L. et al.: ApJ, **629**, 636 (2005)

[3] Benson A.J. et al.: MNRAS, **599**, 38 (2003)
[4] Crawford C.S. et al: MNRAS, **306**, 857 (1999)
[5] Bower R.G. et al: MNRAS, **351**, 63 (2004)
[6] Dijkstra M. et al: submitted to ApJ (astro-ph/0510409) (2005)
[7] Edge A.C.: MNRAS, **328**, 762 (2001)
[8] Edge A.C. et al: MNRAS, **337**, 49 (2002)
[9] Pettini M. et al: ApJ, **554**, 981 (2001)
[10] Steidel C.C. et al.: ApJ, **532**, 170 (2000)
[11] Wilman R.J. et al: Nature, **436**, 227 (2005)

# Part IX

# The High Redshift Universe

# Integral Field Spectroscopy of Galaxies in the Distant Universe

A. M. Swinbank

Dept. of Physics and Astronomy, University of Durham a.m.swinbank@dur.ac.uk

## 1 Introduction

Understanding the processes by which galaxies evolve requires searching for trends in the properties of both local galaxies as well as the distant galaxy population. Many studies concentrate on the statistical properties of galaxies derived from large galaxy redshift surveys, however in order to understand how galaxies evolve, it is also necessary to study individual galaxies; allowing us to understand their intrinsic properties, such as distribution of star-formation, dynamics and chemical enrichment. Combined with large redshift surveys, we can therefore build up a complete picture of galaxy evolution. Of course, at high redshift this becomes increasingly difficult since galaxies are fainter and smaller and therefore obtaining high signal-to-noise imaging or spectroscopy requires significant investments of time, even on eight or ten meter class telescopes. Indeed, even with long integrations, the limit of the sky background can limit faint object spectroscopy.

One way to overcome this problem is to use galaxy clusters as natural telescopes (8),(2),(9),(4). Galaxy cluster lensing has two effects: (i) the total signal is amplified so that we can study objects which would otherwise be too faint to observe in detail; and (ii) the galaxy image is not only amplified, it is also stretched. As a result, the component parts of the galaxy can be spatially resolved, making it possible to study the galaxy even though an unlensed image of a galaxy a $z$ ~1 is smaller than 0.4″ (FWHM). Thus, gravitational lensing provides us with galaxies which are bright enough and extended enough that we can compare them with starburst galaxies in the local Universe.

## 2 Resolved Spectroscopy of $z$ ~1 Luminous Arcs

Above $z$=1 some of the fundamental questions regarding the relationships between the total mass of a galaxy and the baryonic mass locked up in stars remain unanswered. In local rotationally-supported spiral galaxies, this relation is best described by the Tully-Fisher relation which is an empirical correlation between the “peak” rotational velocity (or line width) and the absolute magnitude of a spiral galaxy. Together these two parameters define a plane of structural parameters for spiral galaxies which may reflect the way in which they were initially formed, perhaps suggesting the presence of self regulating processes of star formation in galactic disks. Observing how this correlation evolves over a substantial history of the Universe thus provides a powerful test of the evolution of galaxy mass (12),(6). A simple test of how galaxies build up their stellar mass is to determine whether the galaxy's stellar mass has built-up in lock-step with the mass of the dark matter halo (the hierarchical galaxy formation model), or whether the galaxies' stellar mass grows within a pre-existing dark matter halo. In simplified outline, the first model predicts that the correlation between peak rotation velocity and stellar mass should evolve little, while the second predicts that the stellar mass corresponding to a fixed rotation velocity should decrease with increasing redshift.

However, at high redshift (i.e. $z$ >1) the small angular size of galaxies means that obtaining spatially resolved rotation curves is a challenge (e.g. a typical local spiral has a scale length of 4 kpc, which corresponds to only $0.5''$ at $z \sim 1$). Fortunately, the lensing magnification produced by the deep potential well of galaxy clusters causes the image of the background galaxy to be magnified and stretched. For a typical $z$=1 lensed galaxy, $0.6''$ on the sky corresponds to only ~1 kpc in the source frame). This allows us to both target galaxies which would otherwise be too faint, and also spatially resolve the kinematics on spatial scales far greater than otherwise possible. Moreover, by targeting these galaxies with Integral Field Spectroscopy, we are able to cleanly decouple the spatial and spectral resolution which are often mixed in traditional longslit observations. We can thus identify which galaxies have regular velocity fields comparable to local spirals (and are therefore suitable for this test).

To this end, we have mapped the two-dimensional velocity fields of a sample of six high redshift, gravitationally lensed galaxies with the GMOS IFU. Using a detailed mass model, we are able to correct for the distortion introduced by the lensing potential and reconstruct the source frame morphology and velocity fields. We use the results to investigate the relationship between the stellar and dark matter components of the galaxy (through the Tully Fisher (TF) relation) and how they compare to disk galaxies in the local Universe (Fig 2.)

The position of the galaxies on the TF relation in both the $B$ and $I$-bands shows good agreement with local data(7),(5),(10). This data is in agreement with existing intermediate and high redshift studies(11)(12)(6). Overall, our observations suggest a 0.5±0.3mag of brightening in the $B$-band TF from the local ($z = 0$) correlation, whilst in the $I$-band we place a limit of $< 0.1$mag between our $z = 1$ sample and the local $z = 0$ correlation.

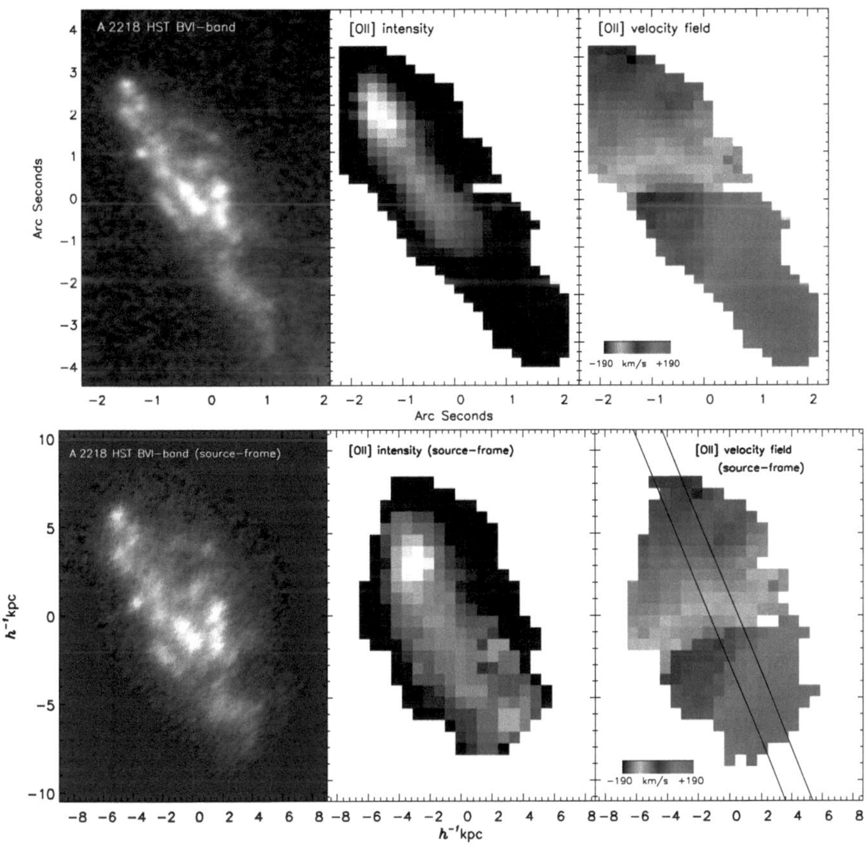


**Fig. 1. Top**: Sky-plane *HST* and IFU observations of Abell 2218 arc#289. *Left:* Combined *HST* WFPC2 $B_{450}$,$V_{606}$ and $I_{814}$-band drizzled images of A 2218 arc289. *Middle:* The [O II] $\lambda$3727 emission map of the arc measured from our IFU observations. The distribution of [O II] emission agrees well with the UV flux seen in the left panel. *Right:* The velocity field of the of the galaxy derived from the [O II] emission. The scale is marked in arcseconds and North is up and East is left. **Bottom:** The reconstructed image of the arc corrected for lens magnification using the mass model of Smith et al. (2003). *Left:* the reconstructed image of the galaxy based on *HST* imaging. *Middle:* The reconstructed [O II] emission line map. *Right:* The velocity map of the galaxy in the source frame (blue-shifted region to upper left). The solid lines show the asymptotic major axis cross section from which a one-dimensional rotation curve was extracted. The luminosity weighted magnification of the source is $4.92 \pm 0.15$, but varies from $\sim 5.6$ to $\sim 4.9$ from the northern to southern end of the arc. The scale shows the size of the galaxy in the source frame. Without a lens, at $z = 1$, $1''$ corresponds to 7.7kpc.

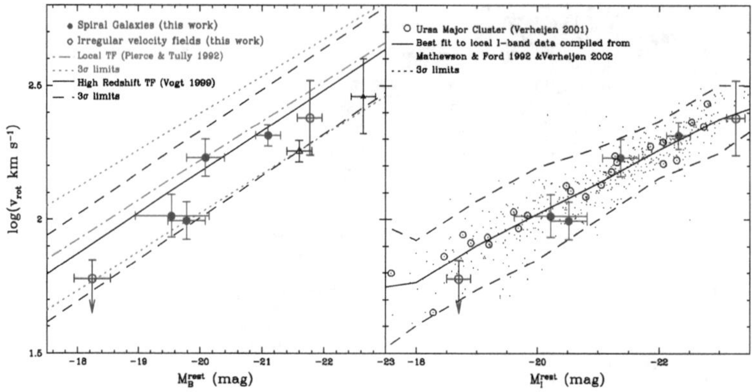


**Fig. 2.** Left: The arcs from our survey on the Tully-Fisher relation in rest frame $B$-band compared to high redshift sample from (12). For comparison we show the low redshift local fit from (7). The solid points show the galaxies which have stable (disk-like) kinematics, but we also indicate the two galaxies for which the dynamics are not well defined by open circles. The solid triangles shows field galaxies (L451) at $z = 1.34$ from van Dokkum & Stanford (2001) and the open triangle shows the position of the field galaxy CFRS22.1313 from CIRPASS IFU observations from Smith et al. (2004). Right: The rest frame $I$-band Tully-Fisher relation compiled from (5) (solid points) and from the Ursa-Major Cluster (open circles; (10)).

## 2.1 z=5 Gravitationally Lensed Galaxies

At higher redshift ($z > 2$), we can turn our attention to one of the most pressing questions which currently plagues models of galaxy formation. The problem with the models is not to understand why galaxies form (this is due to the cooling and condensing of gas in dark matter halos), but to understand why such a small fraction of baryons cool to form stars. Galaxy formation models which only include cooling predict that more than 50% of baryons should form stars, yet a census of the baryons in the local Universe show that less than 10% are locked up in stars, the rest is in a hot diffuse state, similar to that in the inter-cluster medium (Balogh et al. 1999). To account for this puzzling inefficiency requires some form of feedback – a method of expelling gas from galaxies, preventing them from forming stars, and hence regulating galaxy formation (1), (9).

The key to resolving this issue is to identify these features in spatially resolved out-flows around distant proto-galaxies. By comparing the velocity field of the outflow with that of the host galaxy the three dimensional structure of the outflow can be established. However, at these great distances even a massive galaxy only spans $1''$ on the sky and therefore obtaining spatially resolved information is extremely difficult.

One of the most striking cases of gravitational lensing is the (highly magnified) $z$=4.88 galaxy behind the rich lensing cluster RCS0224-002 (3).

Recently, we have used the VLT VIMOS and SINFONI IFUs to study the star forming and kinematical properties in this $z$=4.88 arc. In left-hand panel of Fig. 3 we show the *HST* image of the cluster core and mark the components of the arc $A$, $B$ and $C$. As can be seen from the *HST* image, the lensed galaxy (or arc) is over 12″ in length and therefore is an ideal candidate for integral field spectroscopy. The arc is multiply imaged, with component $A$ appearing to comprise a dense knot surrounded by a halo of diffuse material (a foreground object is also superposed). The morphology of components $B$ and $C$ mirror those of $A$.

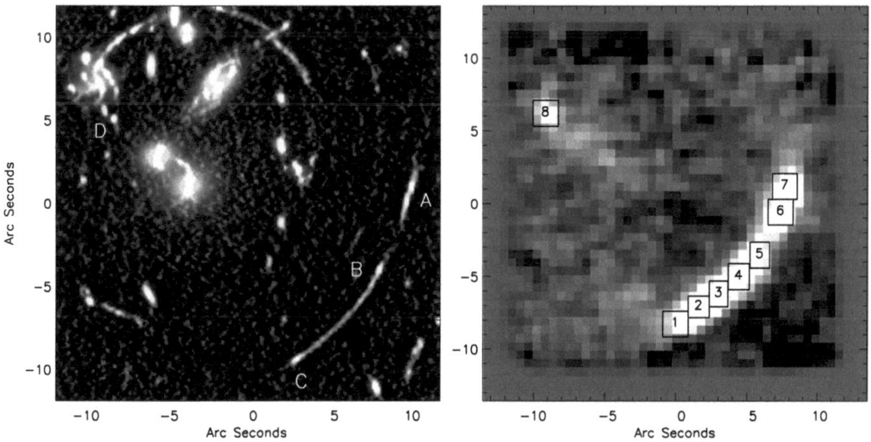


**Fig. 3.** *Left: HST VI* image of the RCS0224-002 cluster core showing the central cluster galaxies as well as the multiple arcs and arclets. The multiply imaged $z$=4.88 arc is labelled A, B and C, whilst the radial counter-image is labelled D. *Right:* Wavelength collapsed (white light) image around the Ly$\alpha$ emission from the $z$=4.88 arc made by collapsing the datacube between 7138 and 7188Å. The $z$=4.88 arc can clearly be been in the Ly$\alpha$ image. We also note that there appears to be another strong Ly$\alpha$ emitter to the North East which is the counter image of the arc.

In Fig 3 we also show the projected the datacube between 7138Å and 7188Å so as to map the spatial distribution of the emission. It is clear that the Ly$\alpha$ emission line morphology traces the that seen in the imaging, with the densest knots in the *HST* image being the brightest in Ly$\alpha$.

An advantage of the spatial size of the VIMOS IFU is that we detect an additional counter image of the arc approximately 10″ to the North-East (labelled #8 in the right panel of Fig. 3). This emission line object is at exactly the same redshift as the main arc. The colours of this extra component are also consistent with the main arc (3).

By extracting a series of independent spectra from the IFU datacube we can investigate the dynamics of the galaxy and the nature of any outflowing material. Even the initial dataset allows us to search for spatial variations in the Lyman alpha emission line. As seen in other young galaxies, the Lyman-alpha emission has a characteristic asymmetric (P-Cygni) profile. To examine the structure of the emission line, we compare the emission from the regions marked 1-8 and find that the structure of the line is remarkably constant from region to region. While there is tentative evidence for structure in the red wing of the line, the blue cut-off occurs at constant wavelength (the variations are less than 30km/s). This is particularly important: the individual star forming regions in the underlying galaxy are expected to be moving at relative speeds in excess of 100km/s (our SINFONI observations will map this): if the superwind was localized to these regions, the structure of the Ly$\alpha$ emission would vary significantly. In contrast, the superwind model predicts that the sharp blue edge of the Lyman-alpha emission line (which is formed by resonant absorption in the outflow) will be uncorrelated with the velocity structure of the host galaxy. The lack of structure suggests that the superwind "bubble" is located well outside of the galaxy and is escaping into inter-galactic space.

Whilst these results are in their early stages, we are showing the power of coupling integral field spectroscopy with gravitationally lensed galaxies to spatially resolve and study the internal dynamics and star-formation properties of primeval galaxies. We will also be able to investigate the dynamics of the inter-stellar medium (spatially resolved) through the absorption lines and the metallicity of the gas through the CIV/Ly$\alpha$ emission line ratios. The SINFONI IFU observations will also probe the [O II] emission from this galaxy. The [O II] emission comes from the star-forming regions and therefore reflects the underlying stellar populations which are responsible for driving the superwind in the galaxy. By coupling the dynamics as measured from the Ly$\alpha$ and [O II] emission we can confirm the speed at which the ejecta is traveling and so test how superwinds regulate galaxy formation.

## Acknowledgments

I would like to thank my collaborators, Richard Bower, Ian Smail, Jean-Paul Kneib, Graham Smith, Richard Ellis, Dan Stark and Andy Bunker. AMS is supported by a PPARC fellowship.

## References

[1] R. G. Bower, S. L. Morris, R. Bacon, et al.: MNRAS, 351, 63, 2004
[2] R. Ellis, M. R. Santos, J. Kneib, and K. Kuijken: ApJL, 560, L119, 2001
[3] M. D. Gladders, H. K. C. Yee, and E. Ellingson: ApJ, 123, 1, 2002
[4] J. Kneib, R. S. Ellis, M. R. Santos, and J. Richard: ApJ, 607, 697, 2004
[5] D. S. Mathewson, V. L. Ford, and M. Buchhorn: ApJS, 81, 413, 1992

[6] B. Milvang-Jensen, A. Aragón-Salamanca, G. K. T. Hau, et al.: MNRAS, 339, L1, 2003.
[7] M. J. Pierce, R. B. Tully: ApJ, 387, 47, 1992
[8] I. Smail, A. Dressler, J. Kneib, et al.: ApJ, 469, 508, 1996
[9] A. M. Swinbank, J. Smith, R. G. Bower, et al.: ApJ, 598, 162, 2003
[10] M. A. W. Verheijen: ApJ, 563, 694, 2001
[11] N. P. Vogt, A. C. Phillips, S. M. Faber, et al: ApJL, 479, L121, 1997
[12] N. P. Vogt, A. C. Phillips, (DEEP Collaboration), European Network on the Formation, and Evolution of Galaxies (TMR) Collaboration; AASM, 200, 2002.

# Integral-Field Studies of the High-Redshift Universe

M. J. Jarvis[1], C. van Breukelen[1], B. P. Venemans[2] and R. J. Wilman[3]

[1] Astrophysics, Department of Physics, Keble Road, Oxford, OX 3RH, U.K.
m.jarvis1@physics.ox.ac.uk, cvb@astro.ox.ac.uk
[2] Sterrewacht Leiden, PO Box 9513, 2300 RA Leiden, The Netherlands
venemans@strw.leidenuniv.nl
[3] Department of Physics, University of Durham, Durham DH1 3LE, U.K.
r.j.wilman@durham.ac.uk

**Summary.** We present results from a new method of exploring the distant Universe. We use 3-D spectroscopy to sample a large cosmological volume at a time when the Universe was less than 3 billion years old to investigate the evolution of star-formation activity. Within this study we also discovered a high redshift type-II quasar which would not have been identified with imaging studies alone. This highlights the crucial role that integral-field spectroscopy may play in surveying the distant Universe in the future.

## 1 Hunting for high-redshift galaxies with IFUs

We initiated a pilot project with a deep, nine hour, VIMOS observation centred on the high-redshift radio galaxy MRC0943-242 at a redshift of $z = 2.92$ in April 2003. The aims of this project were to probe the giant-Ly$\alpha$ emitting halo surrounding this source and the distribution of galaxies within the volume probed by the IFU. With its spectral coverage and large field-of-view, VIMOS is currently the best IFU for such studies. We are able to detect all the galaxies with bright emission lines over the whole volume. For Ly$\alpha$ emission this range is $2.3 < z < 4.6$, and for [OII]$\lambda$3727 emission, we probe $0.08 < z < 0.83$. Therefore we can search for emission-line galaxies over a large fraction of cosmic volume along the sightline of the IFU.

The process of detecting and selecting emission line objects from IFU data is difficult and involves multiple steps, therefore we refer the reader to van (10) for full details. However, our selection enabled us to detect 17 emission-line objects over the volume probed with the IFU. These are predominantly single line objects, and for 14 all of the characteristics point to them being hydrogen Ly$\alpha$ emission-line galaxies (two others are [OII] emitters and the third is the type-II quasar discussed later in this article), we will now concentrate on these Ly$\alpha$ emitters. Ly$\alpha$ emission is produced by massive stars photoionizing hydrogen gas. By using some simple assumptions it is possible to estimate the star-formation rate in galaxies which exhibit Ly$\alpha$ emission by measuring the luminosity of the emission line. Although undoubtedly crude, this does at least produce a lower limit for the star-formation activity in distant galaxies. By binning all of the Ly$\alpha$ luminosities in the volume we construct the Ly$\alpha$ emitter luminosity function. Construction of the luminosity function is a non-trivial task for this type of data because those galaxies with bright emission line can be seen to much greater distance in the volume covered in our data, thus the volume probed is a strong function of the luminosity of the emission lines and an accurate sensitivity map of the field is crucial (see (10)).

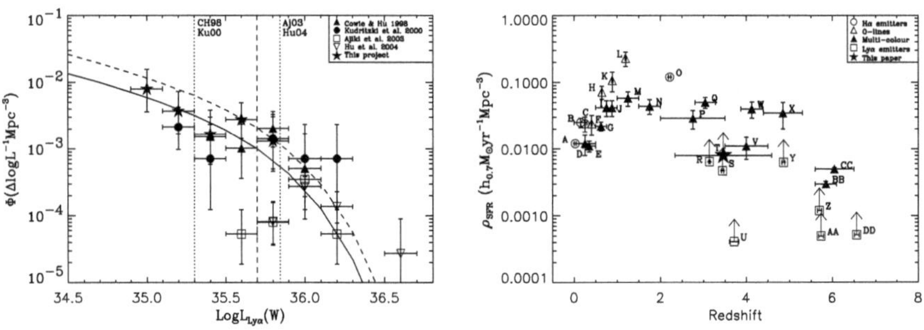


**Fig. 1.** (*left*) The number density of Ly$\alpha$ emitters plotted against the luminosity. The filled symbols mark surveys with an average redshift similar to ours (triangles and circles) and the open symbols stand for surveys at redshift $z = 5.7$ (squares and inverted triangles). Over-plotted are two Schechter luminosity functions: the solid line is the fit to all our data points and the dashed line is the fit to our two highest luminosity data points and those of the surveys at similar redshift with $L > 5 \times 10^{35}$ W (dashed horizontal line) to ensure completeness. The dotted horizontal lines mark the detection limits of the surveys. (*right*) Star-formation rate densities as derived by various types of surveys. The result from our work is denoted by the filled star. The different types of surveys are marked with different symbols: the open circles are H$\alpha$ searches, the open triangles are surveys aimed at oxygen emission lines, the filled triangles are multicolour surveys, and the open squares are Ly$\alpha$ searches (full details of the other surveys can be found in (10)).

Fig. 1 (*left*) shows the Ly$\alpha$ luminosity function derived from this study compared to the luminosity function measured from narrow-band studies and multi-colour selection. One can see that our luminosity function, which probes the redshift range $2.3 < z < 4.6$ extends the work of the narrow-band searches to fainter luminosities where the luminosity function keeps the same Schechter function form up to $z \sim 6$. This implies that there is little evolution in the star-formation rate density over this redshift range, albeit small number statistics preclude strong statements regarding any evolution.

As stated above, knowledge of the luminosity of the Ly$\alpha$ emission line in these galaxies informs on the total star-formation rate. Using typical assumptions of hydrogen recombination the star-formation rate is related to the Ly$\alpha$ luminosity by $SFR = 9.1 \times 10^{-36}(L_{\mathrm{Ly}\alpha}/\mathrm{W})\ \mathrm{M}_\odot\ \mathrm{yr}^{-1}$ [see (10) for details]. By integrating over the Ly$\alpha$ luminosity function we are therefore able to measure the star-formation rate at the redshifts covered by our data. This plot, along with the star-formation rate density derived by other methods, is shown in Fig. 1 (*right*) for $0 < z < 6$. Due to the fact that Ly$\alpha$ can be resonantly scattered and absorbed by neutral hydrogen around the source, the measured SFR from studies using Ly$\alpha$ are hard lower limits. Also, the presence of dust preferentially extinguishes the UV continuum emission, therefore even multi-colour searches are prone to biases which work to reduce the estimated SFR. Therefore, we also show the estimated star-formation rate corrected for obscuration. With this correction in place it is apparent that our IFU search is in line with previous studies conducted in a number of different ways. However, the benefit of using the integral-field approach is that we select sources at all redshifts in our volume in precisely the same way, thus reducing the biases involved in comparing studies at different redshifts from different surveys, which may utilize different techniques.

## 2 Discovery of a type-2 quasar in the IFU deep field

In this section we discuss the way in which our integral-field data has also led to the discovery of two Active Galactic Nuclei (AGN) in the volume probed, in addition to the radio galaxy which was targeted. One of these is a 'normal' unobscured type-I quasar with broad emission lines and an unresolved morphology on optical images at a redshift of $z = 1.79$. However, the other AGN exhibits only narrow-emission lines [Fig. 2 (*left*)] and has a resolved morphology in the optical image [Fig. 2 (*right*)].

These type-II AGNs are relatively difficult to find compared to the type-I counterparts. This is principally due to the fact that type-II AGN look like normal galaxies, and it is only by looking for other signatures of AGN activity, which do not suffer from extinction due to the torus, can they be found, e.g. X-rays from the central engine which penetrate the torus (e.g. (9)), radio emission from powerful jets e.g. (2) or reprocessed dust emission in the mid-infrared from the torus itself (e.g. (7)). However, with the integral-field approach we are sensitive to the bright narrow-emission lines that are characteristic of an obscured AGN, as we obtain the spectrum of any object in the IFU field immediately.

J094531-242831 (hereafter J0945-242) exhibits these bright narrow-emission lines, in the CIV doublet, HeII and CIII], all characteristic of a type-II AGN. The radio map shows that there is no radio emission down to a radio flux limit of 0.15 mJy at 5GHz. At a redshift of $z = 1.65$ this is significantly below the typical luminosity of a radio galaxy, thus we confirm that this is a genuine radio-quiet type-II quasar. The line luminosity ratios of the CIV, HeII and CIII] lines are also consistent with the ratios for radio galaxies, and not the generally lower-luminosity Seyfert-I galaxies and the unobscured quasars (8). Using these line luminosities it is possible to estimate the lower mass limit of the accreting black hole in the centre of this galaxy. We assume the typical line ratios of radio galaxies to convert the HeII luminosity to a line luminosity in [OII], which is correlated with the total bolometric luminosity of the AGN. Under the assumption that the quasar is accreting at its maximum rate, i.e. the Eddington limit, then this bolometric luminosity equates to a black-hole mass of $3 \times 10^8$ $M_\odot$.

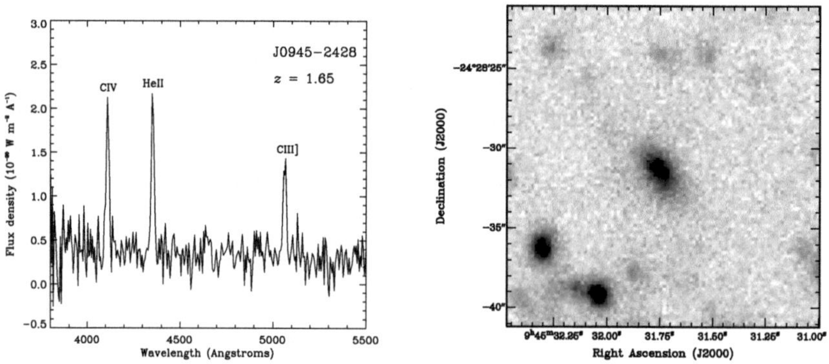


**Fig. 2.** (*left*) The 1-D spectrum of the type-II quasar, J0945-2428 at $z = 1.65$. The spectrum was extracted over the whole galaxy. (*right*) V-band image of J0945-2428, clearly showing that the quasar is resolved.

In the local Universe there is now a well known correlation between the mass of black holes and the luminosity of their host galaxy (see e.g. (5)). The near-infrared $K-$band magnitude of J0945-242 is very faint, with $K = 20.5$. Radio galaxies at $z = 1.65$ typically have host galaxy luminosities of $K \sim 18$ e.g.(3). Thus the host galaxy of J0945-242 appears to be 2.5 mag fainter than that for a typical radio-loud type-II AGN. If this faintness of the host galaxy is caused by extinction from dust then we would expect the blue end of the galaxy spectrum to be fainter, as dust attenuates the blue light more readily than at red wavelengths. However, the host galaxy of J0945-242 is extremely blue, indicative of ongoing star-formation. Therefore, the faintness in the $K-$band light indicates that the host galaxy has a dearth of old, massive stars, which in turn implies that the galaxy is not yet fully formed at $z = 1.65$. Whereas the black hole has already grown, presumably by accretion of matter, close to its final mass due to the fact that the low-redshift black-hole mass function shows that supermassive black holes appear to have a maximum mass of around $10^{10}$ $M_\odot$ (e.g. (6)).

This relatively large black-hole mass associated with a host galaxy approximately a factor of 10 fainter than would be expected from the local relation implies that supermassive black holes at high redshift may essentially be fully grown before the host galaxy has fully formed. This is in qualitative agreement with what we already see in high-redshift radio galaxies, where the small, young, radio sources appear to have extremely bright sub-millimetre luminosities (e.g. (1)). In order to produce these sub-millimetre luminosities, star-formation rates of up to 1000 $M_{\odot}\ yr^{-1}$ are needed, typical of a galaxy undergoing its first major bout of star formation activity.

## 3 Summary

The new method of detecting emission-line galaxies at high redshift along with the serendipitous discovery of an obscured quasar at $z = 1.65$, highlights the way in which relatively wide-area integral-field units on large telescopes could open up a unique window on the Universe. VIMOS is currently the only instrument which has the capability of large spectral coverage coupled with a 1 square arcminute field-of-view. However, future instruments, such as the Multi-Unit Spectroscopic Explorer (MUSE; http://muse.univ-lyon1.fr), will expand the initial work taking place in this field with VIMOS. Furthermore, volumetric surveys with IFUs may begin to find types of object we have yet to discover in traditional surveys, and thus offer a whole new view of the Universe.

Full details of the work presented in this article can be found in (10) and (4).

## References

[1] Archibald E.N., Dunlop J.S., Hughes D.H., Rawlings S., Eales S.A., Ivison R.J., 2001, MNRAS, 323, 417
[2] Jarvis M.J., et al., 2001, MNRAS, 326, 1563
[3] Jarvis M.J., et al., 2001b, MNRAS, 326, 1585
[4] Jarvis M.J., van Breukelen C., Wilman R.J., 2005, MNRAS, 358, 11
[5] Magorrian J., et al., 1998, AJ, 115, 2285
[6] Marconi A., Risaliti G., Gilli R., Hunt L.K., Maiolino R., Salvati M., 2004, MNRAS, 351, 169
[7] Martínez-Sansigre A., Rawlings S., Lacy M., Fadda D., Marleau F.R., Simpson C., Willott C.J., Jarvis M.J., 2005, Nature, 436, 666
[8] McCarthy P.J., 1993, ARAA, 31, 639a
[9] Norman C., et al., 2002, ApJ, 571, 218
[10] van Breukelen C., Jarvis M.J., Venemans B.P., 2005, MNRAS, 359, 895

# 3D Kinematics of High-z Galaxies as Seen Through the Gravitational Telescope

M. Lemoine-Busserolle[1], S. F. Sánchez[2], M. Kissler-Patig[3], R. Pelló[4], J.-P. Kneib[5,6], A. Bunker[7] and T. Contini[4]

[1] Institute of Astronomy, University of Cambridge, Madingley Road, Cambridge, CB3 0HA, UK - lemoine@ast.cam.ac.uk
[2] Centro Astronómico Hispano Alemán (CSIC-MPI), C/Jesus Durban Remon 2-2, E-04004 Almeria - Spain
[3] ESO, Karl-Schwarzschild-Str.2, 85748 Garching - Germany
[4] Laomp - UMR 5572, 14 Avenue E. Belin, F-31400 Toulouse, France
[5] LAM - UMR 6110, Traverse du Siphon-Les trois Lucs, F-13376 Marseille Cedex 12, France
[6] Caltech Astronomy, mail code 105-24, Pasadena, CA 91125, USA
[7] School of Physics, University of Exeter, Stocker Road, Exeter, EX4 4QL, UK

**Summary.** The study of the physical properties of high-redshift galaxies has become one of the major goals of extragalactic astronomy. In particular the mass-assembly histories of galaxies have been the focus of many studies at redshifts 1 to 3. In the purpose of probing the dynamics of intermediate and high-redshift galaxies, we have designed a research program to carry out a near-infrared spectroscopic follow up of spatially resolved distant galaxies. Here, we present the results for A370-A5 (z=1.341), an arc behind the lens cluster Abell 370 (z=0.374), observed in the case of science verification programme of SINFONI/VLT. The natural magnification due to massive galaxy clusters allows to spatially resolve and constrain the dynamics of young star forming galaxies 1 to 3 magnitudes fainter than those selected in blank fields. Thus, the study of lensed galaxies allows to probe a low mass regime of galaxies not accessible in standard observation. In this particular case, we found that the gas distribution and kinematics are consistent with a bipolar outflow with a range of velocities of $v \sim 100 km/s$.

## 1 Introduction

The star formation in the universe may have peaked in the redshift range $1 < z < 2.5$ (7). It is thus crucial to understand the structure of galaxies at this epoch: did most stars form in regular disk galaxies, or is star formation the product of the assembly of galaxies? Do the morphology and kinematics of galaxies at this redshift resemble that of local ones? These are key questions for our understanding of the formation and evolution of galaxies. Many different major projects have been built in order to assess these questions, most of them based on high-resolution HST imaging (8; 5), and long-slit spectroscopic follow-up (2). However, the lack of a two-dimensional mapping of the gas content and kinematics in these galaxies does not give a complete picture of their nature. Now with the advent of NIR integral field spectrographs on

8m-class telescopes, it become possible to perform full 3D spectroscopic studies of galaxies at these redshifts. Clusters of galaxies acting as Gravitational Telescopes (hereafter GTs) constitute a particular and powerful tool in this context of high redshift surveys. The main advantage of such "telescopes" is that they take benefit from the large magnification factor in the core of lensing clusters, close to the critical lines, which typically ranges between 1 and 3 magnitudes. Thus, GTs can be used to access the most distant and faintest population of galaxies.

The plan of the paper is as follows. Section 2 presents a summary of the use of cluster lenses as Gravitational Telescopes. In sect. 3 we describe our observations and data. In sect. 4 we address the analysis and discussion of the physical properties of the arc A370-A5. In sect. 5 we summarise our conclusions. We assume a cosmology with $\Omega_M = 0.3$, $\Omega_\Lambda = 0.7$ and $H_0 = 70$ km s$^{-1}$ Mpc$^{-1}$ throughout. In this case, at $z = 1.341$ 1" = 8.41 kpc.

## 2 Lensing Clusters as Gravitational Telescopes

From the point of view of galaxy evolution studies, GTs can be considered as an alternative way to blank fields to investigate the properties of distant galaxies. For a given limiting magnitude or flux, lensed samples will complement the currently available field surveys towards the faint end of the luminosity function, and towards higher limits in redshift. It is worth noting that only lensing clusters with fairly well constrained mass distributions can be used as efficient GTs for galaxy evolution studies. An interesting property of GTs is that they conserve the surface brightness and the spectral energy distribution (SED) of lensed galaxies. Besides the spatial magnification, sources (or regions of sources) smaller than the seeing will effectively gain in surface brightness. These two effects contribute to increase the interest of GTs in spectroscopic studies. The power of GTs to study the spectrophotometrical and morphological properties of high-$z$ galaxies was emphasized by different authors since the nineties (see Fort & Mellier 1994 for a review). Number of high-$z$ lensed galaxies observed during the nineties were discovered serendipitously, but efficient selection procedures have been defined, based on photometric or lensing criteria. Indeed, GTs have been successfully used during the last ten years to perform detailed studies of distant galaxies at different wavelengths, from UV to submillimeter. A particular and recent application of GTs is the identification and study of extremely distant galaxies, including the search for the first galaxies formed in the early Universe, up to redshifts of the order of $z \sim 10$.

## 3 Data and observations of the arc A370-A5

The lens cluster Abell 370 is the most distant cluster of galaxies in the Abell catalogue(9), with a redshift of $z$=0.374. Its mass distribution, with a bimodal shape, is well known (6; 10; 3). The arclet A5 located at 56" from the center is the most extended one, with a large elongation of 9". In this paper we use near infrared integral field spectroscopy to map the emission line morphology of this arc, with the aim of mapping the star formation through H$\alpha$. The results presented here are part of our science verification programme for the new ES0 instrument SINFONI [8].

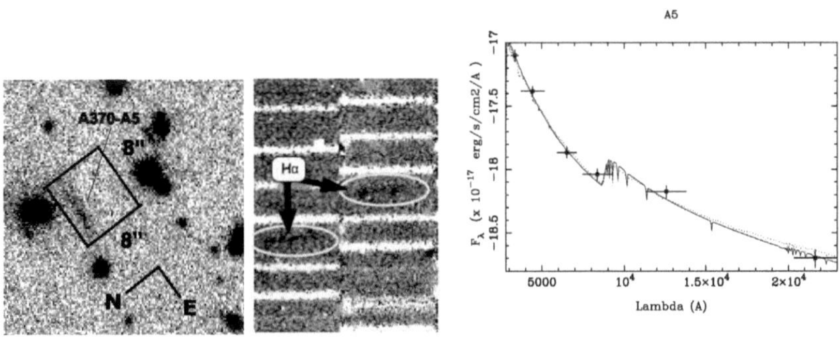


**Fig. 1.** *Left*: CFHT-I band image of A370-A5. The SINFONI FOV used in this study is indicated with a square overlaid to the image. *Center*: Subsection of the 2D SINFONI frames. It shows a detail of two of the 32 slitlets, with the spatial direction along the X-axis and the dispersion direction along the Y-axis. The H$\alpha$ emission is clearly seen as a black stripe. White stripes correspond to residuals after sky-subtraction. *Right*: Photometric SED of A5 corresponding to: $U = 21.51 \pm 0.11$, $B = 22.76 \pm 0.10$, $R = 22.98 \pm 0.11$, $I = 22.61 \pm 0.11$, $J = 21.61 \pm 0.12$, $Ks = 20.79 \pm 0.17$, with error bars including sky noise and zero-point uncertainties. Solid and dashed lines display respectively the best fits obtained with constant star-formation rate and burst models. The best fit correspond to a constant star-formation model of $\sim 500$ Myr, with 1/200 solar metallicity and $E(B-V) = 0$

In addition to our IFS data, we used publicly available images of the arc A370-A5 at different bands: U (F336/WHST, 1), B, R and I (CFHT, 6) and J and Ks (ESO/NTT [9]), in order to derive the UBRI photometry. We correct for the lensing effect both the datacube and the broad-band images. The main effect of the shear is to stretch the image of the galaxy source by a factor of 3.3 in the East-West direction.

[8]http://www.eso.org/science/vltsv/sinfonisv/lensed.html
[9]ESO program 61.A-0296(B)

# 4 Analysis of the physical properties of A370-A5

We used the Hα emission line to determine the distribution and kinematics of the ionized gas by derived from line intensity, the velocity and the velocity dispersion maps from the datacube (see Fig. 2). Similar techniques have been applied to kinematics studies at lower redshifts (4).

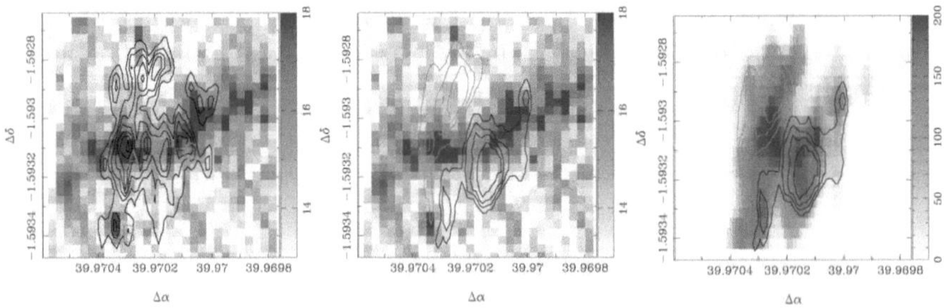


**Fig. 2.** *Left*: Grayscale of the F336W-band image corresponding to the rest-frame UV emission ($\lambda \sim$1435Å) obtained using the WFPC at the HST, once applied the correction from the lensing effect. Images are oriented North-East, and scale is on degrees. The clumpy structure described in the text is clearly seen. Counters show the Hα flux intensity map obtained by fitting the emission line of each spectrum to a single gaussian function. Labels indicate those knots where the Hα emission trace the brightest UV flux. Two Hα extended emission line regions, without any detected continuum contourparts, are clearly seen, extended north-east and south-west from the main axis of the UV continuum emission of the arc. *Center*: Same grayscale image with the Hα velocity map obtained by the fitting procedure overlaid as contours. The approaching velocities are in blue and the receding ones in green. First contours start at $\pm 10$ km/s, with an increasing step of 10 km/s between consecutive contours. *Right*: Grayscale of the Hα velocity dispersion map in km/s obtained by the fitting procedure. The contours show the same velocity map as described in the previous panel.

## 4.1 Morphology, gas distribution and Stellar population

We find the same morphology at any wavelength range covered by our set of broad-band images: a clear clumpy/fragmented morphology, without an underlying smooth component. This may indicate that the galaxy is irregular, without the presence of a well defined shape. The Hα emission is also clumpy, with some of its knots matching the UV emission ones, as indicated in Fig 2 as A, B and C (left panel). This is expected from starforming regions, where the hot stars ionize the surrounding gas. However, there are two emission line regions extending north-east and south-west, out of the continuum emission of the arc (ie., without any detected counterpart in the broad-band images).

The *UBRIJK* Spectral Energy Distribution (SED) of A370-A5 was used to constrain the stellar content of this galaxy. We have explored the metallicity versus extinction parameter space for continuous star-forming (CSF) and single-burst models from the GISSEL code (Bruzual & Charlot 2003), with Salpeter IMF, using a standard $\chi^2$ minimization procedure based on *Hyperz* routines (Bolzonella et al. 2000). In summary, good fits can be obtained with burst models younger than $\sim$ 200 Myr and CSF models, in all cases with negligible dust extinction ($E(B-V) \sim 0$). Although metallicity/age/extinction degeneracy exists for this star-forming galaxy, sub-solar metallicity models seem to be privileged.

## 5 Conclusions

We obtained NIR 3D spectroscopy with SINFONI for the arc A370-A5 behind the lens cluster A370, covering the H-band. This configuration allowed us to sample the H$\alpha$ region at the redshift of the object ($z = 1.341$). Due to the lens effect the image of the galaxy was spatially resolved, which allowed us to infer the spatial distribution and kinematics of the ionized gas. From broad-band archival imaging, we determine the SED from $\lambda_{\rm rest}1300-10000$ Å. The best-fit stellar population is comparatively young (180Myr). The gas distribution (compared with the stellar one) and gas kinematics indicate, most probably, the presence of a bipolar outflow due to galactic wind. A violent starformation (located in the clumpy regions) may be at the origin of this wind. The properties of this galaxy derived from the gas and stellar population may suggest that we are observing it in formation.

*Acknowledgements* We are very grateful to the VLT Obs. for accepting this programme for SINFONI Science Verification Time. This work was supported by the Marie Curie Research Training Network *Euro3D; contract No. HPRN-CT-2002-00305.* J-P. Kneib thanks support from CNRS and Caltech.

## References

[1] Bézecourt, J., Soucail, G., Ellis, R. S., & Kneib, J.-P. 1999, A&A, 351, 433
[2] Erb, D. K., Steidel, C. C., & et al. 2004, APJ, 612, 122
[3] Fort, B. & Mellier, Y. 1994, A&A, 5, 239
[4] García-Lorenzo, B., Sánchez, S. F., & et al. 2005, APJ, 621, 146
[5] Giavalisco, M., Ferguson, H. C., & et al. 2004, APJ L., 600, L93
[6] Kneib, J. P., Mellier, Y., Fort, B., & Mathez, G. 1993, A&A, 273, 367
[7] Madau, P., Pozzetti, L., & Dickinson, M. 1998, APJ, 498, 106
[8] Rix, H.-W., Barden, M., Beckwith, S. V. W., et al. 2004, APJ, 152, 163
[9] Sarazin, C. L., Rood, H. J., & Struble, M. F. 1982, A&A, 108, L7
[10] Smail, I., Dressler, A., Kneib, J.-P., et al. 1996, APJ, 469, 508

# AGN Feedback at High Redshift: Shaping the Most Massive Galaxies?

N. Nesvadba[1], M. D. Lehnert[1] and F. Eisenhauer[1]

Max-Planck-Institut für extraterrestrische Physik, Giessenbachstraße, 85748 Garching bei München, Germany, nicole@mpe.mpg.de

**Summary.** Using the near-infrared integral-field spectrograph SINFONI on the VLT, we studied the spatially-resolved dynamics in the optical emission line gas around the $z = 2.2$ powerful radio galaxy MRC1138-262, a likely progenitor of the massive low-z galaxy population. Velocities and FWHMs of up to 2400 km $s^{-1}$ indicate an outflow driven by energies similar to the energy output of the AGN and fulfilling tight evolutionary constraints set by massive low-z galaxies. Is this the "smoking gun" of high-z AGN feedback? High-z AGN feedback plays a major role in recent cosmological simulations, but observational constraints are still missing.

## 1 Introduction

The old and evolved stellar populations, low gas content, and enhanced $[\alpha/\mathrm{Fe}]$ abundance ratios in nearby massive galaxies likely signal a powerful feedback mechanism in the early Universe. A rapidly increasing number of cosmological simulations suggest that AGN feedback at high redshift played a major role in the evolution of the most massive galaxies (1). Likely more powerful than starburst-driven winds, they should be able to remove significant gas masses from the deep potentials of their massive ($M \sim 10^{12-13}$ $M_\odot$) hosts, perhaps efficiently explaining the characteristics of local massive early type galaxies.

However, relativistic jets are not thought to couple strongly with the ambient ISM of their host (2), due to the clumpy ISM, a rather small working surface, and AGN producing "light jets". In a pilot study to elucidate the impact of high-z AGN feedback on the gas kinematics of the host galaxy, we used an early run with the SINFONI integral-field spectrograph SPIFFI (3) in April 2003 (on UT1 of the VLT) to obtain deep spectroscopy of the $z = 2.2$ powerful radio galaxy MRC1138-262 in the near-infrared H and K bands (rest-frame optical). Observations and data reduction are described in (4). Radio galaxies at high redshift host powerful AGN, and are generally thought to be the likely progenitors of the most massive galaxies seen during a rapid evolutionary phase. This makes them the ideal candidates for studying the impact and cosmological implications of AGN winds.

## 2 Evidence for a Massive AGN-Driven Outflow at $z = 2.2$

[O III]$\lambda\lambda$4959,5007 line emission in the H band is very prominent, with FWHM= 3990 km $s^{-1}$ and complex line profiles. H$\alpha$ and [NII]$\lambda$6583 are blended, making [O III]$\lambda$5007 a more accurate tracer of the gas dynamics. Over radii of $\sim 2.5''$ ($\sim$ 20 kpc), we fitted spectra extracted from $0.75'' \times 0.75''$ apertures. Relative velocities do not show a smooth gradient consistent with rotation, but regions of roughly uniform internal kinematics, and $\sim$ 1000 km $s^{-1}$ velocity relative offsets (Fig. 1). Typical velocities in starburst-driven winds are only a few $\times$ 100 km $s^{-1}$, indicating a more powerful mechanism in MRC1138-262. The total H$\alpha$ off-nuclear flux in the K band corresponds to a minimum ionized gas mass of $4 \times 10^9$ $M_\odot$ for a $H_0 = 70$ km $s^{-1}$, $\Omega_\Lambda = 0.7$ flat cosmology and assuming case B recombination. This implies an outflow rate of $M \sim 400$ $M_\odot$ $yr^{-1}$ for an estimated AGN lifetime of $10^7$ yrs (5).

In analogy with starburst-driven winds, we model the energy injection rate of an energy-conserving bubble which inflates due to its own internal energy, sweeping up, ionizing, and accelerating the observed H$\alpha$ emitting gas, $E \sim 4 \times 10^{45}$ ergs $s^{-1}$. Estimating the energy output of the AGN from the bolometric luminosity $\mathcal{L}_{bol}$ and the kinetic luminosity of the radio jet (following (6)), we find that the kinetic energy of the *observed* outflowing H$\alpha$ emitting gas amounts to $\mathcal{O}(10\%)$ of the input energy, a factor $\sim$ 100 larger than predicted by the simple dentist drill model (7).

## 3 Low-Redshift Constraints and Cosmological Impact

The most massive galaxies at low redshift, mostly spheroids, appear highly evolved and have luminosity-weighted ages consistent with high-redshift formation (8; 9). Perhaps the tightest evolutionary constraint comes from their metal abundance ratios, especially [$\alpha$/Fe], which records the relative contribution of SNIa and SNII supernovae to the chemical enrichment, and constrains the length of the star-formation to $\sim$ 0.5 Gyr (8; 9). If the observed outflow rate of 400 $M_\odot$ $yr^{-1}$ is roughly constant, MRC1138-262 will loose $\mathcal{O}(10\%)$ of its $5 \times 10^{12}$ $M_\odot$ stellar mass (10) in $10^8$ yrs, within the tight limit set by the [$\alpha$/Fe] clock. Over the $10^7$ yrs AGN lifetime, a total of $E_{kin} = \mathcal{O}(10^{61})$ ergs will be injected into the ISM, roughly the binding energy of the host, so that a large fraction of the gas will likely escape from the host potential, while the existing stellar population will evolve passively to become "old, red, and dead" by $z = 0$, in good agreement with observations.

MRC1138-262 is the first HzRG for which we carried out such an analysis, which requires integral-field spectroscopy. Indirect evidence based on longslit spectra suggest this is not a unique source: similar line profiles and luminosities have been observed in other radio galaxies (11). We are currently extending our sample. If the observed outflow rate is roughly constant over the $10^7$ yr AGN lifetime, then a total of $\sim (4-80) \times 10^9$ $M_\odot$ of material will be ejected per $M^*$ galaxy, including $(2-40) \times 10^7$ $M_\odot$ in metals (for solar metallicity). The total ejected energy density is at least similar, if not greater, than in starburst-driven winds. (12) find from chemodynamical evolution models that AGN winds eject more iron than starburst-driven winds.

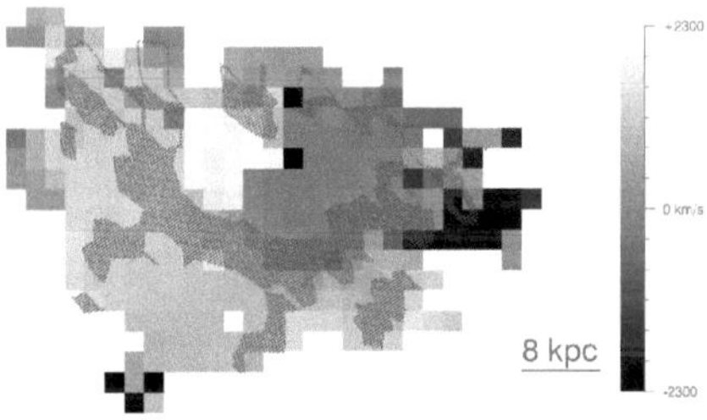


**Fig. 1.** Velocity map of MRC1138-262. Hatched areas show regions with FWHM $> 1400$ km s$^{-1}$.

## 4 Acknowledgements

We would like to thank the SPIFFI commissioning team for carrying out the observations, and ESO staff for helpful and enthusiastic support.

## References

[1] Hopkins, P. F., Hernquist, L., Martini, P. et al. 2005, ApJL, 625, L71
[2] Begelman, M. C., & Cioffi, D. F. 1989, ApJ, 345, L21
[3] Eisenhauer, F. et al., 2000, SPIE,4008,289
[4] Nesvadba, N. et al., in prep.
[5] Martini, P. 2004, Coevolution of Black Holes and Galaxies, 170
[6] Wan, L., Daly, R. A., & Guerra, E. J. 2000, ApJ, 544, 671
[7] Scheuer, P. A. G. 1982, IAUS 97, 163
[8] Thomas, D., Maraston, C., et al. 2005, ApJ, 621, 673
[9] Romano, D., Silva, L., Matteucci, F., & Danese, L. 2002, MNRAS, 334, 444
[10] Pentericci, L., 1999, PhD thesis
[11] McCarthy, P. J., Baum, S. A., & Spinrad, H. 1996, ApJS, 106, 281
[12] Kawata, D., & Gibson, B. K. 2005, MNRAS, 358, L16

# Integral Field Observations of a Distant Cluster

D. Vergani[1*], C. Balkowski[1], V. Cayatte[1], H. Flores[1], F. Hammer[1], S. Mei[2] and J. P. Blakeslee[2]

[1] Observatoire de Paris, Sect. de Meudon, GEPI, 5 place Jules Janssen, F-92195 Meudon Cedex, France; [*] *current address:* INAF–IASF Milano, Istituto di Astrofisica Spaziale e Fisica Cosmica, Italy; daniela@lambrate.inaf.it

[2] Dep. of Physics and Astronomy, Johns Hopkins Univ., Baltimore, MD 21218

## 1 Motivation

Clusters of galaxies are known to be the largest gravitationally bound objects in the Universe. They are excellent tools to explore the distant Universe and to constrain cosmological models. Processes such as tidal interactions, mergers, ram pressure stripping etc. mainly drive the evolution of galaxies (see (3) for a review). In clusters, these environmental mechanisms act on galaxy morphology and kinematics even more efficiently.

In this respect, 3D spectroscopy plays a privileged role to map the resolved kinematics by sampling the whole velocity field. This limits the uncertainties related to the long-slit spectroscopy technique, e.g. slit misalignments from the major axis, warps. In fact, these uncertainties are expected to be most effective in dense environments, where disturbances in morphology and kinematics occur more frequently as by-products of evolutionary processes.

## 2 Target and Dataset

We have used the FLAMES multiple integral field unit (IFU) system of the European Southern Observatory (VLT) to obtain[3] the spatially resolved kinematics of the cluster members in MS0451.6-0305 ($z = 0.5386$). This set-up allow us to observe up to fifteen objects simultaneously with deployable IFUs of a size $2'' \times 3''$ each. Finally, we have determined the velocity field for eight cluster members tracing the redshifted [O II]3727Å emission line.

Available archival data obtained with the Wide Field Channel of the Advanced Camera for Surveys on the Hubble Space Telescope was used to obtain the morphological properties. These images were obtained in the F814W ($I_{814}$) filter for programme GO-9836 (V bandpass at $z \approx 0.5$). We determined the set of structural parameters which reproduces the model for individual objects that best minimizes the residuals with the real image. The internal

[3] *Under the Guaranteed Time of the Obs. Paris-Meudon/GEPI team (PI F. Hammer), programme 71.A-0170A (P.I. V. Cayatte).*

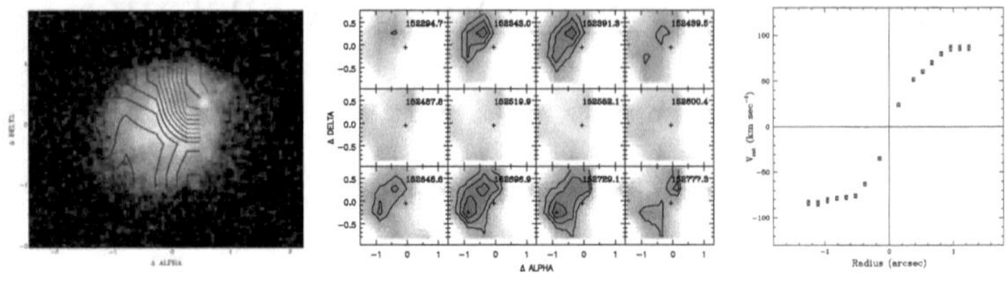


**Fig. 1.** Example of FLAMES/IFU 2D kinematics. (*Left*) HST/ACS image with [O II] velocity field in contours. (*Center*) A contour map of integrated [O II] distribution. (*Right*) An example of a regular rotation curve.

parameters have been used as comparison with those computed from the kinematics.

## 3 Results and Discussions

We find differing behavior among the kinematics and morphology of the eight cluster galaxies observed in MS0451.6-0305 at $z = 0.5386$. The one common element is the kinematic disturbance with the lack of a straight kinematic axis. In the majority of them some faint condensations may be detected. In the vicinity (radius$\leq$7kpc) of the detected cluster members several (up to four) low surface brightness possible companions are observed.

We present an example of our kinematic analysis for one galaxy member MS0451-03 PPP 1912 in Fig. 1. All rotation curves have been computed according to galaxy kinematic peculiarities. One of the most regular rotation curves among the eight is shown in Fig. 1.

In Fig. 2, we display the typical analysis performed on the photometric data-set. For four objects which present a regular structure, the residuals in the modeling process are represented by spiral arms. Two of them have

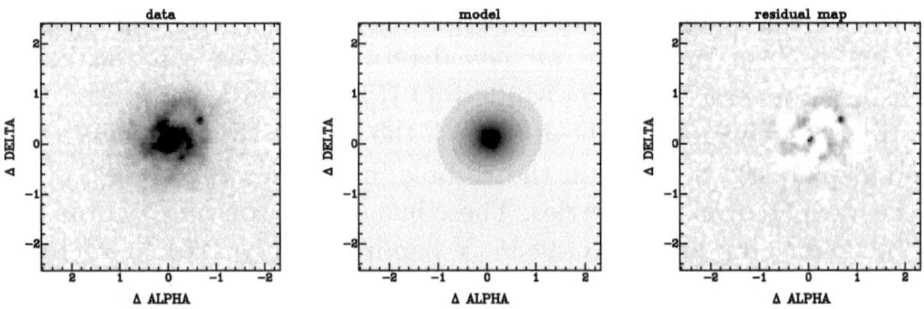


**Fig. 2.** Example of morphological (HST/ACS) analysis for the bulge/Disc decomposition in PPP1912. (*Left*) The original data. (*Center*) The model image. (*Right*) Residuals from the modeling analysis.

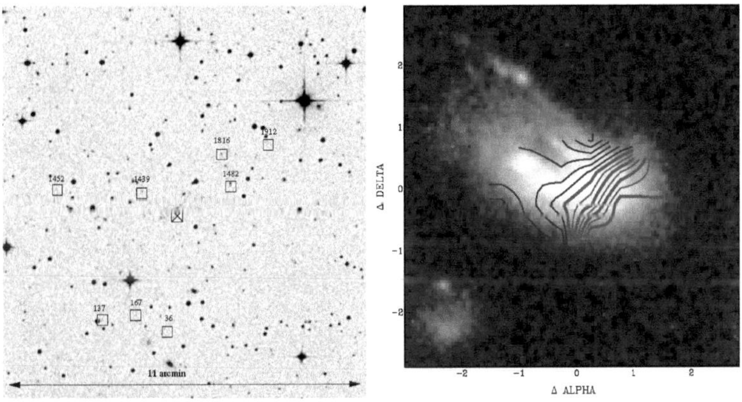


**Fig. 3.** (*Left*) Locations of the eight cluster galaxies observed with FLAMES-IFU according to the notation of (2). (*Right*) A special case of gas stripping/interaction event in the dense environment of MS0451-03.

very peculiar features, such as a one-sided arm. In one of the galaxies (see Fig. 3), we revealed strong evidence for the ram-pressure stripping and/or other mechanisms acting both on the morphology and kinematics. This object very likely resembles C153 in Abell 2125 for which multi-wavelength data are available. The authors of this study have proved that this object at $z = 0.25$ is undergoing a violent stripping event (Keel et al. 2004, STScI-PRC2004-02a). In the Virgo cluster, many other examples with similar characteristics have been studied in detail (e.g. NGC 4654, (5)). Further investigations have been planned on this very peculiar object in MS0451-03 to discriminate the different evolutionary mechanisms. These results will be confirmed by comparing with numerical simulations.

A further step in our *distant cluster* galaxy study is to derive the Tully–Fisher relation and compare with both *local* 3-D kinematic results e.g. (4),(1) and with *distant field* kinematics (Hammer et al., these proceedings). It is not clear in fact whether the evolution in the Tully-Fisher reported by several authors is still significant, within the errors, in the long-slit kinematics when only non-perturbed galaxies are considered (Hammer et al., these proceedings).

These examples of distant velocity fields establish the FLAMES/IFU mode as an efficient tool to study the dynamic evolution of galaxies in cluster environments which avoids the uncertainties inherent in traditional long-slit techniques.

DV acknowledges supports through a Euro3D RTN on Integral Field Spectroscopy (No. HPRN-CT-2002-00305) and the Marie Curie ERG grant (No. MERG-CT-2005-021704), funded by the European Commission. DV also acknowledges ESO support to attend the “Science Perspectives for 3D Spectroscopy” conference.

## References

[1] Chemin L., Balkowski C., Cayatte V., et al., 2005, MNRAS in press, astro-ph/0511417
[2] Ellingson, E., Yee, H. K. C., Abraham, R. G., et al., 1998, ApJS 116, 247
[3] Poggianti, B. M. 2003, Ap&SS, 285, 121
[4] Verheijen, M.A.W., Bershady M.A., Swaters R.A., et al. 2005, In: *"Island Universes - Structure and Evolution of Disk Galaxies"*, astro-ph/0510360
[5] Vollmer, B., 2003, A&A 398, 525

# ESO ASTROPHYSICS SYMPOSIA
European Southern Observatory

Series Editor: Bruno Leibundgut

J. Bergeron, A. Renzini (Eds.),
**From Extrasolar Planets to Cosmology: The VLT Opening Symposium**
Proceedings, 1999. XXVIII, 575 pages. 2000.

A. Weiss, T.G. Abel, V. Hill (Eds.),
**The First Stars**
Proceedings, 1999. XIII, 355 pages. 2000.

A. Fitzsimmons, D. Jewitt, R.M. West (Eds.),
**Minor Bodies in the Outer Solar System**
Proceedings, 1998. XV, 192 pages. 2000.

L. Kaper, E.P.J. van den Heuvel, P.A. Woudt (Eds.),
**Black Holes in Binaries and Galactic Nuclei: Diagnostics, Demography and Formation**
Proceedings, 1999. XXIII, 378 pages. 2001.

G. Setti, J.-P. Swings (Eds.)
**Quasars, AGNs and Related Research Across 2000**
Proceedings, 2000. XVII, 220 pages. 2001.

A.J. Banday, S. Zaroubi, M. Bartelmann (Eds.),
**Mining the Sky**
Proceedings, 2000. XV, 705 pages. 2001.

E. Costa, F. Frontera, J. Hjorth (Eds.),
**Gamma-Ray Bursts in the Afterglow Era**
Proceedings, 2000. XIX, 459 pages. 2001.

S. Cristiani, A. Renzini, R.E. Williams (Eds.),
**Deep Fields**
Proceedings, 2000. XXVI, 379 pages. 2001.

J.F. Alves, M.J. McCaughrean (Eds.),
**The Origins of Stars and Planets: The VLT View**
Proceedings, 2001. XXVII, 515 pages. 2002.

J. Bergeron, G. Monnet (Eds.),
**Scientific Drivers for ESO Future VLT/VLTI Instrumentation**
Proceedings, 2001. XVII, 356 pages. 2002.

M. Gilfanov, R. Sunyaev, E. Churazov (Eds.),
**Lighthouses of the Universe: The Most Luminous Celestial Objects and Their Use for Cosmology**
Proceedings, 2001. XIV, 618 pages. 2002.

R. Bender, A. Renzini (Eds.),
**The Mass of Galaxies at Low and High Redshift**
Proceedings, 2001. XXII, 363 pages. 2003.

W. Hillebrandt, B. Leibundgut (Eds.),
**From Twilight to Highlight: The Physics of Supernovae**
Proceedings, 2002. XVII, 414 pages. 2003.

P.A. Shaver, L. DiLella, A. Giménez (Eds.),
**Astronomy, Cosmology and Fundamental Physics**
Proceedings, 2002. XXI, 501 pages. 2003.

M. Kissler-Patig (Ed.),
**Extragalactic Globular Cluster Systems**
Proceedings, 2002. XVI, 356 pages. 2003.

P.J. Quinn, K.M. Górski (Eds.),
**Toward an International Virtual Observatory**
Proceedings, 2002. XXVII, 341 pages. 2004.

W. Brander, M. Kasper (Eds.),
**Science with Adaptive Optics**
Proceedings, 2003. XX, 387 pages. 2005.

A. Merloni, S. Nayakshin, R.A. Sunyaev (Eds.),
**Growing Black Holes: Accretion in a Cosmological Context**
Proceedings, 2004. XIV, 506 pages. 2005.

L. Stanghellini, J.R. Walsh, N.G. Douglas (Eds.)
**Planetary Nebulae Beyond the Milky Way**
Proceedings, 2004. XVI, 372 pages. 2006.

S. Randich, L. Pasquini (Eds.)
**Chemical Abundances and Mixing in Stars in the Milky Way and its Satellites**
Proceedings, 2004. XXIV, 411 pages. 2006.

A.P. Lobanov, J.A. Zensus, C. Cesarsky, P.J. Diamond (Eds.)
**Exploring the Cosmic Frontier**
Astrophysical Instruments for the 21st Century. XXVI, 270 pages. 2006.

I. Saviane, V.D. Ivanov, J. Borissova (Eds.)
**Groups of Galaxies in the Nearby Universe**
Proceedings, 2005. XX, 390 pages. 2007.

G.W. Pratt, P. Schuecker, H. Boehringer (Eds.)
**Heating versus Cooling in Galaxies and Clusters of Galaxies**
Proceedings, 2006. ??, ?? pages. 2007.

M. Kissler-Patig, R.J. Walsh, M.M. Roth (Eds.)
**Science Perspectives for 3D Spectroscopy**
Proceedings of the ESO Workshop held in Garching, Germany, 10-14 October 2005
Proceedings, 2006. XVIII, 402 pages. 2007.

MIX
Papier aus verantwortungsvollen Quellen
Paper from responsible sources
FSC® C105338

If you have any concerns about our products,
you can contact us on
**ProductSafety@springernature.com**

In case Publisher is established outside the EU,
the EU authorized representative is:
**Springer Nature Customer Service Center GmbH**
**Europaplatz 3, 69115 Heidelberg, Germany**

Printed by Libri Plureos GmbH
in Hamburg, Germany